# Human–Robot Interaction

# Human–Robot Interaction

**Edited by**
**Mansour Rahimi**
Institute of Safety and Systems Management
University of Southern California,

**and**

**Waldemar Karwowski**
Tampere University of Technology,
Tampere, Finland and
Center for Industrial Ergonomics,
University of Louisville

*Taylor & Francis*
*London • Washington, DC*
*1992*

| | |
|---|---|
| **UK** | Taylor & Francis Ltd, 4 John St, London WC1N 2ET |
| **USA** | Taylor & Francis Inc., 1900 Frost Road, Suite 101, Bristol, PA 19007 |

---



**British Library Cataloguing in Publication Data**

Human-robot Interaction
A catalogue record for this book is available from the British Library

ISBN 0-85066-809-3

**Library of Congress Cataloging-in-Publication Data is available**

Cover design by Amanda Barragry
Typeset by Euroset, 2 Dover Close, Alresford, Hampshire SO24 9PG.

Printed in Great Britain by Burgess Science Press, Basingstoke, on paper which has a specified pH value on final paper manufacture of not less than 7.5 and therefore 'acid free'.

# *Contents*

# Preface

The increasing application of robotics technology has resulted in a widespread interest in robotic research and development efforts. The use of industrial robots during the last 10 years has grown steadily in Japan, Europe and the USA (Goto, 1987; Ishitani and Kaya, 1989; Tami, 1989). According to the International Federation of Robotics (Lempiainen, 1990), in 1989 there were 219 667 installed robots in Japan, followed by the USSR (62 339 robots), the USA (36 977 robots) and Germany (22 397 robots). With robot population growing at a rate of 12–27% annually (in 1989), many countries have reached or exceeded the ratio of 10 robots per 10 000 industrial workers. The variety of robot applications has also increased, as more robots are now being used not only in manufacturing (material handling, welding and assembly) but also in service industries, health care and rehabilitation, home use, teleoperation and construction.

Counteracting some early predictions in robot popularity, the ultimate goal of large-scale implementations in unmanned factories and totally autonomous robotic systems has not been achieved. Recent developments in advanced computer-controlled technology pose significant questions with respect to the role and involvement of humans in productivity and safety of these hybrid systems. From a productivity point of view, automation engineers and system designers are involved in designing more efficient and reliable robotic systems. Managers are concerned with implementing an array of organizational structures, job and work designs and technology implementations to increase the efficiency of their robot-based operations. Robot users are also involved with efficient and safe robot installation, programming, monitoring and supervision, trouble-shooting, repair and maintenance. Therefore, in order to succeed, technological knowledge must be integrated with human capabilities within an interactive socio-technical infrastructure.

To some extent, this volume reflects the complexity and the need for attention to current multi-disciplinary problems intrinsic in human–robot interaction. The interactions between humans and robots include a complex array of system constituents and their characteristics such as operator cognitive, motor and perceptual abilities and limitations, robot software, robot hardware characteristics and interfaces, dynamically changing job station layout and environment, organizational and management structures, and safety.

The main goal of this book is to highlight important issues in human–robot interaction, and to encourage more research and application in this new field of scientific endeavour. The book consists of 17 chapters, organized around three main central themes.

**Part I. Human Factors** explores the strengths and weaknesses of humans and robots in an interactive robotic environment.
**Part II. Safety** contains issues related to robot accidents and incidents, human intrusion and protection, robot protection, and total system reliability and integrity.
**Part III. Design and Implementation** explains the need to incorporate human aspects in designing and using robotic systems and the appropriateness of such considerations in different applications.

The contents of this book are intended to be informative and useful for a wide range of scientists: engineers and designers of robotic systems, human factors (ergonomics) specialists, managers and users of robotic systems, and students interested in studying advanced topics in human aspects of robotic systems.

The idea of *Human–Robot Interaction* was initiated by the editors in discussions which followed conferences covering topics of a similar nature. So far, we have been motivated and supported to organize two conferences in the broader field of ergonomics of advanced manufacturing systems. We hope that this book succeeds in incorporating some key discussions in this new field of human factors, and will serve as a catalyst for future research activities with the goal of productive and safe integration of people and their smart steel-collar machines. Most of the challenge lies ahead of us in this germinal field.

Our initial appreciation is to our contributors. Our Taylor & Francis Editor, Dr Robin Mellors has been exceedingly supportive and co-operative in our editing activities. We would also like to thank Mr Glenn Azevedo from the University of Southern California and Mrs Laura Abell, Secretary at the Center for Industrial Ergonomics, University of Louisville, for providing logistic support on this project. While working on this book, the first editor was partially supported by the Faculty Research and Innovation Fund of the Unversity of Southern California, and the second editor was supported by the Fulbright Scholarship at Occupatational Safety Engineering Program, Tampere University of Technology, Tampere, Finland.

MANSOUR RAHIMI
WALDEMAR KARWOWSKI

## References

Goto, M., 1987. Occupational safety and health measures taken for the introduction of robots in the automobile industry. In Noro, K. (Ed.) *Occupational Health and Safety in Automation and Robotics* (London: Taylor & Francis), pp. 399–417.

Ishitani, M. and Kaya, Y., 1989. Robotization in Japanese manufacturing industries. *Technological Forecasting and Social Change,* **35,** 97–131.

Lempiainen, J., 1990. State-of-the-art in world-wide use of industrial robots. *Robotics Society in Finland,* 6–7.

Tami, A., 1989. International comparison of industrial robot penetration. *Technological Forecasting and Social Change,* **34,** 191–210.

# *PART I*

# *HUMAN FACTORS*

# Introduction

Human factors in robotics is the study of principles concerning human behaviour and characteristics for efficient design, evaluation, operation and maintenance of robots. Since people and robots differ considerably in their abilities and limitations to do productive work, the knowledge of these differences should be very useful in designing interactive human–robot workspaces. In order to aid the reader throughout the book, a model for human–robot functional interactions is presented in Figure I.1. This diagram is a representation of the functional characteristics of human–robot interactions at the micro systems level.

This section consists of five chapters. In the first chapter, the techniques currently used in measurement of human and robot performance are reviewed and evaluated. The authors present a strategic model developed to aid in the allocation of tasks between operators and robots. It is suggested that redesigning tasks and robot workstation layouts may be the simplest and most effective approach to provide compatibility between operators and robots.

In the second chapter, the reader is presented with a review of robot teaching and programming methods. Since robot-teach programming is one of the most frequent human–robot interactions, an in-depth understanding of these methods is essential. The authors explain the differences between two programming design concepts: 'showing' and 'telling'. Also, the use of high-level robot programming languages is discussed. In addition, programmer productivity and safety while teaching a robot is explained briefly.

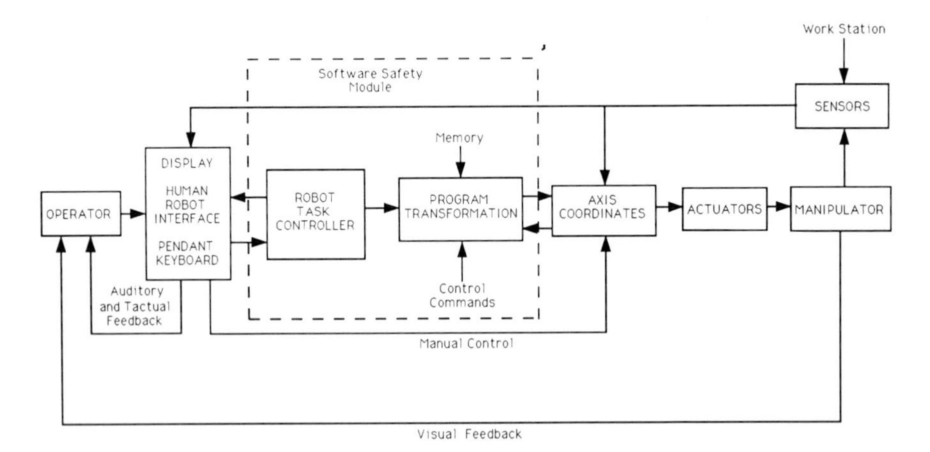


*Figure I.1 A block diagram showing the functional components of a human–robot system.*

Presently, the most frequent tool for teaching a robot is a teach pendant. In the third chapter, the authors address the issue of human control of robot movements using teach pendants. They especially focus on the perceptual and cognitive aspects of human–robot interaction using pendants. It is argued that considerations for operator information processing capabilities in the design of pendants will significantly increase the effectiveness, reliability and safety of robotic systems. In a closely related subject, Chapter four outlines a robot teaching expert system using human-like fuzzy image processing and Robot Time and Motion (RTM). This system is developed to improve the control of robot motions during robot teaching activities.

Off-line programming has distinct advantages over the traditional on-line programming. In the final chapter of this section, the authors present an integrated graphical simulation system for off-line robot programming. The system consists of five components: interactive programming surface, three-dimensional graphics modules, a high-level language compiler, a data interpreter and a combined simulation package. It is indicated that this system relieves robot operators from the routine and sometimes inefficient task of on-line programming, and improves system reliability by avoiding programming mistakes.

# Chapter 1
# Robot and human performance evaluation

**A. M. Genaidy[1] and T. Gupta[2]**

[1]*Occupational Biomechanics Laboratory, Department of Mechanical, Industrial and Nuclear Engineering, University of Cincinnati, Cincinnati, OH 45221-0116, USA*
[2]*Department of Industrial Engineering, Western Michigan University, Kalamazoo, MI 49008, USA*

**Abstract.** This chapter reviews and evaluates the techniques currently used in the measurement of human and robot performances. A strategic model is presented which may assist job designers in the selection process between humans and robots. A case study is presented to illustrate the use of the strategic model.

## *Introduction*

Robots have been successfully implemented in various industrial applications such as painting, welding and handling heavy and fragile parts. Furthermore, programmable assembly systems have effectively utilized robots to automate low-to-medium volume and batch assembly operations which are labour-intensive in nature.

Since robots have the potential to replace humans in many areas, there is a need for developing work methods and measurement techniques to assess robot and human performances. Such techniques will assist in conducting an analysis of the economic feasibility of implementing robots in the workplace.

This chapter reviews and evaluates the techniques currently used in measuring both human and robot performances. A strategic model is presented to provide job designers with a framework for the selection process between humans and robots. A case study is reported to illustrate the use of this strategic model in the selection of humans versus robots.

## *Review of measurement techniques*

Robot and motion time, ROBOT MOST (Maynard Operation Sequence Technique), job and skills analysis and time study have been reported in the literature with the purpose of comparing human and robot capabilities and performances. Table 1.1 briefly summarizes the four techniques. The detail of each technique is given below.

*Table 1.1 A comparison of measurement techniques.*

| Technique | Source | Basis | Comment |
|---|---|---|---|
| RTM | Paul and Nof (1979) | This procedure measures time elements on the basis of MTM-1 system | Based on assumptions of PMTS[a] |
| ROBOT MOST | Wygant (1986) | This procedure measures time elements on the basis of MOST technique | Based on assumptions of PMTS[a] |
| Job and skill analysis | Nof *et al.* (1980) | This technique uses left and right hand analysis to arrive at the estimated time | |
| RMC | Nof *et al.* (1980) | This method is based on the traditional man-machine charts in comparing the relative abilities of robots and humans | |
| Time study | Genaidy *et al.* (1990) | This method estimates time values on the basis of the traditional time study concept | |

PMTS, Predetermined Motion Time Systems.

## Robot and motion time

Paul and Nof (1979) developed a method called Robot Time and Motion (RTM). This method was based on the concept of Methods-Time Measurement (MTM) which was originated by Maynard *et al.* (1948). The RTM technique consists of eight elements categorized as reach, stop on error, stop on force or stop on touch, move, grasp, release, vision and process time delay (Table 1.2). The standard time of the whole operation is the sum of time values assigned to the various elements making up the operation.

The RTM methodology differs from the MTM procedure in that the elements are based, among other things, on the robots' physical parameters, maximum torque, resolution and sensors. On the other hand, MTM takes human variability into account in providing estimates of element times.

Paul and Nof (1979) used RTM and MTM to evaluate robot and human performances for a pump assembly task. It was found that the robot performance required more than eight times the estimated human performance time.

## ROBOT MOST

Wygant (1986) and Wygant and Donaghey (1987) developed ROBOT MOST which is based on the Maynard Operation Sequence Technique known as MOST

*Table 1.2 Description of Robot Time and Motion (RTM) elements.*

| Element | Symbol | Description |
|---|---|---|
| Reach | Rn | Describes the motion of an empty hand to a position. Since a robot may not always reach directly, it may have to move through a series of (n – 1) intermediate points before it reaches the final position. |
| Stop on error | SE | Describes the manipulator being brought to rest within a given position error balance. |
| Stop on force | SF | Describes the manipulator being brought to rest |
| Stop on touch | ST | by force or touch sensing. |
| Move | Mn | Describes the motion of a loaded hand to a position. Mn is identical to Rn except that its time is increased depending on the load. |
| Grasp | GR | Describes closing the fingers. This element is used to grasp an object in a given position. It also describes closing the hand. |
| Release | RE | Describes opening the fingers. |
| Vision | VI | Describes the robot obtaining visual input. It is usually used to identify and locate objects and their features. |
| Process time delay | TI | Specifies unavoidable delays during which the robot must wait. |

*Source:* Paul and Nof, 1979.

(Zandin, 1980). The MOST procedure assumes that the movement of an object can be described as a sequence of basic motions:

1. General Move Sequence (for the movement of an object manually from one location to another freely through air);
2. Controlled Move Sequence (for the movement of an object when it remains in contact with a surface or is attached to another object during movement);
3. Tool Use Sequence (for the use of common hand tools as fastening or loosening, cutting, cleaning, gauging and writing).

The General Move Sequence Model consists of three distinct phases: GET/PUT/RETURN. The Controlled Move Sequence, like the General Move, has three phases: GET/MOVE or ACTUATE/RETURN. The Tool Use Model follows a fixed sequence of subactivities occurring in five main phases: GET OBJECT or TOOL/PLACE OBJECT or TOOL/USE TOOL/ASIDE OBJECT or TOOL/RETURN.

The ROBOT MOST uses the same General Move and Controlled Move Sequence Models developed by Zandin (1980). The Tool Use Sequence Model, however, is not included in ROBOT MOST since a robot does not use hand tools the same way a human operator does. The detail of ROBOT MOST is given in Table 1.3. The time values for ROBOT MOST were developed for the Prab Model E, General Electric P50 and Cincinnati T-3 robots.

*Table 1.3* ROBOT MOST *index values.*

| Basic motion | Description |
|---|---|
| Action distance (A) | The time for the robot action distance is the maximum values of the motions in the *X*, *Y* and *Z* axes for each move. The user selects a specific robot model from the data base and the time for each motion is calculated using the appropriate equation for the individual moves. |
| Body motion (B) | Since the robot moves in the vertical direction simultaneously with the horizontal motions, no time is added for the body motion unless the motion cannot be performed by a robot (e.g. passing through a door or climbing on or off a work platform). In the case where the robot cannot perform the motion, the same index value that is used for the manual body motion will be used for robot body motion. |
| Gain control (G) | To gain control of an object the robot must move to the object, in some cases orient the end-of-arm, and close the gripper. Since the actual path of the robot arm is not known most motions are programmed to a position just clear of the object and then a second motion is made to gain control to ensure that the gripper does not come in contact with the object before it is ready to grasp. The time to close the gripper includes 100 ms for cycle time and the actual close time which varies by robot model. Also, the orientation time will vary depending on the particular model that is being analysed. |
| Place (P) | To place an object the robot must move to the object, orient the end-of-arm as required, and open the gripper. The method of calculating the index value for place is the same as used in gain control. |
| Move controlled (M) | For a robot, this motion is simply a single move or a series of short moves which are calculated in the same way as action distance. |
| Process time (X) | Process time is the same for both manual and robot and sequence models. The analyst enters the process time for each task. This time is used for both the manual and robot time calculations. |
| Align (I) | Since most robot models have the capability of aligning or orienting the object with the desired accuracy during the move controlled activity, no time is allowed in the robot sequence model for align. |

*Source:* Wygant, 1986.

Wygant (1986) compared the actual total time taken by robots and humans for drilling a 1.27 cm deep hole in a casting. The time values estimated by MOST and ROBOT MOST were 0.22 and 0.50 minutes for humans and robots, respectively. Thus the robot took 127% more time than a human to complete the task.

## Job and skill analysis

The concept of job and skill analysis was originated by Nof *et al.* (1980). This approach involves two sequential steps: Robot–Man Charts (RMC); and Job and Skill Analysis (JSA) method. The RMC procedure provides a means of identifying jobs which can be done either by robots or humans. Once it is determined that a robot can perform the job, JSA is conducted.

The RMC technique is a modification of the traditional man–machine charts used in motion and time study. It compares the relative abilities of humans and industrial robots in the following categories (Table 1.4).

1. *Action and manipulation*: manipulation abilities (typical robot structures are polar robot, cylindrical robot, cylindrical moving robot and pick-and-place robot), body dimensions, strength and power, consistency, overload/underload performance, reaction speed and self-diagnosis.
2. *Brain and control*: computational capability, memory, intelligence, reasoning and signal processing.
3. *Energy and utilities*: power requirements, utilities, fatigue/down-time/life expectancy and energy efficiency.
4. *Interface*: sensing and inter-operator communication.
5. *Miscellaneous factors*: environmental constraints, brain–muscle combination, training and social and psychological needs.
6. *Individual differences*.

The JSA method is based on the idea that a task is broken down to elements with their time values and is detailed in Table 1.5. The traditional job and skills analysis utilized for human performance is also given to compare it with that of a robot. In a case study reported by Nof *et al.* (1980), it was found that a more efficient use of an already available camera reduced the task time for the robot by about 30%.

## Time study

Genaidy *et al.* (1990) compared human and robot performances for a simple assembly operation by conducting a time study. The task performed consisted of assembling two cubes connected in the centre by a metal rod. The cubes were continuously fed by two gravitational chutes and placed in a finished part bin after being assembled.

Two workstations, one robotic workstation and one human workstation, were used in this study. Ten male volunteers participated in this study and each person completed 100 work cycles. An equal number of work cycles was completed by two robots.

The mean cycle time for robots was 10.01 s with a standard deviation of 0.276 s and coefficient of variation of 2.76%. The mean cycle time for humans was 2.95 s with a standard deviation of 0.49 s and coefficient of variation of 16.63%. While the mean cycle time for humans was significantly lower than the mean cycle time for robots, the corresponding standard deviation and coefficient of variation were significantly higher.

Results also showed that while the cycle time was virtually the same across all work cycles for the robots, the mean cycle time for human operators decreased initially and then levelled off. The levelling off occurred after the thirtieth cycle.

*Table 1.4 Categories compared in Robot–Man Charts (RMC).*

| Category | Factors | Description of factors |
|---|---|---|
| Physical skills and characteristics | Action and manipulation | Body |
| | | types |
| | | typical range of maximum motion capacity (TRMM) |
| | | Arm |
| | | types |
| | | number |
| | | TRMM |
| | | Wrist |
| | | types |
| | | TRMM |
| | | End effector |
| | | types |
| | | TRMM |
| | Body dimensions | Main body |
| | | Floor area required |
| | Strength and power | Useful arm load |
| | | Power |
| | Consistency | |
| | Overload/underload performance | |
| | Environmental constraints | Ambient temperatures |
| | | Humidity |
| | | Sensitivity |
| Mental and communicative skills | Computational capability | |
| | Memory | |
| | Intelligence | |
| | Reasoning | |
| | Signal processing | |
| | Brain–muscle combination | |
| | Training | |
| | Social and psychological needs | |
| | Inter-operator communication | |
| | Reaction speed | |
| | Self diagnosis | |
| | Individual differences | |
| Energy considerations | Power requirements | |
| | Utilities | |
| | Fatigue | |
| | Downtime | |
| | Life expectancy | |
| | Energy efficiency | |

*Source:* Nof *et al.* (1980).

*Table 1.5 Job and skill analysis utilized for human and robot performances.*

| | Job and Skill Analysis |
|---|---|
| *Human performance* | |
| Element | An operationally definable task component that has clear beginning and end points such as moving an object from point A to point B. |
| Time | Time to perform the given element. |
| Left hand<br>Right hand | Describes which part of the element is performed by which hand (separate descriptions can be used for left and right hand). |
| Vision | Indicates the extent to which vision is used for the performance of the element (e.g. confirm by vision the current position of an object). |
| Other | Indicates the extent to which other senses such as touch, senses kinesthesis, hearing, etc. are used for the performance of an element. |
| Comments | Components of the element where some special precaution has to be exercised (e.g. in joining two parts, the holes in component A must match the holes in component B). |
| *Robot performance* | |
| Element | As above, except each element is a detailed micromotion. |
| Time | The time to perform the given motion. |
| Senses | Each of the major robot senses, namely, vision, touch and force compliance, is treated in the same manner as vision is treated above. |
| Limbs | Each of the major limbs is treated individually. Memory and specific memory requirements needed for the element such as program word length, storage capacity, etc. Program required includes reference information and logic for decision making in the element. |
| Comments | Comments about special requirements such as engineering tolerances, electricity, air pressure and other utilities. |

*Source:* Nof *et al.* (1980).

## *Evaluation of measurement techniques*

Four techniques dealing with the measurement of human and robot performances were reviewed. The outcome of each technique is a set of time values that can be used to compare human and robot productivity. Thus the application of this information in economic analyses is very useful in order to justify robotization (purchase, installation and maintenance of robots in the workplace).

The philosophy behind RTM and ROBOT MOST is based on the concept of Predetermined Motion Time Systems (PMTS). The PMTS are based on two fundamental assumptions: an accurate universal time value can be assigned to each basic motion; and the addition of individual basic motion time values is equal to the time for the whole task. In a survey reported by Genaidy *et al.* (1989), it is not clear whether these assumptions are met. Two opposing views have been presented regarding this issue. One group of investigators supports the validity of PMTS assumptions, while the other group has raised serious doubts with respect to the validity of these assumptions.

The RTM technique requires a detailed analysis of robot motion using the micromotion analysis concept of MTM. This is a very time consuming process. Moreover, RTM does not provide a convenient way to estimate the standard time for human performance using MTM. The ROBOT MOST method is more suitable than the RTM technique since it has an option that prints the time standard for both humans and robots. Also, ROBOT MOST requires less training on the part of the analyst. Although the time study conducted by Genaidy *et al.* (1989) provides a direct measure of standard time which does not suffer from the limitations of PMTS, it should be done for each operation conducted by a robot or a human. The JSA method is a combination of the RTM and conventional work methods techniques. Although a convenient way to improve work methods for robots, it does not provide a framework that can be used systematically for this purpose.

In sum, all techniques developed provide useful means to assess and compare human and robot performances. However, these techniques need to be integrated in order to develop a strategy aiming at placing the right robot and/or human on the right job. In the next section such a strategy is discussed.

## Economic justification of robotization

Once the time it takes the robot to perform an industrial operation is estimated, it is possible to conduct an economic analysis for justification of robotization. Mital and Vinayagamoorthy (1987), Mital and Genaidy (1988), Abdel-Malek (1989) and Genaidy *et al.* (1990) outlined procedures for economic analysis. The reader is referred to these papers for detailed discussion of the subject.

## Strategic model for robot/human evaluation and selection

The main objective of any strategic model within the context of job design and analysis is to match job demands with human and/or robot capabilities, based on cost effectiveness and safety considerations. Job demands include skills, effort, conditions and consistency. Robot and human capabilities can be described in terms of action and manipulation, brain and control, energy and utilities, interface and other factors such as environmental conditions.

Based on the aforementioned discussion, the following strategic model may be used within the context of job design and analysis (Figure 1.1):

1. Determine job demands;
2. Determine robot capabilities (using RMC procedure). This step requires a survey of the various robots available on the market;
3. If job demands do not match robot capabilities, then find another alternative. If they do match, proceed to the next step;

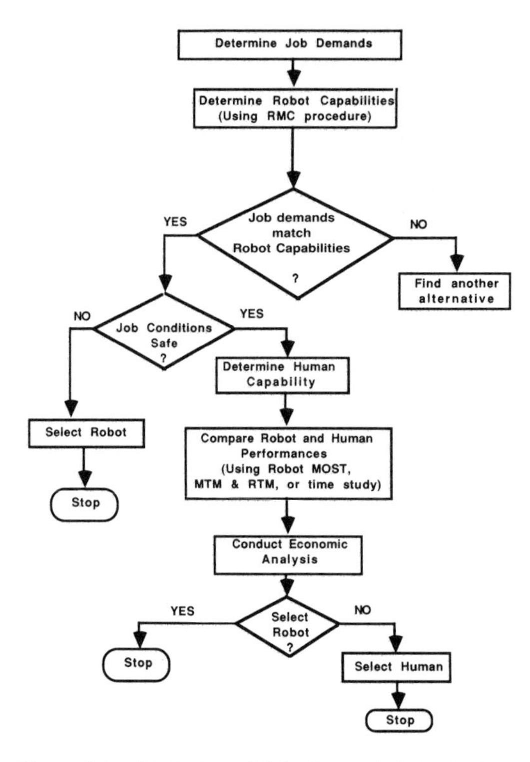


*Figure 1.1 Strategic model for human/robot selection.*

4. If the job conditions are unsafe, select robots. If the job conditions are safe, proceed to the next step;
5. Determine human capabilities;
6. Compare human and robot performances (using ROBOT and ROBOT MOST, MTM and RTM, or time study);
7. Conduct economic analysis; and
8. Depending on the outcome of economic analysis, select robot or human.

The outlined strategic model can be easily implemented on a computer to further its application.

## *Case study*

The case study involves a decision-making process in selecting between robots and humans for a welding operation. This operation is characterized by a hazardous work environment and low labour productivity. These factors, however, are subjective and require detailed analyses to make the right decision.

The strategic model was used for a production system with nine condenser mounting brackets. Welding and brushing were the primary operations involved in the production of each bracket. A combination arc, TIG (tungsten inert gas), and spot welding were determined from the design sheet for each item.

For each of the nine brackets, the various elements of the operation are:

1. The cycle begins with a robot to pick up a mounting bracket and the components to be welded on the bracket and to locate them on the shuttle jig;
2. The shuttle jig then enters the welding robot station (the station consists of two robot arms; the TIG and spot welding units are mounted on the two arms);
3. A robot arm with a TIG unit welds a corner of each major component on the mounting bracket to hold the component in position;
4. Then, the spot weld robot arm goes through a series of spot welds to secure the components on the bracket;
5. After the spot welding is complete, the robot arm drops the bracket on an exit conveyor and the part is ready to leave the station; and
6. Brushing, which follows the welding operation, is performed manually to remove the oxides formed on the surface.

The first four steps of the strategic model were in favour of robotic operation. To compare human and robotic performances, two time studies for the robotic and human operations were conducted. Six-axis PUMA robots were used in the robotic workstations. In each case, three of six degrees of freedom were associated with the wrist.

Each of the nine jobs was performed on the robotic workstation first, and the operation cycle time was recorded for fifty units per job. Subsequently, the same jobs were performed by five professional welders and the cycle times were recorded for 100 units for each of the nine jobs. Fewer cycles for the robotic operations were adequate since robotic learning is minimal compared with human learning (Genaidy *et al.*, 1990).

The mean and standard deviation of cycle times for humans and robots are given in Tables 1.6 and 1.7, respectively. The average cycle time for robots ranged between 14.23 and 43.14 s for the nine jobs. The average cycle time for humans ranged between 23.43 and 73.32 s. There were significant differences at the 5% level between robots and humans in terms of cycle time.

*Table 1.6 Mean and standard deviation of cycle time (s) for five operators.*

| Job No. | Human cycle time (s) | | | | | | | | | |
|---|---|---|---|---|---|---|---|---|---|---|
| | Operator | | | | | | | | | |
| | 1 | | 2 | | 3 | | 4 | | 5 | |
| | m | sd | m | sd | m | sd | m | sd | m | sd |
| 1 | 22.62 | 1.12 | 23.54 | 1.40 | 22.92 | 2.32 | 23.70 | 1.71 | 24.37 | 1.42 |
| 2 | 30.45 | 1.65 | 41.23 | 1.78 | 34.50 | 2.84 | 40.47 | 1.95 | 30.45 | 1.64 |
| 3 | 42.28 | 1.45 | 36.45 | 1.81 | 41.38 | 2.26 | 32.78 | 1.84 | 48.06 | 2.45 |
| 4 | 31.27 | 1.23 | 39.46 | 1.92 | 33.78 | 2.45 | 40.05 | 2.10 | 37.90 | 2.65 |
| 5 | 67.40 | 2.43 | 61.30 | 2.20 | 52.94 | 2.87 | 60.25 | 3.24 | 47.39 | 2.86 |
| 6 | 52.40 | 2.12 | 61.28 | 2.57 | 42.39 | 1.92 | 31.05 | 2.38 | 29.95 | 2.45 |
| 7 | 45.23 | 1.65 | 64.49 | 1.89 | 38.74 | 2.02 | 55.67 | 2.24 | 45.63 | 2.34 |
| 8 | 76.92 | 2.13 | 78.43 | 3.45 | 67.11 | 2.67 | 69.01 | 3.84 | 75.13 | 3.10 |
| 9 | 45.09 | 1.23 | 53.20 | 1.54 | 56.33 | 2.12 | 41.81 | 2.28 | 48.85 | 2.18 |

m, mean; sd, standard deviation.

*Table 1.7 Mean and standard deviation of cycle time (s) for robots and humans.*

| Job no. | Robot cycle time | | Human cycle time | |
|---|---|---|---|---|
| | m | sd | m | sd |
| 1 | 14.23 | 0.33 | 23.43 | 1.82 |
| 2 | 21.47 | 0.33 | 35.42 | 2.13 |
| 3 | 31.36 | 0.42 | 40.19 | 2.03 |
| 4 | 23.38 | 0.39 | 36.49 | 2.76 |
| 5 | 42.73 | 0.32 | 57.85 | 2.94 |
| 6 | 29.30 | 0.43 | 43.41 | 2.67 |
| 7 | 37.92 | 0.44 | 49.95 | 2.10 |
| 8 | 43.14 | 0.37 | 73.32 | 3.26 |
| 9 | 36.32 | 0.37 | 49.05 | 2.04 |

m, mean; sd, standard deviation.

An economic analysis followed the time study. Additional factors were collected (e.g. utilization level, work cell cost, percentage arc-on-time, weld velocity and material savings). The procedure described by Genaidy *et al.* (1990) was used for this purpose. An annual cost saving of $65 000 was estimated for the projected annual demand of 100 000 units of each of the nine brackets.

## Directions for future research

Future research should focus on conducting longitudinal comparisons between humans and robots for existing and new operations. Then, an economic analysis should follow such comparisons. Moreover, research is warranted to develop new work methods and measurement techniques for advanced production systems (Karwowski and Rahimi, 1989).

## References

Abdel-Malek, L. L., 1989, Assessment of the economic feasibility of robotic assembly while conveyor tracking (RACT). *International Journal of Production Research,* **27,** 7, 1209–24.

Genaidy, A. M., Duggal, J. S. and Mital, A., 1990, A comparison of human and robot performances for simple assembly tasks. *The International Journal of Industrial Ergonomics,* **5,** 73–81.

Genaidy, A. M., Obeidat, M. and Mital, A., 1989, The validity of predetermined motion time systems in setting standards for industrial tasks. *The International Journal of Industrial Ergonomics,* **3,** 249–63.

Karwowski, W. and Rahimi, M., 1989, Work design and work measurement: implications for advanced production systems. *The International Journal of Industrial Ergonomics,* **4,** 185–93.

Maynard, H., Stegmerten, G. and Schwab, J., 1948, *Methods-Time Measurement* (New York: McGraw Hill).

Mital, A. and Genaidy, A. M., 1988, 'Automation, robotization in particular, is always economically desirable' – fact or fiction? In Karwowski, W., Parsaei, H. R. and

Wilhelm, M. R. (Eds) *Ergonomics of Hybrid Automated Systems I* (Amsterdam: Elsevier Science Publishers).

Mital, A. and Vinayagamoorthy, R., 1987, Economic feasibility of a robot installation. *The Engineering Economist,* **32,** 3, 173–95.

Nof, S. Y., Knight, J. L. and Salvendy, G., 1980, Effective utilization of industrial robots — a job and skills analysis approach. *AIIE Transactions,* **12,** 3, 216–25.

Paul, R. P. and Nof, S. Y., 1979, Work methods measurement — a comparison between robot and human task performance. *International Journal of Production Research,* **17,** 3, 277–303.

Wygant, R. M., 1986, 'Robots vs Humans in Performing Industrial Tasks: A Comparison of Capabilities and Limitations'. Unpublished PhD Dissertation, University of Houston, Houston, Texas.

Wygant, R. M. and Donaghey, C., 1987, A computer selection model to evaluate robot performance. *Proceedings of the IXth International Conference on Production Research* (Cincinnati, Ohio: University of Cincinnati), pp. 337–41.

Zandin, K. B., 1980, *MOST Work Measurement Systems* (New York: Marcel Dekker).

# *Chapter 2*
# *Human factors in robot teach programming*

**T. Glagowski, H. Pedram and Y. Shamash**

*Department of Electrical Engineering and Computer Science, Washington State University, Pullman, WA 99164–*2752, USA

**Abstract.** In this chapter, different robot teaching/programming methods are explained along with the human factors involved. There are two different classes of robot teaching: teach by showing (show the robot); and programming using a high-level robot programming language (tell the robot). 'Showing' includes methods that guide the robot step by step through the task. 'Telling' exploits high-level language structures for efficient programming, providing the ability to deal with real-time decision making in response to sensor-based information. The major human factors related to teaching methods are required skills, safety considerations and programmer productivity. Human factors are important since they influence the selection of the appropriate teaching method for a given robotics application. Of the two teaching methods, the use of high-level robot programming languages is more important and covers a wider range of applications. In depth investigation of this teaching method is presented. Three typical robot programming languages are reviewed and the characteristics of these languages are explained. Important concepts in robot teaching/programming such as on-line and off-line programming are also discussed.

## *Introduction*

In this chapter we introduce different robot teaching methods considering the human factors involved. The most important human elements in robot programming are safety, productivity, and required human skills. Almost all modern robots are computer-based systems, and as such, they always have human elements within their task performance cycles. A person may interact directly with the hardware and software, conducting a dialogue that drives the function of the system; in all cases people are responsible for the development, support, or maintenance of the system (Pressman, 1987). In the introductory part of this chapter the basic concepts in robot programming are reviewed. The second part is devoted to explaining different robot teaching/programming methods; and the third part, which is in fact a continuation of robot programming, discusses the robot programming languages which are the main elements of computer programming for robots.

## Background

An industrial robot may be defined as a general purpose programmable manipulator (i.e. movable arm) that can be programmed through a sequence of motions in order to perform a useful task. The integration of teleoperators (controlled by remote manipulators) and Numerically Controlled (NC) machines led to the development of industrial robots. The first industrial robot was called Unimate, and was installed in a General Motors plant for unloading a die-casting machine in 1961. The operator used a hand-held controller (pendant) and guided the robot point-by-point through the whole task to record each point. Path teaching by manually guiding a robot through a task and recording the successive points also became a popular robot teaching method. Both these methods are still very much in use. It was not until the 1970s that the use of computers in robot control systems became economically feasible. This led to the use of more sophisticated programming methods and the introduction of textual robot programming languages. In the 1970s the first generation of robot programming languages emerged. Examples are WAVE and AL which were developed at Stanford University, and VAL which was developed by Unimation. In the early 1980s second generation languages began to appear. Examples are AML (by IBM), RAIL (by Automatix), VAL II (by Unimation), later versions of AL, and others. The intent of these developments was to make it easier for the user to program/teach a robot to execute more sophisticated tasks. Along with the second generation languages, proper programming tools and environments were developed to ease the programming task. Third generation languages (still in the research phase) deal with intelligent robots. The user may define a task at a very abstract level and need not go into the details of the task specification. Presently, robots may operate 'intelligently' in a programmed way, but the goal is to operate intelligently on an artificial intelligence basis.

## Programming robots

### *Methods*

A robot is a general purpose programmable manipulator. A task intended for the robot must be 'taught' to it. The process of teaching a robot is also called 'programming' the robot. There are different teaching/programming methods available to meet the requirements of different operator/programmer skills and the complexity of the task, as well as the related productivity and safety issues. As will be explained later, there are two major categories of robot teaching methods: (1) on-line; (2) off-line. Alternative ways of viewing these methods are: (1) teaching by showing; and (2) teaching/programming by using textual programming languages. The former has a simple programming environment in contrast to a variety of different programming environments available for the latter methods. The programming methods and environments are discussed below.

*Human factors*

Each programming method has its own advantages and application domain. Teaching by showing is simpler and requires fewer programming skills; it is limited in applications, and strict safety measures must be applied while programming and testing since the operator is in close proximity to the robot. Using robot programming languages to program a robot requires higher initial investment in apparatus and higher programming skills; these methods cover a broader range of applications, and programming safety measures are less stringent since the programmer does not need to be in the workcell of the robot. Using robot programming languages also allows the programmer to include certain run-time safety features in the application program.

Programmer productivity is one of the main factors influencing the selection of a programming method and the related facilities. For example it is not productive to use a teach-by-showing method for an application that deals with a number of sensors and many decisions made during the robot's operation. A tremendous amount of programming time would be necessary, numerous safety measures would need to be applied and the final product (program) would hardly be adaptable even to small programming changes.

## Elements of a robot system

First some general terms in robotics will be explained.

*Robot cell.* Robot cell is the working environment of a robot including the robot itself and all the objects with which the robot interacts.

*Point.* Point is the basic building block of a robotics task. A point is a sufficient description of a robot manipulator configuration at a given time. There are two alternative ways of representing a point mathematically: (1) by a vector of robot joint positions as read from the robot position sensors. This is a Joint-Level Representation (JLR) of a point; and (2) by the vector of position and orientation with respect to a reference coordinate system. This is called a World Coordinate Representation (WCR) of a point (Shamash *et al.*, 1988).

*Coordinate transformation.* Besides the reference frame (world coordinate system), there may be other frames in the system as well. For example, a vision system gets an object's data with respect to the camera's coordinate system; or the end-effector's configuration may be defined with respect to the robot's base coordinate system. The act of mapping a configuration from one frame to another frame is called coordinate transformation.

*Path.* A set of points defines a path.

*Trajectory.* Successive points in a path may be distanced arbitrarily. A trajectory is the route that the manipulator takes to connect two points. The trajectory may be linear on joints (each joint will take the same amount of time to change its value), straight line, or circular.

*Robotics software.* A robot is a computer-based system, and as such, different levels of programming are required. A distinction should be made between the underlying software (or Robot Operating System) and the application software. The former is always present and manages input and output operations as well as operator interactions. Even when we are using simple teaching methods (teach-by-showing), it is the underlying software that accepts the commands and lets the robot be programmed. Application software is a different concept since it programs the robot for a specific task. Different levels of application software are possible depending on the programming environment that a robotics system provides. Still, it is the robot operating system software that opens the way for the application software to run. Typical elements of a robot system are shown in Figure 2.1.

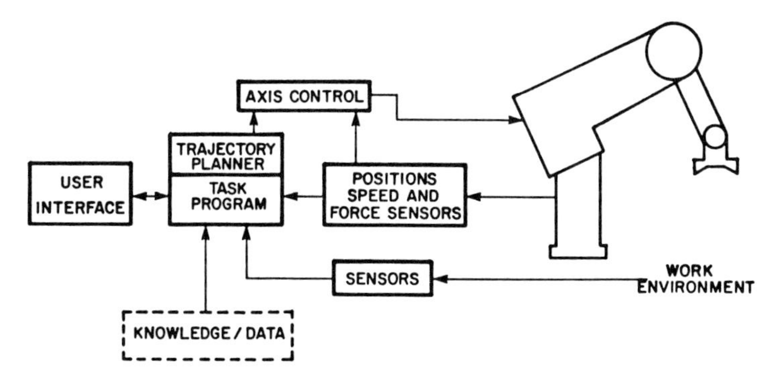


*Figure 2.1 Elements of a robot system.*

The user creates the task application program by showing or by using a textual programming language. The trajectory planner must calculate the successive joint values for the defined task and feed the axis control. The axis control is a feedback control system that gets input from the trajectory planner and receives feedback data from the robot's internal sensors. The task program may contain some decision making protocols such as 'if the force is more than 1 lb, change the move direction'. For decision-making purposes, the task program needs feedback from both the internal and external sensors. Knowledge-based systems, or data from Computer-Aided Design/Computer-Aided Manufacturing (CAD/CAM) systems may be part of the task program as well.

## *Robot teaching*

Teaching a robot is in fact programming it to perform a specific task. A large part of robot programming involves defining a path for the robot to take. There are two main approaches considered in robot teaching: on-line programming and off-line programming. When using on-line methods, the robot itself is used during programming, while the off-line approach allows the user to program the task on a different computer system and download the task application program into the robot's control system.

Depending on the particular application of the robot, different teaching methods may be used. The basic goal in modern robot design is to make the teaching process as user-friendly as possible considering the safety issues.

## On-line programming

On-line programming methods use the robot controller as the programming vehicle, and therefore the robot is not available for production during on-line programming. There are three distinct on-line programming methods, although combinations are possible. These methods are 'lead-through teaching', 'pendant teaching', and teaching via robot programming languages (Shamash *et al.*, 1988).

### *Why on-line programming?*

On-line programming provides direct interaction between human and robot, and appears to be the most natural method of robot teaching. Since the robot may be observed visually during teaching, the on-line operator/programmer has an excellent perception of the robot actions. One should bear in mind that wrong commands to the robot may cause serious damage to the equipment or endanger the safety of the operator (see Gray *et al.*, this volume). Precautions must be considered to avoid undesirable and hazardous situations. Nevertheless, the visual perception and the lower level of required computer programming skills make on-line programming a widely used robot teaching method.

### *Lead-through teaching*

Lead-through teaching is the simplest of all robot teaching methods. To program the robot, the operator simply sets the robot to the free movement mode (in which the servo controls are off), and moves the robot arm manually through a desired path (Figure 2.2). As the operator is moving the robot along the desired path, successive points along this path are stored in the robot controller's memory. During playback, the robot will traverse the taught path by following the stored path points. Some editing capabilities may be available after the task is taught. If the free movement mode for teaching is not available, the operator can use a robot simulator for teach programming and use the stored program to drive the intended robot. The robot simulator is identical to the actual robot except that it is usually smaller in size and has no drives or transmission elements. The simulator communicates with the robot controller. Lead-through teaching is normally used for low precision applications such as spray painting.

For the lead-through method, the operator must be a skilled worker capable of performing the task manually and with high precision. However, the lead-through operator need not have any computer programming skills.

*Figure 2.2 Lead-through teaching.*

*Teach pendant programming*

A teach pendant is a hand-held keyboard which is used as a remote control device for directing a robot arm to different positions required by the the user (Figure 2.3). Its main function is to record the desired points along the path for later playback. The robot system records only the endpoints of paths, leaving the intermediate points to be calculated by some interpolation algorithm during playback. Interpolation may be linear on joint angles, straight line, circular, or a weaving pattern (for welding robots).

A teach pendant may be used as a stand-alone teach programming device or as part of a more sophisticated on-line programming environment. In its simplest form, a teach pendant can drive the robot arm to points of importance and record their coordinates either as a joint vector or as a three-dimensional (3-D) vector in the world coordinate system. During playback the robot repeats the point-to-point movement in a specified sequence. However, teach pendants are usually

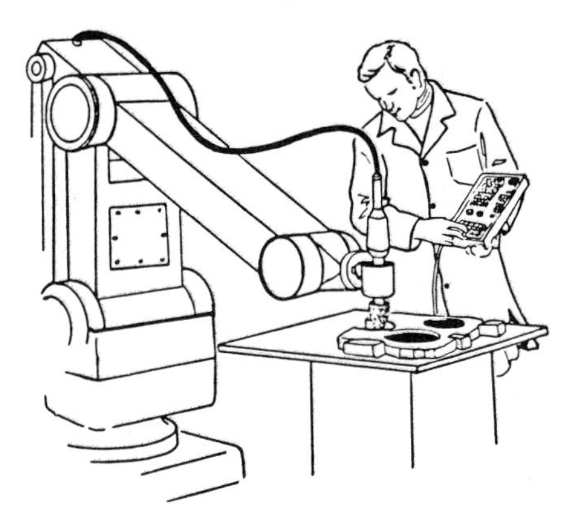

*Figure 2.3 Pendant teaching.*

more powerful than the simple one described and offer additional programming features as well, such as the teach pendant used in the KAREL system (GMFanuc, 1990b), shown in Figure 2.4. This teach pendant is equipped with a keypad and a Liquid Crystal Display (LCD) screen and is capable of: jogging (moving with respect to a defined frame) the robot; teaching positional data (points and paths); testing program execution; recovering from errors; displaying user messages, error messages, prompts and menus; displaying and modifying positional data, program variables, and inputs and outputs; and performing operations on files (copy, delete, transfer, etc.).

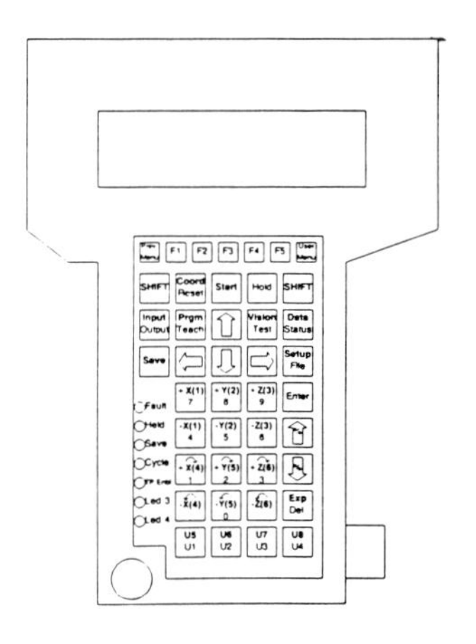


*Figure 2.4 KAREL Teach Pendant (courtesy of GMF Robotics).*

As an example of teach pendant programming, let us consider a palletizing operation. A pallet is a common type of container used in a factory to hold and move parts. Suppose that an operation requires the robot to pick up parts from a conveyor and place them on a pallet with 24 positions, as depicted in Figure 2.5. An operator would lead the robot, step by step, through one cycle of operation and record each move in the robot controller. Additionally, functional data and motion parameters would be entered as the points are programmed. The programming steps for the palletizing example are shown in Figure 2.5.

Two notes of interest can be observed from the preceding example. First, teach pendant programming is fairly simple and minimal operator training is necessary. Second, it is obvious that the number of programming steps can become quite large as the number of repeated subtasks in one cycle of the robot operation increases.

Simple pendant teaching is appropriate for small and simple tasks. More complex tasks require teach pendants with additional programming features such as branching. In addition to the mentioned capabilities, a teach pendant may also include a joystick to move the robot arm to the desired positions.

The teach pendant programmer should be a technician who knows the job (not necessarily skilfully), and has some basic understanding of programming

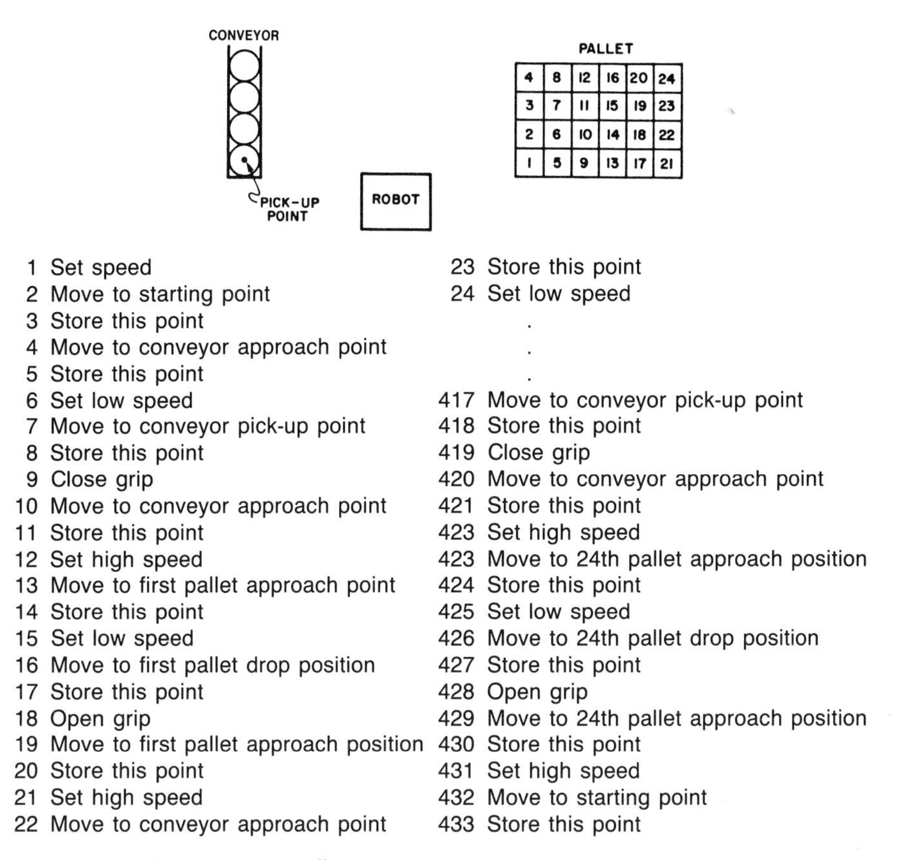


*Figure 2.5 A palletizing operation and the programming steps.*

devices. From job allocation and training perspectives there is a trade-off regarding teach pendant applications. A simple teach pendant may be used by an operator who is familiar with the job and has minimal programming knowledge. More sophisticated teach pendants may be used to program more complex tasks, but the operator then needs to be familiar with the basics of computer programming.

Another issue that influences the operator teaching tasks is the design of the teach pendant. Important design requirements include safety, packaging and reliability. Safety features include having an emergency stop button. Packaging features include having a lightweight unit with a proper hand grip, an easy-to-read display and a proper location and legend of the keys. For a more detailed description of these features see Parsons, this volume.

### *Textual programming languages in the on-line mode*

As mentioned earlier, on-line programming implies using the robot control system along with other peripherals as the programming environment. Using a high-level robotics programming language on the robot control system to program the robot is another on-line programming method. Robotics languages are typically based on Pascal, C, and BASIC (because of ease of use).

In the teach pendant example, we observed the need for more programming constructs for practical purposes. The more sophisticated teach pendants provide some of the desired programming constructs in a low-level language, but there are limitations that are common to pendant-driven robotics systems. These limitations include: the ability to show only one line of the program on the LED display of the pendant at any time; the lack of structured programming constructs and subroutines; and cumbersome editing activities. In order to overcome some of these problems, high-level robotics languages were developed that make robot programming more flexible and more efficient. To demonstrate the flexibility and effectiveness of high-level language programming, another version of the palletizing example in Figure 2.5 is presented here using AML (a robotics language explained later). The efficiency and readability of the program is noticeable (see Figure 2.6).

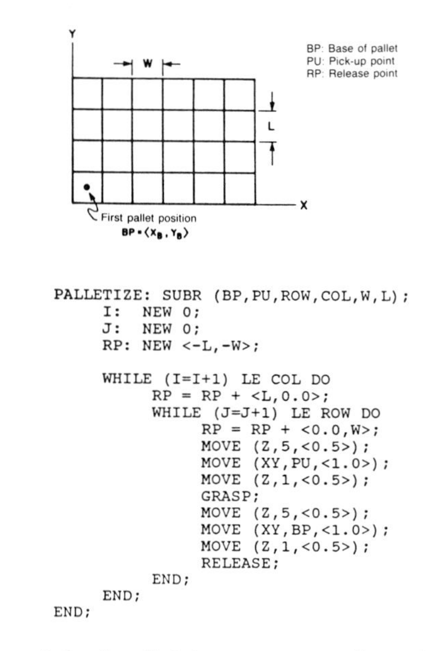


```
PALLETIZE: SUBR (BP,PU,ROW,COL,W,L);
     I:  NEW 0;
     J:  NEW 0;
     RP: NEW <-L,-W>;

     WHILE (I=I+1) LE COL DO
          RP = RP + <L,0.0>;
          WHILE (J=J+1) LE ROW DO
               RP = RP + <0.0,W>;
               MOVE (Z,5,<0.5>);
               MOVE (XY,PU,<1.0>);
               MOVE (Z,1,<0.5>);
               GRASP;
               MOVE (Z,5,<0.5>);
               MOVE (XY,BP,<1.0>);
               MOVE (Z,1,<0.5>);
               RELEASE;
          END;
     END;
END;
```

*Figure 2.6 A palletizing operation: pallet and program.*

The minimum configuration for a high-level on-line programming system includes a computer as the robot control system, a programmer's console and mass storage. Adequate software tools for editing, interpretation and debugging purposes are also necessary. A teach pendant may be used as part of this on-line programming environment. From this perspective, the teach pendant complements the primary programming system by providing direct system control within the robot's workcell.

One important drawback of using high-level programming languages is the skill gap which typically exists between the shop-floor technician and the computer programmer. It could be frustrating to train the shop-floor operator to use a high-level language, and a computer programmer may not have an adequate sense of the real task to be performed. Despite this drawback, high-level programming languages provide undeniable advantages that should not be ignored.

Although high-level structured languages are used in on-line programming, they are more advantageous when used in an off-line programming environment as discussed in the next section.

## Off-line programming

Off-line programming suggests developing robotics control programs away from the robot system and perhaps on a different computer. The idea has been around for years but it gained importance recently due to the integration of simulation packages, Computer-Aided Design (CAD), and computer graphics into one programming environment. Now, a programmer can create robotics application programs on a general purpose computer by, (1) getting the parts specifications from CAD-generated data, (2) generating a procedural description of the task using a high-level robotics language (generate robot's path data), (3) simulating the task, and (4) using computer graphics packages to visualize an animation of the task on the screen.

### *Why off-line programming?*

There are several reasons for off-line programming:

*Reduction in robot downtime.* Plant production time is extremely valuable and competes with robot programming time. Using off-line programming, if an engineering change is made, the robot's programs may be developed off-line and loaded into the robot's system. Also, with on-line programming the entire robot workcell must be completely designed and installed before the robot moves can be programmed. Off-line programming makes programming possible even before the robot's workcell is completely installed;

*Using common available general purpose computers.* Common available computer systems such as Unix-based or MS-DOS-based personal computers may be used for off-line programming developments. Besides the obvious economic advantage, using commonly available computer systems eliminates the need to train programmers for a specialized robotics programming environment;

*Ease of programming.* Off-line programming considerably reduces the time required to create a program and makes programming changes much easier; in addition, the programmed task may be visualized on the computer screen before operating the real robot. Also, developing a program in the plant environment creates inconveniences which do not allow for a productive working

environment for the programmer. Fortunately, off-line programming eliminates these distractions, thereby greatly increasing productivity;
*Use of CAD systems.* The availability of CAD systems means that computers can acquire design data specifications and derive procedures to implement a working program for the robot. The cost effectiveness of using personal computers with CAD systems is a decisive factor in the popularity of off-line programming; and
*Visualization.* With the aid of computer graphics and the CAD representation of robots, machine tools and different objects, it is possible to generate an animated simulation of the robot's tasks. This simulation of positional data makes it easy to generate robotics programs that are accurate. Visualization can also aid in designing a robot's workcell layout. By experimentation, the designer can find the optimized location for the robot(s) and other equipment in the workcell.

### *Problems with off-line programming*

Off-line programming may seem to be the ideal solution for robotics applications. However, like all engineering solutions, there are associated difficulties that need to be stressed. They can be categorized as:

*Differences between simulated task and real robot operation.* This is the most serious problem associated with off-line programming and it is caused by the differences between the data stored in the data base and the actual components of the robot's cell. Robot parameters drift with time and realignment of the robot or replacement of parts can change the CAD model of the robot to some extent. There are two possible solutions to this problem: (1) Perform a periodic calibration of the robot system; and (2) use a combination of both off-line and on-line programming (i.e. allow some data to be defined on-line);
*Communication and networking.* Communication is a two-way path (Cunningham, 1988). That is, the application program developed off-line must be loaded into the robot's controller memory, and the modified program from the shop floor must be loaded back into the off-line system in order to keep the latest version and maintain consistency. If this process involves one robot system, a single communications port can handle the task. In the case of a multiple robot system, a hierarchical network based on a mainframe computer system and several layers of workstations may become necessary;
*Open architecture for the robot.* A software vendor must understand a robot's architecture to be able to write a simulation package for that robot. If the robot architecture is proprietary, the amount of information that can be released to the software vendor is always a problem. Some manufacturers provide their own simulation packages for their robots to avoid this problem; and
*Too many robotics programming languages.* Almost every robot manufacturer has its own programming language. If different robots are being used in a plant, which is very likely, there will be a burden on the programmers to be familiar with several robot languages. One solution would be to have standard

structurerd programming languages for robotics (analogous to general computer programming languages such as C, Pascal, or FORTRAN), and then postprocessors may be used to translate programs into specific robotics languages. Little work has been done on language standardization,. but it is fortunate that most of the robotics programming languages are at least based on standard computer languages, thereby reducing the learning curve.

*Elements of off-line programming*

The elements of an off-line programming system are demonstrated in Figure 2.7 which illustrates the components of a typical off-line programming station.

The off-line programming system consists of a programming environment which allows a user to develop robot programs, a CAD system and parts data base to provide task data, and simulation and computer graphics packages to facilitate simulation and visualization of the programmed task. All these components fit into a common general purpose computer system such as an IBM PC-AT at the personal computer level or a DEC VAX at the mini computer level.

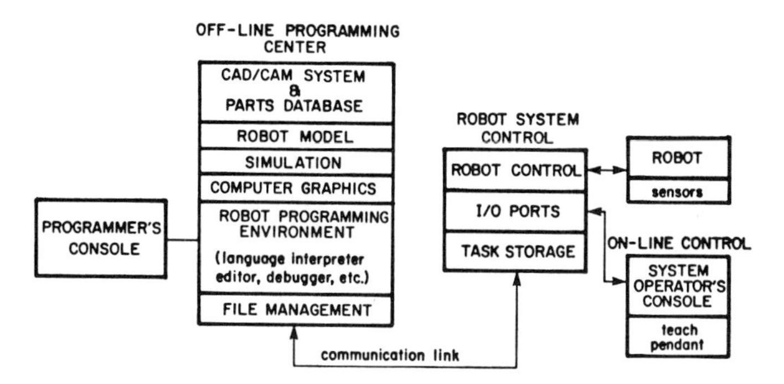


*Figure 2.7 Off-line programming system.*

There are a number of packages that provide a complete off-line programming environment such as IGRIP from GMF, Robographics from Computervision, and Positioner Layout and Cell Evaluation (PLACE) from McDonnell-Douglas. As an example the PLACE system will be explained briefly (Groover *et al.*, 1986). The PLACE graphics simulation package consists of four modules:

1. PLACE. This module is used to construct a three-dimensional model of the robot workcell in the CAD/CAM data base and to evaluate the operation of the cell;
2. BUILD. This module is used to construct models of the individual robots that might be used in the cell;
3. COMMAND. This module is used to create and debug programs off-line that would be downloaded to the robot: and
4. ADJUST. To compensate for the differences between the computer graphics model of the workcell and the actual workcell, the ADJUST module is used to calibrate the cell.

Figure 2.8 demonstrates the relationship between the different modules in the PLACE system.

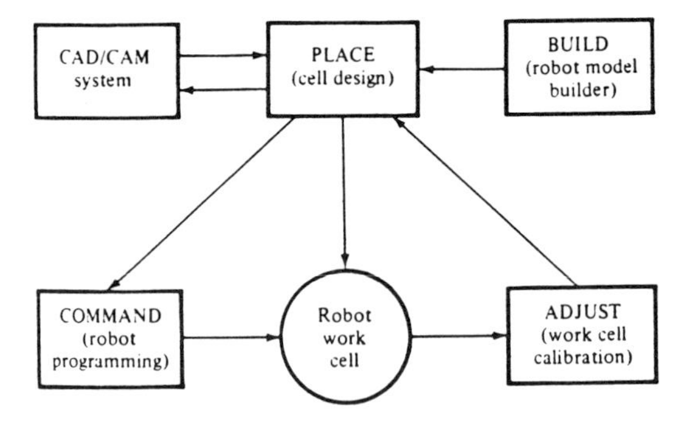


*Figure 2.8 The PLACE System (from Groover et al.,1986).*

## Trade-offs and human factors

There are trade-offs in using the two different types of teaching methods. The on-line methods of using a teach pendant and lead-through teaching are both simple, easy to use and efficient for simple tasks. Lead-through teaching, however, is limited to simple continuous path tasks such as spray painting, whereas teach pendant programming can include more complicated tasks such as pick-and-place, machine loading/unloading and spot welding. Using a teach pendant provides a more accurate point-to-point programming of the robot than lead-through teaching, but on the other hand, the operator must have some understanding of the basics of computer programming.

Although lead-through and pendant teaching are efficient methods for small or simple tasks, there is a growing number of applications that require complex commands such as branching and subroutines. These kinds of commands are usually embedded in a high-level programming language. Consequently, programming in a high-level language has emerged as another method in on-line programming. Obviously, this method of programming is more demanding and more advanced computer programming skills are required.

Off-line programming provides an alternative method in robot programming. The programmer uses a computer system other than the robot controller to program a robotics task. The programmer has a number of programming tools at his/her disposal, such as CAD/CAM tools, knowledge-based systems, etc. It is possible to program very sophisticated tasks, and then to simulate and observe the results on a computer screen. This factor is essential in making radical changes in a robot task without being worried about the potential damages and hazards in the workcell. The power of simulation gives more confidence to the programmer and makes programming changes easier. Off-line programming requires high-level computer programming skills, but it releases the robot programmer from the hazards and physical limitations of the shop floor and provides powerful computer tools to facilitate robot programming. There is always a gap between the simulated task and the real operation of the robot. This gap may be corrected using calibration techniques and/or trial and error.

From a human factors point of view, there are three items to be considered; required skills, safety, and productivity. The required operator training increases as we move from lead-through to pendant teaching to high-level language programming to off-line programming. The concepts of computer programming and software development must be mastered for more complex robot programming methods. Safety considerations, on the other hand, are less disruptive with more sophisticated computer programming methods, because computer programming, especially the off-line mode, requires little or no direct interaction with the robot. Obviously there should be safety considerations when running the application program for the first time. One can summarize programming skill requirements for on-line programming as follows:

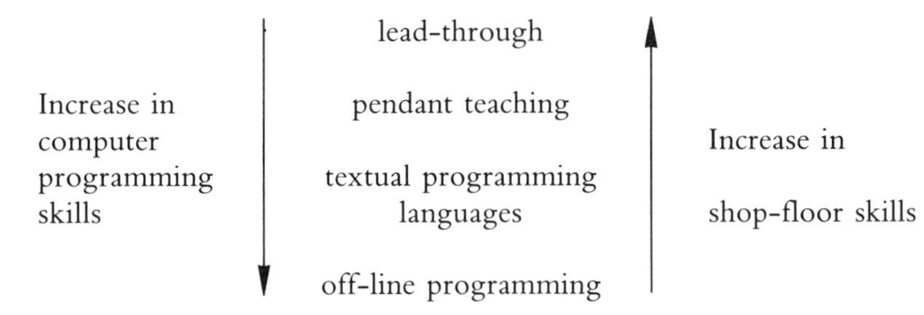


Generally, productivity may be defined as the ratio of 'output units' to 'input units'. The direct input to a programming system is the product of skill and time (skill*time) factor, and the direct output is a program or certain number of Lines Of Code (LOC). Both skill*time and LOC are quantifiable measures, but there are other factors that influence the weight of input measures. The fixed investment for programming tools and safety measures (system overhead) must be considered, and the quality of the developed code is an important factor. This quality may be measured in terms of flexibility—capability to handle different run-time conditions—, ease of program modification and code reusability.

The fixed investment (overhead) may appear as a constant factor added to the input, and the quality of code may be considered as a factor scaling the number of lines. The 'optimum quality' is application dependent. For example, in a simple pick-and-place task, pendant teaching is appropriate because it requires relatively less skill*time and low overhead. Due to the fact that the program consists of a few limited steps, it is easy to make modifications. Other quality measures are irrelevant for this simple task. A complex assembly task may involve several subtasks such as parts locating, parts handling, parts mating and parts joining. Each of these subtasks requires a sophisticated subprogram, so the overall developed program will become large and expensive. In this case, software quality becomes a very important factor. Clearly, this is a case for using an off-line programming method, one that preferably is coupled with CAD data, and has simulation capabilities.

## *Robot programming languages*

Robot programming languages are the most powerful means by which a robot programmer can express the intended operation of a robot. All of the progress in modern robot applications would not have been possible without using these languages. Due to their importance and their impact on robot teaching, this entire section will be dedicated to robot programming languages. Their purpose, characteristics and related design considerations will be discussed. Human factors considerations such as ease of use, and the effect of computer programming environment (hardware, software) will also be considered. Examples of typical robot programming languages are provided in this section.

### What is a robot programming language?

A working definition of a robot programming language as given by the NATO workshop on robot programming languages (Volz, 1988) is that: 'A robot programming language is a means by which programmers can express the intended operation of a robot and associated activities'. The 'associated activities' refers to interfacing a number of system elements together such as users, robot control systems, sensors, CAD/CAM systems, application programs, knowledge systems, and so on.

Programmability of the robot was the central idea for creating and developing robot programming languages. Use of a teach pendant is an excellent method of teaching robots fairly simple tasks such as pick-and-place. However, when the job is more complex and different responses to a number of sensors are required (decision making), or external event synchronization must be considered, the teach pendant system becomes inadequate. Using high-level structured programming languages allows the system to combine several functional capabilities such as manipulation of the robot arm, sensing the external world through sensors, limited intelligent behaviour in a programmed way, and the interaction between the robot controller and available data bases.

There are more advantages to using high-level robot programming languages, such as: (1) robots may be integrated in a manufacturing system and no longer be considered as separate devices acting entirely on their own; (2) higher levels of abstraction are possible to express robot operations such as 'put peg A in hole B', or 'open the door'; (3) handling programming changes (fix-ups) is much easier, i.e. using sensory feedback allows the programmer to consider corrections due to unpredictable factors such as uncertainty about part positions and tolerances; (4) the robot system acquires the ability to record information and to process it for later use; (5) reusability of code both within and between applications is achievable.

The central role of programmability had been recognized in university research laboratories long before the industry embraced the idea. The first robot programming languages were developed in academic research in the early 1970s. In the late 1970s, the robotics industry had the first generation of language-based robot controllers. These languages had simple control structures (IF GOTO), integer arithmetic, limited interrupt handling for applications, keyboard/CRT (cathode ray tube) input-output, and a relatively slow interpretive execution. Second generation robot languages were introduced in the mid 1980s. The main characteristics of these languages are: (1) complete control structures; (2) varied data types with full arithmetic support; (3) powerful interrupt handling with process control support; (4) full Input/Output (I/O) including teach pendant and secondary storage; and (5) faster application program execution (Ward and Stoddard, 1985).

Programming environments play an important role in the productivity and reliability of programming. Software engineers have come to realize that programming environments and programming languages go hand in hand. The same must be considered true for robot programming. As a result, different programming environments have been developed for different robot systems.

Standardization is another issue in robot programming. Currently, there are as many robot programming languages as there are companies in the robot business. If in a production line there are different robots from different companies, the task facing the programmer will be very difficult. The need for standardized languages is obvious, but several questions need to be answered before any popular standard robot programming language emerges. In the mean time, standardization will probably be achieved by product popularity, i.e. the more robot systems a company sells, the more likely its programming method is to become the *de facto* standard (Moore, 1985).

## Characteristics of robot programming languages

Second generation robot programming languages are considered here because they are currently dominant in the programming of robots. The major improvements introduced by second generation robot programming languages over the first generation ones were mainly in the areas of program control structure, sensor interaction capabilities and communication with other sytems. Let us first consider why this upgrade was necessary.

In the past, the robot was a stand-alone system performing simple tasks. However, a modern robot is usually considered as part of an integrated automated system. Consequently, the robot must interact with other controllers in its environment, i.e. gathering information through sensors or receiving/sending commands or data from/to other parts of the system. The robot controller must handle complexities of real-time manipulator control, sensory feedback, general input-output, and man-machine and machine-to-machine communications. If one compares the requirements for the programming of robot systems with those of writing operating systems, it is clear that the robot programmer needs at least the language capabilities of the operating system programmer. In other words, a flexible and powerful manipulator language should present features similar to those of a high-level computer programming language.

In general, a robot programming language should present the following capabilities: (1) a high-level programming language construct with both control and data structures; (2) motion statements; (3) process I/O; (4) user interrupt handling; (5) special data structures specific to robotics applications; and (6) communication capabilities.

## Robot programming language design

Traditionally, there have been two approaches to robotics language design. One approach is to design a dedicated programming language for robotics applications by focusing on the robot control primitives, and then adding the necessary language constructs to support the robotics tasks. The advantage of this approach is that the language may be tailored to specific robot applications so that elegant and concise robot programming is possible. A disadvantage of this approach is the need to create a complete programming environment including an operating system, a language interpreter/compiler, an editor and other tools not directly related to robotics. A further disadvantage is that the language and system are unique and put a burden on the programmer in the form of a learning curve. Another approach is to choose a general-purpose language and extend it for robotics applications which eliminates the disadvantages of the first approach. But this form of solution has not gained popularity in industry, partly because of compatibility problems with the older versions, and extra training for a new language.

There are several examples of languages specific to a special kind of robot or to a company, such as VAL II from Unimation to control PUMA robots (Shimano *et al.*, 1984), RAIL from Automatix (Automatix Inc, 1988), AML from IBM to control IBM 7565 and IBM 7535 robots (Taylor *et al.*, 1982), KAREL from GMF (GMFanuc, 1990a,b) and so on. However, there are a few examples of popular robotics languages designed strictly by extending a general purpose language. One notable example is Robot Control C Library (RCCL), an extension of the popular C language, which was designed at Purdue University (Paul and Hayward, 1986). It seems that the robot industry is more comfortable with robot specific languages, in contrast with the academic research community which is leaning more toward general purpose language extensions.

There are two main factors that influence the robotics industry to choose customized languages. First, is its ease of use: the robotics industry targets a wide range of users and does not require all of them to be experienced computer programmers. The industrial robot programming language should have flexibility to meet different levels of programming skills such as pendant teaching, simple programming employing an easy subset of the language, and sophisticated robot programming using full features of the robot language. The second factor is ‘interpretation v. compilation’. Interpretive languages are preferred because they allow for easy changes without the necessity of recompilation, and are capable of executing only parts of the program. Ease of use and interpretation are two factors common to most of the industrial robot programming languages.

An extended general purpose programming language does not provide these benefits, although it helps standardization.

## Survey of three robot programming languages

This section is intended for those who are familiar with computer programming languages.

In this section the industrial programming languages KAREL (developed by GMF Robotics), AML (developed by IBM) and an academic based language, AL (developed at Stanford University), are explained briefly. Due to the fact that many industrial robot programming languages are influenced by the features introduced in AL, this section will start by introducing it.

### *AL*

AL is a manipulator programming language developed at Stanford University in the 1970s. The importance of this language is that it paved the way for the development of later industrial robot programming languages. Portions of the material presented in this section are reprinted from Goldman (1985) with permission.

#### AL programming system

In the AL User's Manual (Mujtaba and Goldman, 1981), AL is introduced as a compiled language, i.e. a language that needs to be converted to a lower level language for execution by a compiler. Later it was reported (Goldman, 1985) that a new programming environment had been developed for AL which includes a syntax-directed editor, an interpreter, a debugger and other related programming tools. The editor is the main interface between the user and the AL system. The user's AL program is entered and modified by the editor. It is also through the editor that the user communicates to the debugger and then to the interpreter. The editor is syntax-directed and guarantees programs to be syntactically correct (though the programs may be incomplete). It is through the debugging facilities of the editor that the user can run the program, try out various statements, examine and modify variable values and set and clear breakpoints.

#### AL programming language

AL is an ALGOL-like language extended to handle the problems of manipulator control. These extensions include primitive data types to describe objects in a three-dimensional world, statements to specify the real-time motion of various devices and the ability to execute statements in parallel. In this section a brief description of the AL language is provided. For a more complete treatment see Mujtaba and Goldman (1981).

*Basic data types.* The basic data types in AL were chosen to facilitate working in the three dimensions of the real world. Scalars are floating point numbers; they correspond to reals in other programming languages. Vectors are a 3-tuple specifying $(X,Y,Z)$ values, which represent either a translation or a location with respect to some coordinate system. Rotations are a $3\times3$ matrix representing either an orientation or a rotation about an axis. A rotation, or 'rot', is constructed from a vector specifying the axis of rotation, and a scalar giving the angle of rotation. Frames are used to represent local coordinate systems. They consist of a vector specifying the location of the origin and a rotation specifying the orientation of the axes. 'Transes' are used to transform frames and vectors from one coordinate system to another, and like frames they consist of a vector and a rotation. Events are used to coordinate processes run in parallel. Arrays are available for all of the above data types. Detailed explanation of the concepts of frames and transformations may be found in Paul (1982, Chapter 1).

AL allows physical dimensions to be associated with variables. Some of the known dimensions include time, distance, angle, force, torque, velocity and angular velocity. New dimensions may be defined if desired. When parsing a program the system will check for consistent usage of dimensioned quantities in expressions and assignments.

*Basic control structures.* As mentioned above, AL is an extension of ALGOL. As such, it is block-structured: a program consists of a sequence of statements between a BEGIN-END pair, and variables must be declared before their use. AL has the traditional ALGOL control structures. These include

```
IF ⟨boolean condition⟩ THEN ⟨statement_1⟩ ELSE ⟨statement_2⟩
FOR ⟨scalar variable⟩←⟨exp⟩ STEP ⟨exp⟩ UNTIL ⟨exp⟩ DO ⟨statement⟩
WHILE ⟨boolean condition⟩ DO ⟨statement⟩
DO ⟨statement⟩ UNTIL ⟨boolean condition⟩
```

along with procedures, CASE statements and a statement to pause for a given time duration. Other familiar features include comments and macros.

*Input/Output.* AL has some input/output facilities. There is a PRINT statement, which can accommodate all the standard data types. The program can request the input as either a scalar number or a boolean value. It is also possible to have the program wait until the user enters a prompt character.

*Affixment.* The spatial relationships between the various features of an object may be modelled by use of the AFFIX statement. AFFIX takes two frames and establishes a transformation between them. Whenever either of them is subsequently changed, the other will be updated to maintain the relationship that existed between them when they were affixed. Thus when an object is moved, all of its features will move with it. Variations of the affixment statement allow the value of the trans defining the relationship to be stated explicitly, or computed from the frames' current positions. The relationship can be broken by the UNFIX statement.

*Motion control.* There is, naturally, a variety of statements dealing with the control of external devices (e.g. manipulators). These statements can specify the device to be controlled, possibly along with a number of modifying clauses describing the desired motion. The MOVE statement is used to move an arm. Possible clauses include the destination location (this must be specified), any intermediate (VIA) points the motion should pass through, the speed with which the motion will take place, any forces to be applied, the stiffness characteristics of the arm, various conditions to monitor, and a host of other variables. The OPEN and CLOSE statements apply to grippers. Other devices (e.g. an electric socket driver) are controlled with the OPERATE statement. There is also a special CENTER statement for grasping objects which causes both the hand and arm to move together. The STOP statement allows a motion to be terminated before it has been completed.

*A palletizing example in AL.* In the following example, the arm picks up castings from one place and puts them on a pallet in six rows of four (Figure 2.9). The castings come in batches of 50, but it is not known ahead of time how many batches there will be. The program is easy to read and comments are provided.

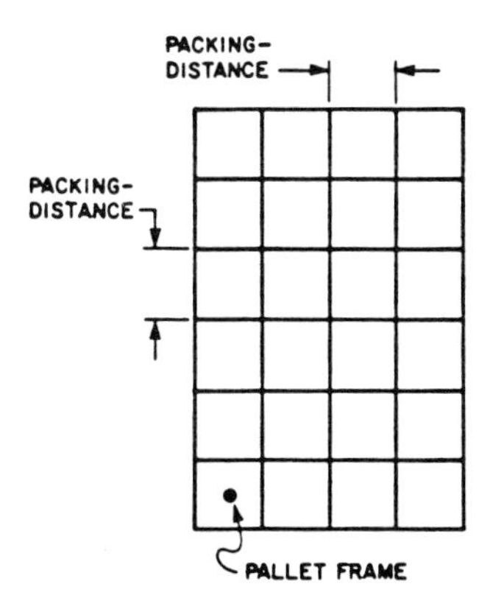


*Figure 2.9 Pallet of the AL palletizing example.*

```
Example—AL palletizing program
  BEGIN "sort casting"
    {variable declarations}
    FRAME pickup,garbage_bin,pallet;
    SCALAR pallet_row,pallet_column;
    DISTANCE SCALAR packing_distance;
    SCALAR ok,more_batches,casting_number;

    {initialization}
    packing_distance←4*inches;
    pallet_row←1; pallet_column←0;
    casting←pickup;

    {main routine}
    OPEN hand TO 3*inches;
    MOVE arm TO pickup DIRECTLY;
    CENTER arm;
```

```
{if the fingers get closer than 1.5 inches together, there is no batch}
IF (hand<1.5*inches) THEN more_batches←FALSE
                      ELSE more_batches←TRUE;
WHILE more-batches DO
  BEGIN "sort 50 castings"
  FOR casting_number←1 STEP 1 UNTIL 50 DO
    BEGIN "sort castings in hand"
    ok←FALSE;
    AFFIX casting TO arm RIGIDLY;
    {raise the casting up three inches in the z direction.}
     {zhat is the direction in the world coordinate system.}
    MOVE casting TO pickup + 3*zhat*inches
    IF pallet_column = 4
      THEN BEGIN pallet_column←0;
                 pallet_row←pallet_row + 1;
           END
      ELSE pallet_column←pallet_column + 1;

    MOVE casting TO pallet
      VECTOR (pallet_column*packing_distance,
              pallet_row*packing_distance,0*inches)
      {vector represents current position's offset with
       respect to pallet FRAME}
      WITH APPROACH = 3*zhat*inches;
      {move the final three inches in the z direction in
       the approach mode (lower speed)}
    UNFIX casting FROM arm;
    OPEN hand TO 3*inches;
    IF (pallet_column = 4) AND (pallet_row = 6)
      THEN BEGIN "pallet full"
      pallet_column←0; pallet_row←1;

      {code to remove this pallet and get a new pallet}

           END "pallet full"

    MOVE arm TO pickup;
    casting←pickup;
    CENTER arm;
    END "sort casting in hand"
  IF (hand<1.5*inches) THEN more_batches←FALSE;
  END "sort 50 castings"

 MOVE arm TO park;
END "sort castings";
```

### *KAREL*

#### KAREL programming system

KAREL is the language used in the KAREL programming system to control robots and was developed by GMF Robotics (Ward, 1985; GMFanuc, 1990a,b). The KAREL development environment consists of a manipulator, a controller and system software. KAREL can direct robot motions, control and communicate with related equipment and interact with an operator.

The KAREL system software includes a language translator, motion controlled routines to calculate trajectories, a file system which is used for file manipulation in secondary storage, a syntax-directed editor which provides the means for editing new or existing KAREL files, system variables that deal with permanently defined parameters declared as part of the KAREL system software, software to control the CRT/KeyBoard or teach pendant screens, and the KAREL Command Language (KCL). This software system is designed to provide a program development environment. Programs are edited using a line-oriented editor that is familiar with the KAREL language syntax. This tool provides commands for program entry and insertion, deletion and replacement of text. During text entry the editor can optionally check for syntax errors within a program. The KAREL system uses a translator–interpreter approach which is a compromise between the compiled and interpretive approaches. The programs must first be translated to an intermediate pseudo-code (a one-pass procedure), then the pseudo-code is ready for execution by the interpreter.

The KAREL Command Language (KCL), used for on-line program development, is the operational language of a KAREL controller. KCL on-line programming facilities include commands to operate the on-line editor and translator, a family of commands to permit the operator to run, abort, pause and resume programs, along with the possibility of single-stepping through a program and setting break-points. KCL has commands for data manipulation, file manipulation, positional teaching, diagnostics, utilities, help, status displays and robot calibration.

#### KAREL programming language

KAREL is a structured high-level programming language based on Pascal. The language is extended to include robotics data structures and robotics general functions.

*Data types.* Besides the usual data types (integer, real boolean, string), KAREL supports structured data types for robotics applications. These data types are positional data, paths and vectors.

Positional data are used to define the location and orientation of an object or the robot's Tool Centre Point (TCP). A tool is attached to the robot's end effector. In KAREL, positional data are presented as a location in 3-D cartesian space $(x,y,z)$ and an orientation with angles $(w,p,r)$, where $w$, $p$ and $r$ are the yaw, pitch, and roll values of orientation (in degrees) with respect to the user reference. In Figure 2.10, $d$ represents the direction of the robot's end-effector and may be obtained by rotating $\boldsymbol{k}$ (unit vector in the $z$ direction) in three directions:

$$d = \text{Rotate}(z,r).\text{Rotate}(y,p).\text{Rotate}(x,w)\ .\ \boldsymbol{k}.$$

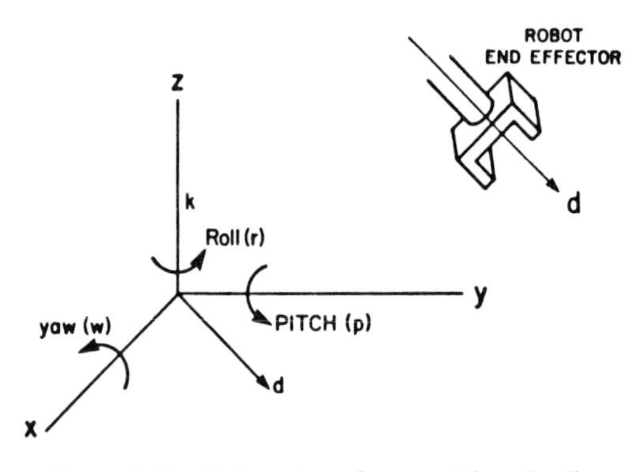


*Figure 2.10 Orientation of yaw, pitch and roll.*

Paths are arrays of positions with each element of the path being one positional datum. Vectors are composed of three elements. The numeric values of these elements are similar to the three-component cartesian vectors used in mathematics. Here are some examples of the structured data types.

Example—use of positional data

```
VAR
  next-pos: POSITION
BEGIN
  next-pos = POS (100,-400.25,0.5,10,-45,30,'n')
  MOVE TO next-pos
END
```

Example—use of path data

```
MOVE TO main-path [1]   {one point only}
MOVE ALONG main-path   {traverse the whole path}
```

Example—use of vector data

```
pos-offset = VEC (100.5,-200,30)
MOVE RELATIVE pos-offset
```

*Basic control structures.* A program consists of a sequence of statements between a BEGIN-END pair. Variables must be declared before their use. Most Pascal-like programming constructs such as FOR, REPEAT, WHILE, IF-THEN-ELSE, and SELECT (case/multiple conditioning) are provided.

*Input/Output (I/O).* KAREL provides both digital and analogue I/O lines for process control. The digital lines may be grouped together to form one input or output channel. The following example shows different ways to control input and output lines.

Example–control of Input/Output lines (Ward, 1985)

| | |
|---|---|
| DOUT[3] = ON | Turns on digital output 3 |
| AOUT[2] = 346 | Sets second analogue output to 346 |
| GOUT[3] = 10 | Sets value 10 on digital output lines defined as group 3 |
| IF DIN[5] THEN. . . | Tests input 5 to be ON, if ON, then performs specified function |

*Monitors and interrupts.* KAREL provides UNTIL and WHEN 'clauses' as a means for interrupting. UNTIL is used in a MOVE statement to specify a local condition handler. WHEN establishes a monitor and is used to specify a condition/action pair in a local and global condition handler. In the following example, the UNTIL clause forces a condition on the MOVE statement, and the move is terminated when either the TCP reaches far-pos or the condition in the UNTIL clause is satisfied. EVAL evaluates the following expression.

Example—using MOVE and UNTIL (local)

```
WRITE ('Enter force scale: ')
READ (f_scale)
MOVE TO far_pos,
  UNTIL AIN [force]>EVAL (10 * f_scale)
ENDMOVE
```

The next example uses a local condition handler to signal the robot to execute the routine start_weld 200 ms before reaching node[2] in the path 'main-path'.

Example—using WHEN (local)

```
MOVE ALONG main-path,
  WHEN TIME 200 BEFORER NODE[2] DO
  start-weld
ENDMOVE
```

The example below enables a global monitor. If the monitor turns true (the digital input safety gate turns on) anywhere in the program execution, then the interrupt to shutdown the system will occur.

Example—using WHEN (global)

```
WHEN [1] DIN[Safety-gate] DO shutdown
```

KAREL has a large set of built-in functions not discussed in this section. Interested readers may refer to GMFanuc (1990a).

### *AML*

The experience of the Automation Research Project at the IBM research facility in Yorktown Heights, New York, led to the creation of a second generation research robot system. This robotic system included an interactive programming environment and a programming language called AML (A Manufacturing Language). This system is used to control the IBM 7565 robot, a cartesian robot with a maximum configuration of six joints plus a gripper. Joints one, two and three are linear, while joints four, five and six are rotational. The gripper is also considered linear in terms of the opening between its fingers (see Figure 2.11). A subset of AML is available on IBM personal computers for programming the IBM 7535 Manufacturing System Robot. The central idea in designing AML was to provide a powerful language with simple subsets, so that programmers with a wide range of experience can use it. An interpreter implements the base language.

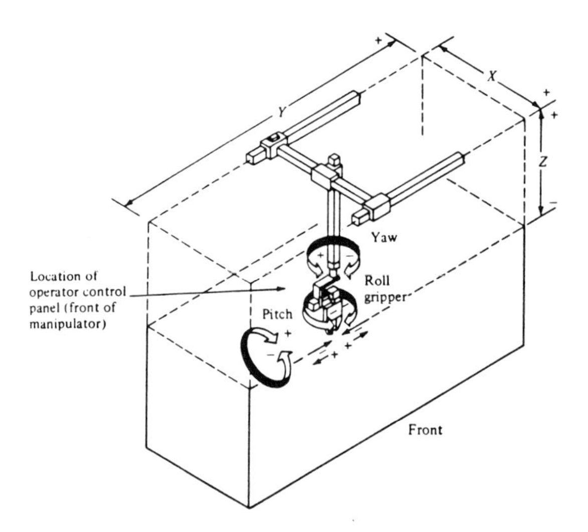


*Figure 2.11 The IBM 7565 robot (courtesy of International Business Machines).*

#### AML programming language

AML is a structured robot programming language designed with the philosophy of 'functional transparency', in the sense that AML subroutines can be written and then executed exactly like built-in system commands. This property allows for the development of functional hierarchies for different user classes. It is an expression-oriented rather than a statement-oriented language. This means that every legal construct of the language produces a value that can be used as part of some other expression. In this section a brief description of the AML language is provided.

*Data types.* The basic data types are divided into scalars and aggregates. Scalars include integer, real and string data types. An aggregate is an ordered set whose elements can include scalars and/or other aggregates. Examples of aggregates are (Taylor *et al.*, 1982):

```
⟨1, 2, 3, 4⟩
⟨'FEEDER_3', 455,⟨5,25,14.75,2.375⟩⟩
⟨'A', 'B', 'C'⟩
⟨ ⟩                     null (empty) aggregate
```

Sample aggregate operations are:

```
⟨1, 2, 3⟩#⟨4,⟨5,6⟩⟩ produces ⟨1,2,3,4,⟨5,6⟩⟩

3 of 'X' evaluates to ⟨'X', 'X', 'X'⟩
If xx has the associated value of ⟨1,⟨2,3⟩,⟨4,5,6⟩⟩
Then:  xx(1)    evaluates to 1
       xx(2)    evaluates to ⟨2,3⟩
       xx(3,2)  evaluates to 5
```

*Basic control structures.* The control structure of AML includes the normal constructs of structured programming. The IF-THEN-ELSE, WHILE-DO and REPEAT-UNTIL constructs are provided. Also compound expressions can appear between a BEGIN-END pair. In the following example, a user function (subroutine), ASK_YESNO, is inserted into the IF condition. Due to the expression-oriented nature of AML, the function ASK_YESNO will be executed as soon as it is encountered by the interpreter and returns a 'true' or 'false' value. The other function, CALIBRATE_PALLET_COORDINATES, is also a user-defined function.

Example—compound expressions in a BEGIN-END pair

```
IF ASK_YESNO ('Calibrate the pallets?','YES') THEN
  BEGIN
  CALIBRATE_PALLET_COORDINATES (PALLET_1);
  CALIBRATE_PALLET_COORDINATES (PALLET_2);
  END;
```

Subroutines may be used or called from within any expression. Although all subroutines are called the same way, they may exist either as user-written AML code or as primitive elements of the base system.

Example—calling system subroutines:

```
MOVE (GRIPPER,3.0)        --open gripper three inches
SENSIO (P1_SHUTTLE,1)     --start shuttle out
FREEZE                    --freeze robot motion
```

Example—calling user-defined subroutines:

```
ASK_YESNO ('Calibrate the pallets?','YES')

--sends 'Calibrate the pallets?' message to the screen and
  if the user responds 'YES', it returns 'TRUE', otherwise
  it returns false.

CALIBRATE_PALLET_COORDINATES (PALLET_1)
```

Example—subroutine to advance an index (Taylor, 1982). Again we go to the pallet example (Figure 2.12). Here is a subroutine to advance the pallet index and calculate the coordinates of the next pallet position (pallet goal). P is a data structure of type PALLET—user defined—which has the structure:

```
PALLET: ⟨⟨rowindex,colindex⟩,⟨maxrow,maxcol⟩,
        ⟨rowspace,colspace⟩,⟨pallet coords⟩,⟨pallet goal⟩⟩
```

The subroutine is as follows:

```
INDEX_PALLET: SUBR(!P)              --! for pass by reference
  IF P(1,1) LT P(2,1) THEN          --is current row full?
     P(1,1) = P(1,1) + 1            --yes, increment row index
  ELSE IF P(1,2) LT P(2,2) THEN --row full, another column
                                      to go?
  P(1) = ⟨1,P⟨1,2⟩ + 1⟩             --yes, reset row, increment
                                      column

       ELSE RETURN ('EXHAUSTED');

  --Here, P(1) = ⟨current row index, current column index⟩
  --Compute an updated coordinate for PALLET_GOAL value by
  --multiplying the index by distance, and adding the
  --coordinates of the pallet.

    P(5) = (P(1) – 1)*P(3)#⟨0⟩ + P(4);
    RETURN ('OK');                  --return success
    END;
```

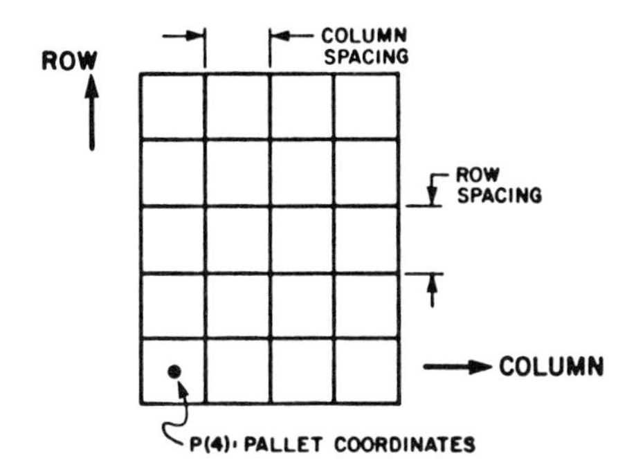


*Figure 2.12 Pallet of the AML palletizing example.*

Example—calling a user-made subroutine (in this case, INDEX_PALLET)

```
IF INDEX_PALLET (PALLET_2) EQ 'EXHAUSTED' THEN
    {statements here}
    END;
```

This subroutine is used as an IF condition. It will either return 'OK' or 'EXHAUSTED'. In any case, the INDEX_PALLET will be executed and the indices and coordinates will be updated.

*AML commands.* AML commands are simply predefined subroutines that define the semantic functions for robotics, mathematical calculations, I/O, and so on. No syntactic distinction is made between these routines and any other subroutine in the system. AML commands can be classified into several categories: fundamental routines, calculational subroutines, robot and sensor I/O commands, system interface commands and data processing commands. Here, we only discuss the robot and sensor I/O commands. Other commands are similar to what were introduced for the other languages. The robot and sensor I/O command group contains robot motion commands, robot state control commands and I/O commands. Examples of robot motion commands are MOVE, AMOVE and DMOVE. The MOVE command returns control once the motion is completed. The AMOVE command returns control as soon as the motion is started, thus allowing the user to overlap computation with motion. The DMOVE command is used to move relative to the current position. The syntax for all three move commands is similar and, except for the first two arguments, the rest of the arguments are optional. The syntax for the MOVE command is shown below:

MOVE (joints, goals,monits,⟨spd,accel,decel,settle⟩)

The IBM 7565 robot is cartesian, and joints one, two and three correspond to the *x*,*y* and *z* coordinates respectively. Joints four, five and six are rotational and define the roll, pitch, and yaw angles of the wrist. Examples are:

```
MOVE (1,10.2);              --move joint 1 to 10.2 (x direction)
MOVE (⟨1,2⟩,0);             --move joints 1 and 2 to 0.0
MOVE (⟨4,5⟩,⟨30,90⟩);       --move joints ⟨4,5⟩ to ⟨30,90⟩ degrees
```

Speed, acceleration, and deceleration may be controlled inside the MOVE command, or with separate commands (SPEED, ACCEL, DECEL). A set of sensor monitor numbers (monit) may be specified with the motion command. If any of these monitors trip before motion is complete, the trajectory is halted at that point and the motion enters its settling phase (exhausting the required time). An optional settling time can control the motion command.

Robot state control commands include STARTUP, SHUTDOWN and FREEZE. The DEFIO command defines logical input or output channels that may be sensed by SENSIO commands or monitored by the MONITOR command periodically for program interruption.

## *Summary*

In the last section, different philosophies regarding languages were discussed, and three specific languages were described briefly. Obviously, one major issue is the languages' ability to cover the main functions required by robotics applications. Another equally important issue is the ease of use from the user's point of view. This issue takes both the programming environment and the programming language into account.

AL is an academic product. It provides an interactive programming environment and a language powerful enough for most robotics applications and research. AL is based on ALGOL and is extended for robotics tasks. Although the language has numerous capabilities and has greatly influenced most of the languages that followed, it is not easy to master.

KAREL is an industrial language. It is part of a complete robot control system that facilitates operator activities as well as programming efforts. KAREL provides an interactive programming environment to develop robotics programs. The KAREL language is a Pascal-like language which is intended for programmers with different levels of skills. It is possible to work with only a simple subset of the complete language if desired. AML is a language used for assembly tasks. The designers first developed a new programming language and then extended it to cover robotics programming requirements. The main philosophy of AML is a functional transparency. There is no syntactic difference between system functions and user-oriented functions. This allows for a hierarchically functional structure which facilitates programmers with different levels of knowledge and skills. In brief, experienced programmers can use the rich language constructs along with system functions, while the non-experienced programmers may use user-defined functions and a subset of language constructs.

There are characteristics common to all second generation robot programming languages including necessary data types, standard language constructs and sufficient I/O handling. These languages are all suitable for programming complex tasks. Table 2.1 compares the specific characteristics of the three languages presented in this section. A more comprehensive comparison of a broad range of languages is presented in Bonner and Shin (1982).

*Table 2.1 A comparison of three programming languages.*

| | Readability[a] | Ease of extension[b] | Range of users[c] | Programming environment[d] | Off-line support |
|---|---|---|---|---|---|
| KAREL | 1 | 1 | 1 | 1 | yes |
| AML | 3 | 1 | 1 | 1 | yes |
| AL | 2 | 2 | 2 | 1 | no |

[a]Readability. Second generation languages read very much like English but the degree of understandability is different: 1, Pascal-like understandability; 2, Pascal-like but includes a number of language dependent functions; 3, contains a number of implicit calls.
[b]Ease of extension: 1, user functions are treated exactly like system functions; 2, user subroutines are easy to add; 3, limitations on user subroutines.
[c]Range of users: machine operators, maintenance personnel, application package users, specific application programmers, application package writers: 1, all classes of users can use the language; 2, restricted to some user classes.
[d]Programming environment: 1, includes all the necessary tools that construct a programming environment such as editor, interpreter and debugger.

One last note about safety should be mentioned here. It has been shown that a majority of accidents occur during teaching, testing and maintenance (Helander, 1988). Using computer programming, especially in the off-line mode, is safer because programming is done away from the robot. Furthermore, if a simulator is available most of the dangerous moves may be detected before testing, making the testing process safer.

There are many safety measures to be considered during the automatic mode as well. Two important categories are guarding and safety sensor systems. Guarding includes fencing, physical stops, interlock guard and trip devices that are installed in the robotics cell environment. Safety sensor systems may include 'approaching robot' and 'robot runaway' detections.

Since reliability is the key feature in any safety system, dependence on software may only be allowed in non-critical safety measures; e.g. software should not be used as the only means to detect intruders very near the robot. In the case of safety sensor systems, software may be used to detect 'perimeter penetration around the workstation', or 'intruder detection within the workstation'. In both cases a warning signal may be activated. Software must not be used to detect intruders very near the robot. Run-time variable checking is also possible in software in order to activate an alarm or stop the robot in the case of out of boundary conditions. Examples are joint values, end-effector position and weight of the load. There are statements or commands in most robot programming languages that enable the system to check the variables periodically, and to react to undesirable situations. In the discussed languages, the following statements are provided;

| | |
|---|---|
| AL: | ON ⟨condition⟩ DO ⟨statement⟩ |
| KAREL: | WHEN ⟨condition⟩ DO ⟨statement⟩ |
| AML: | MONITOR command |

The three statements operate in the same way. They monitor a sensor periodically, and if the condition turns 'true', the program will be interrupted to execute the DO part.

In short, there are three sources for the implementation of safety measures: the system hardware; the robot operating system; and the application software. The emergency and critical safety measures must be very simple and provided through independent hardware. Some basic constraints such as 'approaching the robot' and singularity conditions may be checked by the robot operating system. The application software can also take care of some non-critical application-dependent constraints such as weight of load or zone identification. It is the responsibility of the software engineer to analyse the safety problem, identify the software implementable safety measures, and verify the correctness and reliability of these measures.

## References


Automatix Inc, 1988, RAIL programming language. (Billerica, Massachusetts: Automatix).

Bonner, S. and Shin, K. G., 1982, A comparative study of robot languages, *IEEE Computer*, December, 82–96.

Boren, R. R., 1985, Graphics simulation and programming for robotics workcell design. *Robotics Age*, August, 30–3.

Buckley, S. J. and Collins, G. F., 1985, A structured robot programming language. In Nof, S. Y. (Ed.) *Handbook of Industrial Robotics* (New York: Wiley).

Carter, S., 1987, Off-line robot programming: the state-of-the-art. *The Industrial Robot*, December, 213–5.

Cooke, I. E. and Heys, J. D., 1985, Flexible programming of the reflex robot. *The Industrial Robot*, June, 117–9.

Cunningham, C., 1988, On the move to off-line programming. *Manufacturing Engineering*, October, 77–9.

Deisenroth, M. P., 1985, Robot teaching. In Nof, S. Y. (Ed.) *Handbook of Industrial Robotics* (New York: Wiley), pp. 352–65.

GMFanuc, 1990a, *KAREL Reference Manual* (Auburn Hills, Michigan: GMF Robotics).

GMFanuc, 1990b, Enhanced KAREL Operational Manual (Auburn Hills, Michigan: GMF Robotics).

Goldman, R., 1985, *Design of an Interactive Manipulator Programming Environment* (Ann Arbor, Michigan: UMI Research Press).

Groover, M. P., Weiss, M., Nagel, R. N. and Odrey, N. G., 1986, *INDUSTRIAL ROBOTICS: Technology, Programming, and Applications, Part 3 (Robot Programming and Languages)* (New York: McGraw-Hill).

Helander, M. G., 1988, Ergonomics, workplace design. In Dorf, R. (Ed.) *International Encyclopedia of Robotics Applications and Automation* (New York: Wiley), pp. 477–87.

Koren, Y., 1986, *Robotics for Engineers*, Chapter 7 (Programming) (New York: McGraw-Hill), pp. 209–27.

Larson, T. M. and Coppola, A., 1984, Flexible language and control system eases robot programming. *Electronics* , June 14, 156–9.

Mason, J. E., 1986, Designing the robot teach pendant. *Robotics Engineering*, November, 23–5.

Moore, G., 1985, Robot programming: the language of labour? *IEE Electronics and Power*, July, 499–502.

Mujtaba, S. and Goldman, R., 1981, *AL User's Manual*, 3rd Edn (Stanford University: Stanford Artificial Intelligence Laboratory).

Paul, R. P., 1982, *Robot Manipulators: Mathematics, Programming, and Control* (Cambridge, Massachusetts: MIT Press).

Paul, R. P. and Hayward, V., 1986, Robot manipulation control under UNIX RCCL: a robot control 'C' library. *The International Journal of Robotics Research,* **5,** 4, 94–111.

Pressman, R. S., 1987, *Software Engineering: A Practitioner's Approach*, 2nd Edn (New York: McGraw-Hill), pp. 50–3.

Sandhu, H. S. and Shildt, H., 1985, ROBOTALK: a new language to control the rhino robot. *Robotics Age*, September, 15–19.

Schreiber, R. R., 1984, How to teach a robot. *Robotics Today*, June.

Shamash, Y., Yang, Y. and Roth, Z., 1988, Teaching a robot. In Dorf, R. (Ed.) *International Encyclopedia of Robotics Applications and Automation* (New York: Wiley), pp. 1689–1701.

Shimano, B. E., Geschke, C. C. and Spalding III, C. H., 1984, VAL II: A New Robot Control System for Automatic Manufacturing, IEEE 1984 International Conference on Robotics and Automation (Silver Spring, Maryland: IEEE Computer Society Press), pp. 278–91.

Shroer, B. J. and Teoh, W., 1986, A graphical simulation tool with off-line robot programming. *Simulation*, August, 63–7.

Snyder, W. E., 1985, INDUSTRIAL ROBOTS: Computer Interfacing and Control. In *Robot Programming Languages* (Englewood Cliffs, N.J.: Prentice-Hall), pp. 200–314.

Taylor, R. H., Summers, P. D. and Meyer, J. M., 1982, AML: a manufacturing language. *The International Journal of Robotics Research,* **1,** 3, 19–41.

Volz, R. A., 1988, Report of the Robot Programming Language Workshop Group: NATO Workshop on Robot Programming Languages. *IEEE Journal of Robotics and Automation,* **4,** 1, 86–90.

Ward, M. R. and Stoddard, K. A., 1985, KAREL: a programming language for the factory floor. *Robotics Age*, September, 10–14.

# Chapter 3
# Human control of robot motion: orientation, perception and compatibility

**S.V. Gray, J. R. Wilson and C. S. Syan**

*Institute for Occupational Ergonomics, Department of Production Engineering and Production Management, University of Nottingham, Nottingham, NG7 2RD UK*

**Abstract.** This chapter focuses on the effectiveness and reliability of hand-held teach pendants for controlling robot movements. Though there is a shift in emphasis toward remote off-line programming, the current hand-held teach pendants appear to remain a dominant device for robot teach programming. A review of the human factors issues of operator perception and cognition of robot movements is presented. Robot safety, on-line displays and robot movement characteristics have been studied. A series of experiments has been conducted to study operators' understanding of robot arm configuration, orientation and compatibility between teach controls and robot movements. The results indicate that the ability to comprehend changes in robot status is largely governed by the perceptual orientation of the operator. Some pendant control design recommendations are suggested.

## Introduction

The introduction of robots to many industries has led to improved product quality, increased productivity and the removal of people from hazardous or monotonous work. However, it has become apparent that robots themselves present a new type of hazard for personnel unfamiliar with their movement flexibility and speed of motion. This has led to strict regulations on guarding to protect workers from being struck by the robot during its normal operation. Unfortunately these regulations do not resolve the problem in specific situations which may require or benefit from the operator being in close proximity to the robot arm whilst it is moving (i.e. during programming or maintenance operations). Both of these tasks may involve control of the robot arm via a remote hand-held control device known as a teach pendant.

Although the use of the teach pendant necessarily enforces speed restrictions on robot motion, there still exists the likely hazard whereby errors in control input using the teach pendant could cause unexpected robot movement. This may result in damage to the robot or equipment, or worse, personal injury. At the very least, such unreliability in control reduces work effectiveness. Lack of standardization in teach pendants has allowed the development of a variety of control designs. This could increase the likelihood of control input errors being made, particularly if the operator is required to control more than one robot system.

This chapter is concerned with the control of robot movement by people using teach pendants. Moreover, although physical design criteria for teach controls and robot safety developments are germane to the chapter, we mainly examine issues relevant to human–robot interaction at the perceptual and cognitive level. The tasks undertaken and the understanding required for controlling a robot are explained. An overview of a series of experiments is presented, through which we hope to increase our knowledge of human–robot interaction in motion control.

## *Previous work*

The focus of this chapter is on the effectiveness and reliability of human control of robots and not necessarily on safety issues *per se*. However, much can be learned about how people behave around robots from the accident prevention and safety literature, as well as from previous work on teach controls.

### Robot safety

Reports on accidents involving robots are few in number, and those that are available each cover only a small number of incidents. Hence it is difficult to obtain a true picture of the overall hazards associated with the use of robots in industry. Moreover, much of the literature appears to be somewhat incestuous. The reports that exist are useful though, for highlighting the circumstances which can lead to accidents and hence may need further investigation. Information available from France, Japan, Sweden and the USA, amongst others, seems to indicate potential risk for all personnel interacting with robots. Statistics differ as to the type of work found to be most hazardous. This may be due to differences in worker-group classification, differences between countries in the nature of the interaction involved for worker groups or in safety policies, or differences in accident data assessment. Almost all reports, though, specify an actual or potential risk to those involved in programming, and specifically teaching, robots (JISHA, 1983; NIOSH, 1984; Carlsson, 1985; Nicolaisen, 1985; Sugimoto and Kawaguchi, 1985; Vautrin and Deisvaldi, 1986; Jiang and Gainer, 1987).

### Programmability and safety improvements

A robot safeguard is any system or device that provides protection to personnel from exposure to hazards associated with robots (MTTA, 1982). Approaches to robot safeguarding generally fall into three categories; physical barriers, presence-sensing devices and 'intelligent' detection and control systems. Possible operational effects of invoking these systems may be described as:

1. auditory/visual warning;
2. slow robot motion;
3. freeze the robot in its current position;
4. prevent gripper movement (release of objects);
5. move the robot arm away (collision avoidance); and
6. complete system shutdown.

All these safeguards have at least some disadvantages, especially where we wish to permit close access and good visibility for teach control. Limit stops or other kinds of limitation on the machine's movement would provide safe areas. This may be sufficient for those not directly involved with the machine, e.g. management staff or visitors to the plant, but it generally will not be so for people intimately involved in robot operations. Current safeguarding measures can initiate shutdown of normal operation if any person enters the robot work area. Disadvantages of this are that downtime is costly and may extend to other machinery within the system, and that operation start-up may be complex and time consuming. The development of presence-sensing devices and sensor-driven intelligent robot control systems may, in the future, allow access to the work area without needing to stop robot operation. If this is so, there may be some advantage to using another system in addition which would provide the operator with advance information of robot movement.

## On-line display of robot movement

During early work at Nottingham University, attempts were made to specify and develop a predictor display of robot position and movement (Wilson *et al.*, 1988). The robot control program is accessed either directly or from a file of commands and, through an emulator, drives a graphic simulation (of greater or lesser fidelity) of the robot, its movement, and its environment. In use as a predictor, the graphics model is advanced in time from the actual robot. The form of the display—in terms of type of view, level of complexity, degree of integration of information over time and amount of time advance—should be alterable by the people who use it. They, having viewed the display, would then either work on the robot or in its environment, or would avoid the robot's sphere of movement for the moment.

The technical implications of such a system are that control programs should be accessible, converted to drive an emulator, and a suitable range of graphics built to depict these data; that there is negligible time delay in the processing involved; and that all this can be done with a system of limited cost. The human implications are that the information displayed is seen, understood and acted upon by relevant personnel, that they are motivated to use the system, feel they can rely upon it and can plan their actions accordingly.

Assessment of the system's potential for safety improvements proved inconclusive at best and so development is now to be channelled into provision of on-line graphics for program testing and proving. A rudimentary display could then perhaps be incorporated into teach pendants of the future.

Of interest in the current context of human–robot interaction will be the data from a series of human factors experiments presently being analysed. As well as addressing most of the questions which might be raised about use of predictor systems (presentation of time information, human ability at elapsed time estimation etc.), experiments also examined perceptual and comprehension issues. These included: the effect of differences in display abstraction (solid, wire etc.) or display dimensions [two-dimensional/three-dimensional (2-D/3-D)]; the extent to which observers can orient themselves and how orientation can be aided; and a look at possible mental models of robots.

From analyses conducted so far, clear evidence has been obtained that subjects comprehend a 3-D wire frame graphics display of robots much better with hidden lines removed, but that time pressure decreases performance for both hidden lines in and hidden lines out. These results might be expected but we are providing evidence on the extent of the differences in performance, useful in cost trade-offs. In addition, certain configurations cause considerable perceptual ambiguity and any such mimic display would require incorporation of further cueing information. For a plan-view only of the robot model, subjects consistently overestimate the potential floor space swept by the robot (useful for safety), and comprehension and estimation grow more accurate when a scaled grid is also provided. This display work is discussed no further here, but its findings will be integrated with those from the research described later in this chapter to feed into improved teach control systems.

## Reaction times and perception of reach envelopes

Safety regulations worldwide require that during programming and maintenance tasks the speed of robot motion must be restricted. However, the recommendation for what is regarded as a safe maximum speed is not yet standardized. In a review of current recommendations, Etherton *et al.* (1988) reported that they range between 15 cm (6 in)/s (National Safety Council, 1985) and 30 cm (12 in)/s (Van Deest, 1984). The variation in recommended motion speeds prompted these researchers to assess human reaction times to unexpected robot motion at different motion speeds. They concluded that there was no evidence to recommend changing the American National Standards Institute (ANSI) standard of 25 cm/s as overrun distances at this speed were small (a mean of 7.77 cm). In a later experiment (Etherton and Sneckenberger, 1990) they recommended that a robot speed of 25 cm/s or less would not be harmful provided that the operator was fully attentive to the robot and was within close reach of a motion disabling switch. Interestingly, they found that at higher motion speeds (30 and 45 cm/s) subjects' reaction times were faster, implying that this was a reflexive action unimpaired by the likely decision cost of unnecessarily stopping the robot which had been more evident at slower speeds.

Helander *et al.* (1987) developed a mathematical model relating robot speed and human reaction time to risk of injury. This model has enabled them to take account of psychological factors (perception and decision making) as well as motor response time.

Taking a different approach to this problem, Rahimi and Karwowski (1990) measured human judgement of that robot speed of movement they considered safe for people working immediately outside the robot danger zone. They found that, on average, subjects selected speeds of 40.7 cm/s for a large robot and 64.2 cm/s for a small robot.

If personnel are required to enter into the robot work area during its normal operation cycle, a potential hazard may arise when the robot is in a condition known as 'waiting'. This occurs when the robot remains stationary for a period of time but has not actually ended its movement cycle. It may be a conditional stop in which the robot waits for an input signal from other machinery before continuing its movements or it may be a deceptive stop in which a certain position is servo-controlled for a fixed period of time. Both cases are potentially hazardous to a person with access to the robot area during operation, as unexpected robot movement may occur. Human judgement of 'safe wait' periods has been experimentally assessed by Nagamachi (1986), who examined five discrete wait periods (0–4 s) at five motion speeds. The results showed that subjects considered it safe to pass under the robot arm when the 'wait' period was longer than 2 s. Shorter wait periods were considered more or less dangerous depending on the speed of robot motion. Karwowski (1989) examined a different aspect of this: the minimum time of robot inactivity that would be perceived by subjects as indicative of program cycle completion. The results showed a significant difference between the subjects that had previously witnessed a simulated accident (who waited an average of 20.2 s before they decided it was safe to approach the robot) compared with subjects who did not see the accident (who waited an average of 17.1 s).

Other experimental work has evaluated human perception of safe areas within the robot's work envelope. Nagamachi (1986) measured the minimum approachable distance to the robot that people considered safe. It was found that subjects ventured closer to the robot arm when movement was at a slower speed: the closest approach at 14 cm/s was 1.5 cm, whereas the median approach distance at 22 cm/s was 22.5 cm. Karwowski *et al.* (1988) found that approach distance was influenced by prior exposure to a simulated operator–robot accident. The average approach distance for those who had seen the accident was 20.9 cm, whereas for those who had not seen the accident it was 15.3 cm. The angle of approach to the robot also seemed to make a difference to the distance considered safe. Subjects ventured much closer to the robot from the front position (7.4 cm) compared to the side positions (35.7 cm and 29.9 cm). Karwowski *et al.* concluded that in general subjects tended to underestimate the true reach of the robot arm and would thus tend to 'invade' the robot's work envelope.

### Teach control and teach pendants

Recent studies have shown that robot users complain of the lack of standardization among designs of teach pendants for different robots (Cousins, 1988) and that problems may arise when operators learn to use one teach pendant design and then transfer to others (Edwards, 1984; Helander and Karwan, 1988). The studies show that teach pendants vary on a whole range of aspects, e.g. size, weight, layout, number of keys, display type, labelling and method of motion control (Parsons, 1986; Cousins, 1988). The nature of the programming task invariably requires the programmer to be in extremely close proximity to the robot arm in order to position the tool precisely. A negative transfer of training effect may produce control errors, especially in hurried or emergency situations (Helander and Karwan, 1988). An analysis of the standards on robot safety currently available shows that the recommendations for teach pendant design are extremely limited. In the UK the Health and Safety Executive (1989) has published guidelines for industrial robot safety. These refer mainly to safeguarding and operational procedures. However, it is recommended that the robot control pendant should be ergonomically designed: 'The application of ergonomic principles to the design of teach pendants can improve safety by simplifying tasks and reducing the scope for human errors'. No more definite or specific guidelines are given as to what the design should be like.

The Japanese standards (JISHA, 1985) recommend that slow speeds are used when programming, and to aid correct movement control they suggest that the direction of each axis is marked upon the robot itself and in a similar way on the teach pendant controls.

Both of these standards only briefly acknowledge the importance of teach pendant design. In the USA a comprehensive set of proposals is devoted to robot teach pendant design (ANSI/RIA, 1988). The proposed standard calls for smaller, lighter pendants using the simplest design for their functional requirements, which personnel with a minimum of training would be capable of operating. On motion control design it is recommended that labelling should be unambiguously displayed for positive and negative directions. Furthermore, the controls should be designed such that '. . . actuation of a control corresponds to the expected control–movement direction and be oriented so that the control motions are compatible with the movements of the [robot]' (Paragraph 5.3.1). However, there is no guidance as to which control design is optimal for this objective.

Some research has already been carried out to evaluate the use of different motion controls. In Canada, Ghosh and Lemay (1985) assessed the ease of use of the Unimate pushbutton teach pendant. A control task was set up which involved the use of all three modes of programming (joint, world and tool), and this was repeated 30 times. The main finding was that the subjects learnt how to use the controls correctly with practice, but that population stereotype expectations had caused some difficulty. For example, as a result of the orientation of the tool, there was a specific position where the minus 'Z' button would move the tool vertically upwards contrary to expectations.

Research from the Health and Safety Executive (1988, personal communication) showed that, in joint mode, when the operator viewed the robot from in front, the joystick pendant produced virtually no errors, compared to 25% errors using the pushbutton pendant. However, in all other human–robot orientations neither pendant design was considered satisfactory for accurate control (25%–44% errors).

These experiments have led to mixed opinions as to the most suitable design for robot motion control hardware and software. This may possibly be due to differences in experimental design and interpretation of results. A common observation though, is that population stereotype expectations and the results of changing the orientation of the robot *vis-à-vis* the operator can adversely affect use of robot motion controls.

## *Control by teaching: the system and the task*

### Hardware and control modes

Before we can start to understand the task of teach controlling robots we must be aware of the main variations in robot systems. Robots themselves can be divided into four types, classified by their physical configurations (Engleberger, 1980; Hall and Hall, 1985).

#### *Polar coordinate configuration*

Also known as 'spherical coordinate', because the workspace within which it can move its arm is a partial sphere. The robot has a rotary base and a pivot that can be used to raise and lower a telescoping arm. One of the most familiar robots, the Unimate model 2000 series, was designed around this configuration.

#### *Cylindrical coordinate configuration*

In this configuration, also known as the 'post-type', the robot body is a vertical column that swivels about a vertical axis. The arm consists of several orthogonal slides which allow the arm to be moved up or down and in or out with respect to the body.

#### *Jointed arm configuration*

The jointed arm configuration is similar in appearance to the human arm; it is also often called the anthropomorphic or articulated type. The arm consists of several straight members connected by joints which are analogous to the human shoulder, elbow and wrist. The robot arm is mounted to a base which can be rotated to provide the robot with the capacity to work within a quasi-spherical space.

#### *Cartesian coordinate configuration*

A robot which is constructed around this configuration (also called rectilinear) consists of three orthogonal slides. The three slides are parallel to the *X, Y* and

*Z* axes of the cartesian coordinate system. By appropriate movements of these slides, the robot is capable of moving its arm to any point within its 3-D rectangularly shaped workspace.

These four robot types have different numbers of linear and rotating axes. Robots in general can have up to six degrees of freedom (or types of movement). These are:

1. Vertical traverse: up and down motions of the arm, caused by pivoting the entire arm about a horizontal axis or moving the arm along a vertical slide;
2. Radial traverse: extension and retraction of the arm (in and out movement);
3. Rotational traverse: rotation about the vertical axis (right or left swivel of the robot arm);
4. Wrist pitch: up and down movement of the wrist;
5. Wrist yaw: right and left swivel of the wrist; and
6. Wrist roll: rotation of the wrist.

Remote teach pendants are used to control and subsequently to program robots by leading them through movements in one of three modes. Control can be in JOINT mode, whereby individual, specified parts of the robot are moved in a certain (rotational) direction about their pivot points. Alternatively, robots can be moved with reference to a set of cartesian coordinates, the actual parts moved depending upon the robot control program; this may be done with reference to coordinates at the robot control axis (WORLD mode) or at the gripper (TOOL mode). The different operational requirements of these three modes, in terms of understanding what the current orientation of the robot is and what movements are required to achieve the task goal, contribute in part to the complexities of using robot teach pendants.

Teach pendants vary on a large number of counts, amongst them overall size and weight, labelling and coding, control type and action. Possibly the most fundamental difference is between pendants with joysticks and pendants with buttons or keys. Work discussed earlier, for instance from HSE (personal communication), looked at differences between these pendant types in terms of human performance. Our research has also involved such comparison.

## The robot control task

At one level the task of controlling a robot, moving it along a specific path using the teach pendant, might be represented by recourse to 'standard' information processing models. Within models of the Human Information Processor (e.g. Wickens, 1984) we can see how the task requires human sensing, perception, decision making and controlled action with feedback assistance, utilizing attention and long-term memory resources and being 'controlled' through working memory. During our work we have argued the central role of perceptual processes in the task of robot control. A person's perception is based upon their knowledge, experiences and expectations, combined with current information from the environment. In our understanding of the robot movement control task, then, we have used the idea of the operators drawing upon their knowledge, and upon their representation of that knowledge, of the robot, task and environment. This is very close to the notion of operator mental models. Although the operator mental model concept has been criticized due to overuse or even abuse (see Wilson and Rutherford, 1989, for a review), it could provide a powerful unifying framework within which to represent such tasks as robot motion control. Work has been underway to develop methods with which to identify mental models of manufacturing processes such as robot systems (Wilson and Rutherford, 1988). However, many criticisms can be levelled at methods reported in the literature (see Rutherford and Wilson, 1991). This, and the still barely framed theoretical status of the notion, has meant that we have not used it here.

The task of robot control is represented in Figure 3.1. This representation integrates events in, or states of, the robot's environment with tasks (actions or plans/decisions) carried out by the operator. Human resources of knowledge and cognitive processes for manipulating and applying such knowledge will be utilized in completion of the tasks; the relevant aspects of such human knowledge and processes are drawn out in 'exploded form' in the representation. Thus all elements in the diagram connected by dotted lines with no arrows are to be regarded as within the central core of knowledge and processing capabilities.

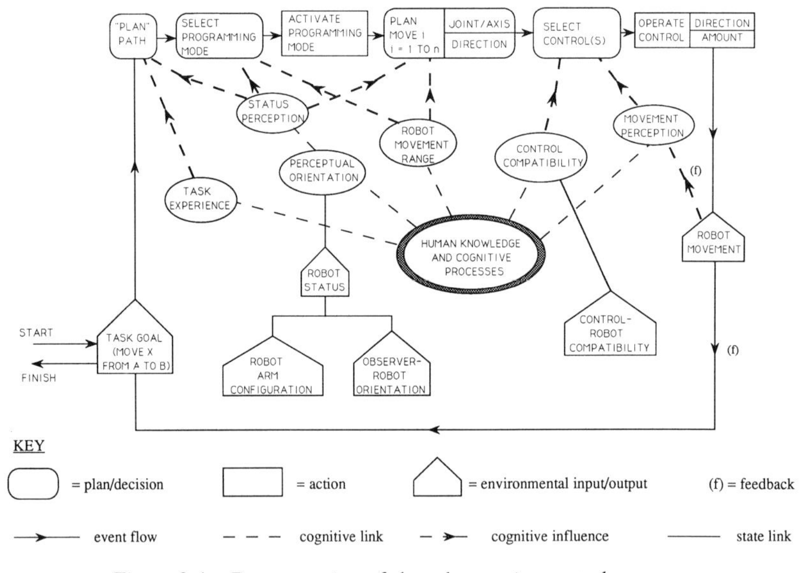


*Figure 3.1 Representation of the robot motion control process.*

The actual robot configuration and actual orientation of the observer to the robot, the robot status and *the perceptual orientation of the observer*, will influence a certain perception or comprehension of the robot's current status. Together with the knowledge of the task goal and previous task experience this will enable 'planning' of the 'best'—most efficient or easiest—movement path to achieve that goal; such a planned path may be more or less well determined at this stage. The operator must then select and activate the most appropriate programming mode, influenced again by perception of current robot status as well as task goal. Status perception and knowledge of movement limitations will then enable planning of the first move (and later, subsequent moves) in terms of the robot joint or axis and direction of movement. Appropriate control(s) are selected and a movement is made by operating the control a certain amount in a certain direction. This again is influenced by the observer's perception of control–robot compatibility and any stereotypical expectations. Knowledge of the actual robot movement will feedback via the operator's perception into control, or next move planning, or even selection of programming mode and path. Consequences of robot movement will be assessed against the task goal; this will take place periodically, continually or terminally.

To some extent our approach might be regarded as carrying out a cognitive task analysis. Rahimi and Azevedo (1990) report something similar; despite some analysis of operator errors, though, what they present is actually more of a task description. Using a representation of the task, such as Figure 3.1, which is admittedly only one possible version and is somewhat simplistic, we can nevertheless identify the key aspects of human–robot interaction which must be investigated to understand human control of robots. These are:

1. how observers identify appropriate robot parts;
2. effects of physical person–robot orientation and the observer's perceptual orientation on understanding of the robot's position, and actual and potential movements;
3. effects of actual robot configuration on the observer's perception of the robot and comprehension of its status and potential;
4. effect of control–robot movement compatibility, and perception of this, on selection and direction of operation of controls;
5. overall ability to control the robot; and
6. differences, if any, in all the above between JOINT and WORLD programming modes and between pushbutton (keys) and joystick teach controls.

It was to investigate such questions that the research described in the rest of this chapter was set up.

## *Experiments in human understanding and control of robot systems*

In the light of assumptions about the robot control task such as that represented in Figure 3.1 an experimental programme has been devised. This was divided into three general stages with several experiments in each. To some extent the experiments form a progressive series, each adding to our knowledge of the task, and allowing us both better to define our task representation and also to develop the next experiments in the series.

Experimental stage 1 evaluated naive subjects' comprehension of the range of robot movement (degrees of freedom, movement restrictions and robot state) and their ability to identify individual joints of the robot. Experimental stage 2 evaluated the subjects' expectations of control–motion compatibility and considered how these are affected by perceptual orientation. Experimental stage

3 considered the role of motion feedback in control selection, using different control designs for subjects with different perceptual orientations in both JOINT and WORLD programming modes.

A Unimate PUMA 560 (Mk 1) anthropomorphic robot was used for all the experiments. This robot was chosen because it has a wide range of movement capabilities which would enable examination of some of the more complicated perceptual problems presented to the robot programmer. The experiments were carried out over a period of 18 months and used a total of 96 undergraduate production engineering students as experimental subjects. None of the subjects had any experience in robot programming or took part in more than one experiment. Naive subjects were used for two reasons. First, it was considered that subjects would not be willing to participate in the entire research programme, necessitating a larger subject population than was available amongst experienced personnel. Second, and more importantly, it was possible to identify the perceptual problems that the task presented to naive subjects as well as their expectations of control–motion compatibility, without the influence of prior training.

In the first two experimental stages, subjects responded to predetermined robot movements and made control selections, therefore receiving no feedback from actual robot motion. This provided a good deal of information on the circumstances that produce 'misperceptions' of robot movement as well as on 'expected' control–motion relationships for two different control designs (joystick and toggle switches).

In the third experimental stage the different motion control designs were interfaced to the robot controller and this enabled examination of the effect of motion feedback on the control task. These last experiments were carried out using both the JOINT and WORLD programming modes. In the interests of clarity only the results of the JOINT mode experiments will be discussed in this chapter.

## Experimental stage 1: comprehension of robot movement

Stage 1 was aimed at assessing subjects' comprehension of the robot movement range and of changes in robot status (as defined by the robot-arm configuration and observer–robot orientation). The robot was programmed to move to specific locations denoting certain configurations (e.g. normal position, rotation and arm 'flip' changes as shown in Figure 3.2). The subject was seated directly in front of the robot, well outside the robot's working envelope. With the robot initially starting from each configuration position shown in Figure 3.2, the subject was shown a series of individual moves in JOINT mode and asked to describe each joint. It was found that the subjects used 'human-like' terms to describe the individual robot joints (e.g. body, arm, elbow, wrist) and could easily recognize the three major joints but had great difficulty in distinguishing the wrist joints. Two types of description were used for direction of joint movement: either as clockwise/anticlockwise rotation of the joint, or as left/right and up/down movement of the joint.

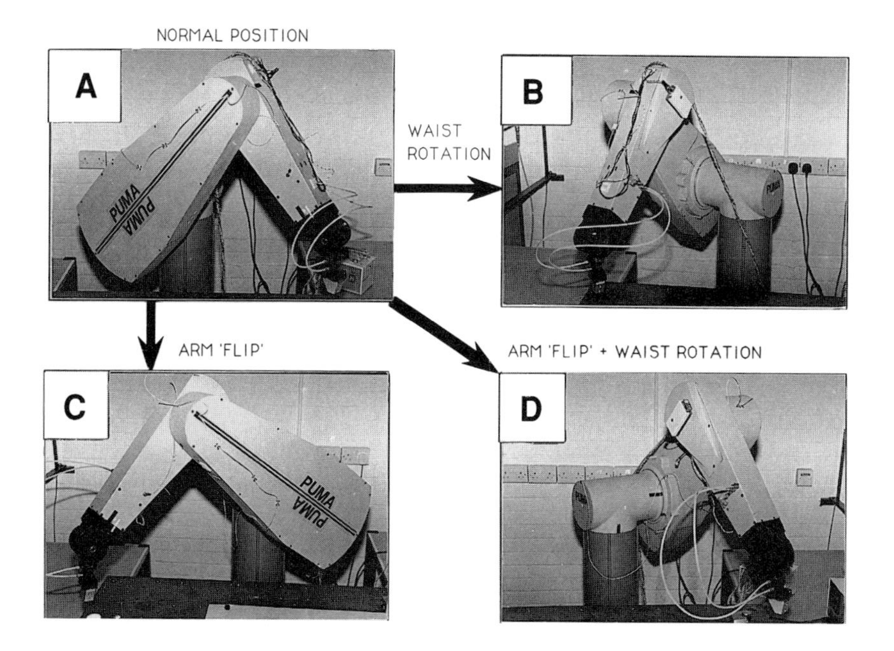


*Figure 3.2 The PUMA 560 anthropomorphic robot shown in four arm-configuration positions (A,B,C,D). The arrows indicate how each new configuration is derived from the normal position.*

Further experimentation found that changes in robot-arm configuration produced inconsistencies in motion descriptions for some subjects (Gray and Wilson, 1989). These subjects did not appear to recognize the configuration change that had occurred and consistently described robot moves 'as-currently-viewed'. For example, a clockwise rotation of joint 2 as viewed in the normal position, at location A (Figure 3.2), would be described as moving anticlockwise when viewed in the rotated configuration at location B. Other subjects were able to recognize configuration changes and described actual robot moves consistently, regardless of the arm configuration.

These early experiments indicated that subjects differ in the way in which they recognize and relate to robot movement. Some subjects use a robot-centred reference point which 'moves with' the robot when its configuration changes, others use an observer-centred perceptual reference which remains consistent to themselves regardless of changes in robot configuration or observer–robot orientation. The classification of subjects in this manner defined what we then termed their 'perceptual orientation' (see Figure 3.1).

## Experimental stage 2: control–motion compatibility

Stage 2 of the experimental programme examined the strategies used by subjects to select the controls appropriate to each robot movement. Two control designs were compared (joystick and toggle switches) in each of two programming modes (JOINT mode and WORLD mode). The experimental results showed that a strong stereotype association between perceived motion and control selection was developed by all subjects (Gray *et al.*, 1989). For example, a 'positive' control would be selected for movements described as right, forwards, up or clockwise. For the subjects using the observer-centred perceptual reference discussed above, this resulted in their selecting different controls for the same actual robot movement when the robot was viewed in different configurations.

## Experimental stage 3: control of motion

Stage 3 of the experiments examined the role of motion feedback on control selection performance. At this point the experimental method differed from the previous experiments. The robot was programmed to move to each of the start locations in Figure 3.2 a number of times. From each position, subjects were given an instruction indicating what movement of the robot they were required to initiate. To avoid any bias or ambiguity associated with the experimenter's descriptions, instructions were written and contained joint numbers (1, 2 or 3) and direction arrows (↑, ↓, ← or →) (Gray *et al.*, 1990). As noted earlier, what follows refers to JOINT mode only although findings are available for WORLD mode also. The results showed that changes in robot configuration caused initial confusion, particularly for the subjects previously tested as employing an observer-centred perceptual reference, in a similar way to the previous experiments. However, subjects learned from the motion feedback and did not make errors on subsequent moves within the same configuration (e.g. if the '2↑' move was achieved using the '+ve' control, then '3↑' would also be '+ve').

An interesting observation during this experiment was that different strategies were used by the subjects to comprehend the control–motion relationship. The subjects who had been identified as 'robot-centred' recognized when a configuration change had occurred, and then had to perform a mental transformation of the robot image to decide what actual move was required. As a result, these subjects took on average twice as long to make their control selections than the 'observer-centred' group (24.5 s and 12.8 s respectively). Some of the subjects in the observer-centred group said that they could not figure out why the control–motion relationship was constantly changing and were forced to 'guess' which direction would be appropriate. Other subjects in this group, however, realized that certain configurations caused a control–motion reversal and used cues on the robot (could they see or not see wires, screws etc.) to identify when reversals had occurred.

A significant difference was found in the performance rates with the different teach controls, with 7% more errors made using the joystick control, across all tasks ($P<0.05$). The majority of the subjects (73%) had said that they preferred using the toggle switches as it was easier to remember the control–motion relationship than with the joystick.

### Interpretations from experimental findings

The experimental work has provided some information on the perceptual/cognitive requirements needed to make correct control selections in robot motion control. This information has made it possible to identify the conditions of the control task that may lead to control errors. In general, it was found that there were two specific problem areas:

1. The observers' comprehension of current robot status: this involves recognition of the robot-arm configuration, taking into account the orientation of observer to the robot. For some observers there can be a 'misperception' of the arm movement(s) required and, in consequence, incorrect control selection; and
2. The compatibility between the teach controls and robot motion and the observers' assumptions about this relationship. For example, if the control–motion relationship is illogical, the observer may have greater difficulty in correctly selecting the control that will move the robot in the required (planned) manner.

It was found that all of the subjects expected a logical relationship between the controls and perceived motion in accordance with population stereotype associations. However, for the control of some movements, the control–motion relationship was incompatible with these expectations. For example, to move joint 1 to the 'left' or 'anticlockwise', the correct control on the Unimate pendant is '1 + ', not '1 − ' as expected. Further work will examine the control–motion relationship for other robot systems and assess how these deviate from user expectations.

The experiments indicated that the ability to comprehend changes in robot status is largely governed by the perceptual orientation of the observer. Fewer 'misperceptions' are made if the observer maintains a robot-centred reference system, and predictable misperceptions are made by observers who use observer-centred reference systems. An obvious practical solution would ensure that all observers comprehend and make use of a robot-centred perspective. This would involve some amount of training and/or cueing aids within the robot system, for example robot joint labelling or feedback of robot status via the teach controls.

## *Conclusions*

Despite some moves to off-line programming, the teach control and programming of robots and similar or related equipment will remain a vital industrial task for some considerable time. That being so, teach control will continue to throw up human factors problems and issues of considerable interest, akin to similar issues in the general realm of remote control. In this chapter we have looked at the problems associated with reliable robot control via a teach pendant. The current work suggests that improved effectiveness of performance and safety will be enabled through better consideration of cognitive factors, specifically perceptual ones. Issues such as how people orientate themselves perceptually *vis-à-vis* the robot, how they understand current robot arm configuration, and their compatibility expectations about use of the controls for certain robot movements must be examined. By doing so, and by identifying ways in which a programmer's knowledge and understanding can be improved or aided, ergonomists can make an important contribution to robot control system development.

Our research into such perceptual issues and into mental models of robots is continuing and, together with engineering colleagues, the human factors group will be applying their findings within a technical programme. Universal teach controls, task level and object level control and programming will be the focus of such technical research, with application to both robots and an extended range of manufacturing systems as our goal.

## Acknowledgements

Some of the work reported here was carried out under grants from the Applications of Computers to Manufacturing Engineering Directorate of the Science and Engineering Research Council (grant GR/D/30242) and the Joint Research Committee of the Economic and Social Research Council and the Science and Engineering Research Council (grant GR/F/07705).

## References

ANSI/RIA R15.02–1988, *Proposed American National Standard of Human Engineering Design Criteria for Hand Held Control Pendants* (New York: American National Standards Institute).

Carlsson, J., 1985, Robot accidents in Sweden. In Bonney, M. C. and Young, Y. F. (Eds) *Robot Safety* (Bedford: IFS Publications Ltd), pp. 49–64.

Cousins, S. A., 1988, Development of human engineering design standard for robot teach pendants. In Karwowski, W., Parsaei, H. R. and Wilhelm, M. R. (Eds) *Ergonomics of Hybrid Automated Systems I* (Amsterdam: Elsevier), pp. 429–36.

Edwards, M., 1984, Robots in industry: an overview. *Applied Ergonomics,* **15,** 1, 45–53.

Engleberger, J. F., 1980, *Robotics in Practice* (Amersham: Avebury Publishing Co).

Etherton, J. and Sneckenberger, J. E., 1990, A robot safety experiment varying robot speed and contrast with human decision cost. *Applied Ergonomics,* **21,** 3, 231–63.

Etherton, J., Beauchamp, Y., Nunez, G. and Ahluwalia, R., 1988, Human response to unexpected robot movements at selected slow speeds. In Karwowski, W., Parsaei, H. R. and Wilhelm, M. R. (Eds) *Ergonomics of Hybrid Automated Systems I* (Amsterdam: Elsevier), pp. 381–9.

Ghosh, K. and Lemay, C., 1985, Man/machine interactions in robotics and their effect on the safety of the workplace. *Proceedings Robots 9,* **2,** 19.1–19.8.

Gray, S. V. and Wilson, J. R., 1989, Teach pendant design for industrial robots: understanding human perception of robot movement. In Megaw, E. D. (Ed.) *Contemporary Ergonomics 1989* (London: Taylor & Francis), pp. 278–83.

Gray, S. V., Syan, C. S. and Wilson, J. R., 1990, Task difficulties in robot motion control. In Karwowski, W. and Rahimi, M. (Eds) *Ergonomics of Hybrid Automated Systems II* (Amsterdam: Elsevier), pp. 825–32.

Gray, S. V., Syan, C. S. and Wilson, J. R., 1991, Experimental evaluation of control selection using different teach pendant designs. In Pridham, M. S. and O'Brien, C. (Eds) *Production Research: Approaching the 21st Century* (London: Taylor & Francis), pp. 413–19 (in press).
Hall, E. L. and Hall, B. C., 1985, *Robotics a User Friendly Introduction* (New York: CBS College Publishing).
Health and Safety Executive, 1989, HS(G) series, HS/G 43, *Industrial Robot Safety* (London: HMSO Publications).
Health and Safety Executive Research and Laboratory Services Division, 1988, *An experimental evaluation of robot teach controls,* personal communication.
Helander, M. G. and Karwan, M. H., 1988, Methods for field evaluation of safety in a robotics workplace. In Karwowski, W., Parsaei, H. R. and Wilhelm, M. R. (Eds) *Ergonomics of Hybrid Automated Systems I* (Amsterdam: Elsevier), pp. 403–410.
Helander, M. G., Karwan, M. H. and Etherton, J., 1987, A model of human reaction time to dangerous robot arm movements. *Proceedings of the Human Factors Society 31st Annual Meeting*, pp. 191–5.
Jiang, B. C. and Gainer, C. A., 1987, A cause and effect analysis of robot accidents. *Journal of Occupational Accidents,* **9**, 1, 27–45.
JISHA, 1983, *General Code for Safety of Industrial Robots,* JIS B8433 (Tokyo: Japan Industrial Safety and Health Association).
JISHA, 1985, *An Interpretation of the Technical Guidance on Safety Standards in the Use, etc. of Industrial Robots* (Tokyo: Japan Industrial Safety and Health Association).
Karwowski, W., 1989, Human–robot interaction: safety aspects of hybrid automated systems. In Adams, A. S. and Hall, R. R. (Eds) *Ergonomics International 88: Proceedings of the 10th Congress of the International Ergonomics Association* (London: Taylor & Francis), pp. 642–4.
Karwowski, W. T., Parsaei, H. R., Nash, D. L. and Rahimi, M., 1988, Human perception of the work envelope of an industrial robot. In Karwowski, W., Parsaei, H. R. and Wilhelm, M. R. (Eds) *Ergonomics of Hybrid Automated Systems I* (Amsterdam: Elsevier), pp. 421–8.
MTTA, 1982, *Safeguarding Industrial Robots: Part 1, Basic Principles* (London: Machine Tool Trades Association).
Nagamachi, M., 1986, Human Factors of industrial robots and robot safety management in Japan. *Applied Ergonomics,* **17**,1, 9–18.
National Safety Council, 1985, *Robots,* Data sheet 1–717–85 (Chicago: National Safety Council).
Nicolaisen, P., 1985, Occupational safety and industrial robots. In Bonney, M. C. and Yong, Y. F. (Eds) *Robot Safety* (Bedford: IFS Publications Ltd and New York: Springer), pp. 33–48.
NIOSH, 1984, *Request for Assistance in Preventing the Injury of Workers by Robots.* Report by the US Department of Health & Human Services, National Institute for Occupational Safety and Health, December.
Parsons, H. M., 1986, Data base of industrial human–robot interfaces. *Proceedings International Society for Optical Engineering, Vol. 726, Intelligent Robots and Computer Vision,* pp. 503–8.
Rahimi, M. and Azevedo, G., 1990, A task analysis of industrial robot teach programming. In Karwowski, W. and Rahimi, M. (Eds) *Ergonomics of Hybrid Automated Systems II.* (Amsterdam: Elsevier), pp. 841–8.
Rahimi, M. and Karwowski, W., 1990, Human perception of robot safe speed and idle time. *Behaviour and Information Technology,* **9**, 5, 381–9.
Rutherford, A. and Wilson, J. R., 1991, Searching for the mental model in human–machine interaction? In Rogers, Y., Rutherford, A. and Bibby, P. (Eds) *Models in the Mind: Perspectives, Theory and Application* (London: Academic Press) (in press).
Sugimoto, N. and Kawaguchi, K., 1985, Fault tree analysis of hazards created by robots.

In Bonney, M. C. and Yong, Y. F. (Eds) *Robot Safety* (Bedford: IFS Publications and New York: Springer), pp. 83–98.

Van Deest, R., 1984, Robotics safety a potential crisis. *Professional Safety,* January, 40–2.

Vautrin, J. P. and Deisvaldi, D., 1986, Manipulating industrial robots in France — effects on health, safety and working conditions: results of the INRS-CRAM survey. *Journal of Occupational Accidents,* **8,** 1–12.

Wickens, C. D., 1984, *Engineering Psychology and Human Performance* (Columbus, Ohio: Charles E. Merrill).

Wilson, J. R. and Rutherford, A., 1988, Methods to identify mental models of manufacturing processes. In Adams, A. S. and Hall, R. R. (Eds) *Ergonomics International 88: Proceedings of the 10th Congress of the International Ergonomics Association* (London: Taylor & Francis), pp. 357–9.

Wilson, J. R. and Rutherford, A., 1989, Mental models: theory and application in human factors. *Human Factors,* **31,** 6, 617–34.

Wilson, J. R., Corlett, E. N., Istance, H., Rutherford, A. and Gray, S. V., 1988, *Improvements to the Programmability and Safety of Robots.* Internal report of the Department of Production Engineering and Production Management, University of Nottingham, UK.

# *Chapter 4*
# *Ergonomics of human–robot motion control*

**S. Y. Lee[1] and M. Nagamachi[2]**

[1]*Department of Industrial Engineering, Korea University, 1, 5-ka, Anam-dong, Sungbuk-ku, Seoul 136, Korea and* [2]*Department of Industrial and Systems Engineering, Hiroshima University, Shitami, Saijo-Cho, Higash Hiroshima-City 724, Japan.*

**Abstract.** The Teaching Expert System/World Coordinate Sytem (TES/WCS) was proposed to improve robot motion control. First, the precision coordinate reading for getting inherent data about position and posture of an object was performed through an integrated image and fuzzy processing. Second, singularity and parameter limitation problems in getting the motion data about position and posture of the robot arm in macro-motion mode were solved by a proposed geometric algorithm. Third, unnecessary robot motion could be reduced by the Robot Time and Motion (RTM) method with the Multi-Geometric Straight Line Motion (MGSLM) method in micro-motion mode. The results demonstrated a reduction in Root Mean Square (RMS) values of error at coordinate reading, and a reduction in mean teaching task time to task order.

## *Introduction*

Today's robots perform tasks in hostile environments such as space, deep ocean, or nuclear power plants. As the field of robot applications expands, the relationship between humans and robots becomes very important. Unfortunately, the field of human–robot interface technology has not progressed much compared with other areas of robot engineering (Tsuji and Ejiri, 1984). One of the important issues in human–robot interface technology is the allocation of tasks to the human and the robot. The human–robot system can be defined as the system which combines the generality of the human and the autonomy of the robot (Sheridan and Ferrell, 1967).

In this paper, an integrated technique is proposed for human–robot system performance. Its goal is to maximize the environmental adaptability of the robot, and to assure that the task is performed effectively through the cooperation of the human and the robot. Task requirements are those that are difficult for the human or the robot to perform alone, from the remote site considering the requirements for macro-judgement and manipulation (Croker and Lyman, 1983). Also, an attempt is made in this paper to use some aspects of robot intelligence for image processing. An advanced teleoperator system is proposed, in which the major parts of the task are executed by a computer program with minimal assistance from the human operator (Takase and Wakamatsu, 1984).

In terms of error recovery, once an error occurs during a task operation in the automatic mode of the robot, it is converted into the manual mode in which a human takes part, without stopping the process. After recovering the error through the manual mode, the robot operation should be converted back into the automatic mode. So, the system performance is enhanced by the continual resumption of a given task. Therefore, a Teaching Expert System/World Coordinate System (TES/WCS) was constructed for the automatic mode and an Error Recovery Expert System/World Coordinate System (ERES/WCS) for the manual mode, applying a WCS designed by the functional analysis of the robot joints (Lee *et al.*, 1989).

## *Characteristics of the TES/WCS*

The TES/WCS was designed to gain inherent data of the task object, and, using an expert system, to obtain the motion data of the robot (Lee *et al.*, 1988a). The inherent data, i.e. the data about the position and the orientation of the task object, provide information such as size and shape of the object. The motion data relate to the position and the posture of the robot for moving the robot hand to the task object. This approach, called human–robot ergonomics, integrates the motion control ability of humans using traditional ergonomics, and the motion control methods of robots using the concept of robot ergonomics.

The TES/WCS was constructed to improve robot motion control by combining the inherent data of the task object and the motion data of the robot. A TES/CCS applying the Cartesian Coordinate System (CCS) was designed by the movement analysis of robot joints (Lee *et al.*, 1988a). The following design problems were found during TES/CCS development. First, it was difficult for the human operator to know the relationship between the robot and the task objects in the workspace. Since the system did not give the criteria of the layout of the task object, it was difficult to solve vagueness of the coordinate readings in relation to the inherent data of the task object. Second, the $X$, $Y$ and $Z$ coordinates and pitch of the robot hand must be considered simultaneously in calculating the motion data of the robot moving to the task object. Third, when moving to the task object using micro-motion, there is an unnecessary robot motion caused by rectangular movement of each axis which needs to be accounted for when calculating the motion data of the robot. Therefore, in order to achieve the goals of human–robot ergonomics, it was necessary to improve the TES/CCS by using the TES/WCS for the teaching task as follows.

First, the integrated image and fuzzy processing was developed to gather the inherent data efficiently. This system was composed of image processing to solve the inaccuracies of readings about the position and posture of the task objects (Tsuji and Ejiri, 1984), and the fuzzy processing module to solve the vagueness of such readings. Second, an algorithm based on geometric analysis was formulated to solve singularity and parameter limitation problems in calculating the macro-motion data of the robot and to correct the robot movements, as well as to reduce the time of operation. Third, the Multi-Geometric Straight Line Motion (MGSLM) method, moving the robot along a straight line with a slant, and the Robot Time and Motion (RTM) method, analysing the robot motions according to the task method of the robot, were combined in order to decrease the amount and time of operation required for position control. Figure 4.1 represents the global background and flow of this study.

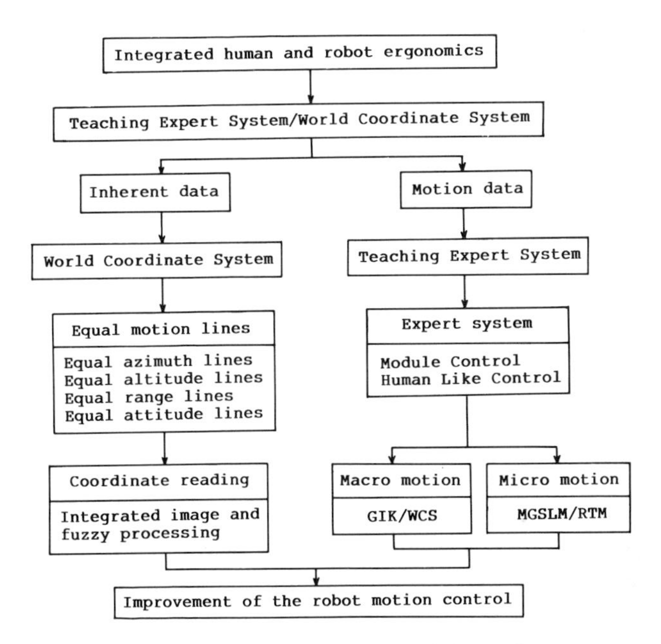


*Figure 4.1 Overall flowchart of this study.*

# Construction of the TES/WCS

## Design of the WCS

Equal motion lines about the robot workspace designed in the WCS were composed of equal azimuth lines, equal altitude lines, equal range lines and equal posture lines (Lee *et al.*, 1988b). The design of the WCS, using three equal motion lines among these four equal motion lines as shown in Figure 4.2, provides the following advantages. First, it enables the human operator to make coordiante reading easy in case of the robot moving to the task object. This is done by a motion control program, i.e. by establishing the criterion of each coordinate through equal motion lines. Second, unnecessary robot motion can be eliminated because the waist joint can be rotated without the need to rotate shoulder and elbow joints by installing the position of task objects using the concept of equal motion lines. Third, the singularity problem can be solved because singularity

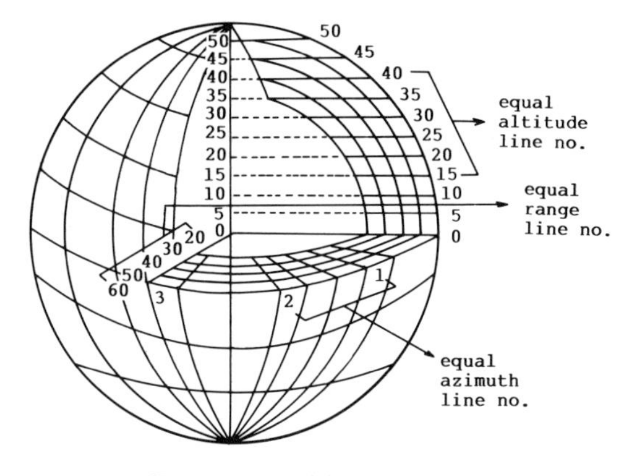


*Figure 4.2 Globe of the WCS.*

is readily recognized by the WCS. Finally, because elbow joint angle in the robot workspace always has constant value according to the given pitch and range, the joint parameter limitation problem can also be solved.

### Design of the TES

The Module Control (MC) combines the inherent data of the task object with the motion data from the expert system. The Geometric Inverse Kinematics/World Coordinate System (GIK/WCS) is used for this combination (Lee *et al.*, 1988c). Figure 4.3 shows the structure of the TES/WCS, where the system indicated with the double line is the main focus of this study. The vagueness of the inherent data and the subjective judgement about it are revised by the MC method through the integrated image and fuzzy processing system. Motion data is derived from an algorithm based on geometric analysis. Also, if the human operator inputs the inherent data of the task object, by considering the Tool Centre Point (TCP) of the robot hand, these data can be changed to motion data through an algorithm of geometric analysis, movement rules of MGSLM/RTM, and heuristic joint selection rules of Unit Contol (UC) and Micro Unit Control (MUC). Robot movements at the joints execute motion data.

## *Improvement of the robot motion control*

In this study, in order to improve the robot motion control efficiently, image processing was used for the calculation of position and posture of the task object according to a criterion of equal motion lines of the WCS. In addition, fuzzy processing was designed to reduce the inaccuracy of the coordinate reading of the task object.

An algorithm based on geometric analysis was formulated to solve the

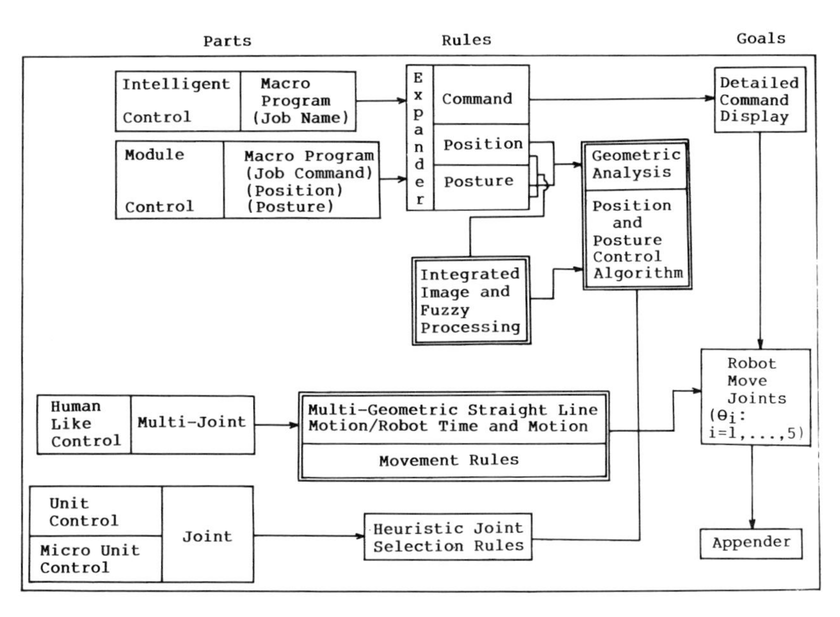


*Figure 4.3 Structure of the TES/WCS.*

singularity and the parameter limitation problems. This algorithm allowed for the calculation of the motion data and for reduction of the inaccuracies of robot movements. It also helped to reduce the time of the operation by a human operator in the micro-motion mode.

In order to calculate the robot motion data by the Human-Like Control (HLC) system, the MGSLM method, i.e. moving the robot along a straight line with a slant, and the RTM method, i.e. analysing robot motions according to the task method, were combined to decrease the scope and the time of operation required in the micro-motion mode. As mentioned above, the robot motion control was improved through the efficient combination of the inherent data of task object and the motion data of a robot as a part of the integrated human–robot ergonomics.

## *Coordinate reading for calculation of inherent data*

The coordinate reading of the position and posture of the task object is composed of image processing, reducing the errors of reading through fuzzy processing, and transforming these values into WCS values based on the shoulder joint of the robot as shown in Figure 4.4.

### Distance

To measure the distance from the robot hand to the task object, the task object is photographed by camera II attached to the robot hand from some point O

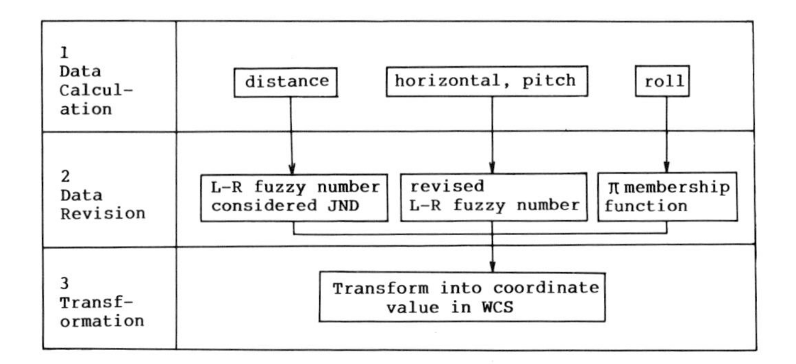


*Figure 4.4 Flowchart of the integrated image and fuzzy processing.*

(Figure 4.5), and its area is calculated as a pixel number using an image processing technique. Also, the robot hand is moved to some point A between point O and the task object. The task object is photographed and its area is calculated again (Ballard and Brown, 1982). The following notation is used to illustrate this approach (see Figure 4.5): $S_0$, the pixel area of the task object when it is photographed at point O; $S_1$, the pixel area of the task object when it is photographed at point A; $d_0$, the distance from point O to the task object; $d_1$, the distance from point A to the task object; and $d_2$, the distance from point O to point A. The value of $d_1$ is obtained from $S_0$, $S_1$, $d_2$ using the following equation:

$$1/d_0^2 : 1/d_1^2 = S_0 : S_1. \tag{1}$$

The value of $d_0$ is replaced with $(d_1 + d_2)$ and $d_1$ is calculated. Because the distance value calculated in the above data processing is not an accurate value regarding the task object, data revision processing is needed. The above distance value can be measured, since point A is located between point O and the task

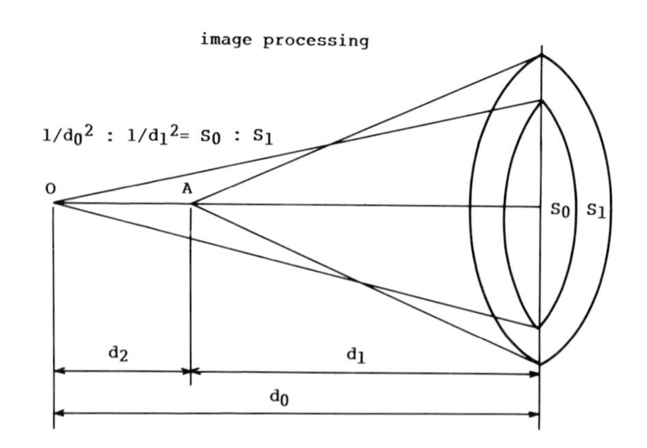


*Figure 4.5 Relationship between distances and areas of similar figure.*

object. In other words, the shape of the task object at point O displayed on a monitor is similar to that of point A, but varies when the robot hand does not move along a straight line. Therefore, an error of the distance value caused by a variation of the assumption is revised by using a L-R fuzzy number (Zimmermann, 1985). This makes a small competition group of the L-R fuzzy number and the random sampling range, and improves accuracy in estimating the distance value.

The degree of ambiguity was defined by considering the degree of variety compared with a similar figure. The degrees of vagueness were set at 0, 0.5 and 1 values according to a pre-perceived degree of grip of the task object. This was done because the object areas vary according to the picture angle (Park, 1988). The revised distance value of the robot hand was transformed into the value in WCS based on its shoulder joint angle. This allowed the motion data for the robot to be obtained. The distance value ($\rho$) transformed on the WCS was defined by the following equation:

$$\rho=[Z^2+\{d\cos p' \pm(\rho'^2 \tfrac{2}{3} Z'^2)^{1/2}\}^2]^{1/2}. \tag{2}$$

$\rho'$ is the distance from shoulder axis to robot hand.

## Horizontal and pitch

In order to calculate the horizontal value and pitch value, equal azimuth lines and equal pitch lines were drawn on a monitor for image processing according to the equal motion lines on the WCS. Nine blocks were made in total. These were referred to as left-up, left-middle, left-down, middle-up, middle-middle, middle-down, right-up, right-middle and right-down, respectively (see Figure 4.6). The human operator indicated the major part of an image for the task object on the nine blocks. Next, the area of the image was calculated by using an image

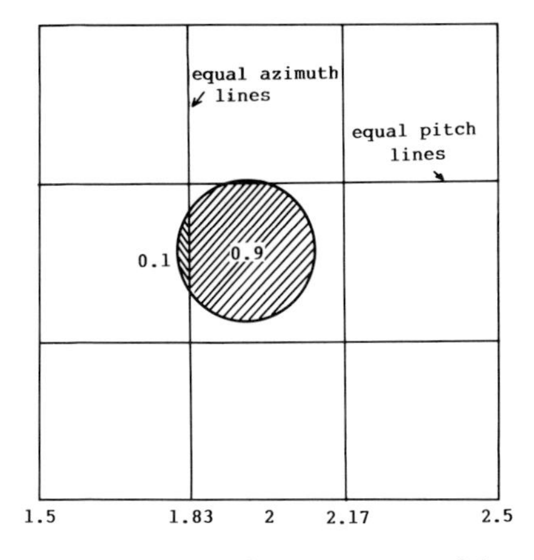


*Figure 4.6* *Monitor to indicate major part of the task object.*

processing technique, and its value was used to represent the membership value in the data revision processing.

The fuzzy set theory for correction of the horizontal value and the pitch value was used, and the revised L-R fuzzy number was utilized (Zimmermann, 1985). Because the position of the task object can be displayed by the membership value on a monitor, the symmetric and linear membership functions should be developed for the above revised L-R fuzzy number. The revised L-R fuzzy number accurately shows the position of the task object, revising the horizontal value and the pitch value. The following equation was used as the reference function of the revised L-R fuzzy number:

$$\mu(x) = \mathrm{L}(x) = \mathrm{R}(x) = 1/X. \tag{3}$$

Also, the membership functions were defined as the following:

$$\mathrm{L}(10/30X + \alpha + 10) = (30X + \alpha + 10)/10, \quad -0.167 \leq X \leq 0; \tag{4}$$

$$\mathrm{R}(10/-30X + \beta + 10) = (-30X + \beta + 10)/10, \quad 0 \leq X \leq 0.167. \tag{5}$$

The vagueness values used were set on 0, 0.5 and 1 (Park, 1988). The decision about the obscure values was made based on similarity and according to the block shape of the object. In case of a block, the centre point of the object is presented by the membership value which is the area ratio in criterion to equal motion lines by the WCS. The centre of the circle, however, is not well presented by the membership value. Therefore, the ambiguity is used for the revision of the error caused by the above assumption.

According to the vagueness system used in this study, $\alpha$ and $\beta$ were defined as follows:

$$\alpha = \beta = 0, \quad V = 0; \tag{6}$$

$$\alpha = \beta = 0.7, \quad V = 0.5; \tag{7}$$

$$\alpha = \beta = 1.4, \quad V = 1. \tag{8}$$

The competition group of the membership function which considers the vagueness of the horizontal value is presented as the equation for the critical value $C$ (Park, 1988). In other words, the interval of the horizontal value which could be chosen was presented as follows:

$$\text{In case of } V = 0, \quad C_g = \{[x] \mid \mu_h(x) = C\}; \tag{9}$$

$$\text{In case of } V = 0.5, \quad C_g = \{[x, x \pm 0.048] \mid \mu_h(x) = C\}; \tag{10}$$

$$\text{In case of } V = 1, \quad C_g = \{[x, x \pm 0.096] \mid \mu_h(x) = C\}. \tag{11}$$

That is, when $V = 0$, the competition group becomes a single point:

$$\mu_h(x) = C. \tag{12}$$

When $V = 0.5$ and $V = 1$, the competition group was formed by the symmetry (L-R type) of the $X$ value, i.e. both sides of the $X$ value. The data revision step about the horizontal module number '2' is shown in Figure 4.7. The data revision step about the pitch value is similar to the horizontal value. Therefore, in transformation processing on the WCS, the horizontal value ($\theta$) and the pitch value ($p$) were transformed as follows (the detailed contents about $\alpha$ are omitted here):

$$\theta = \theta' + \alpha; \tag{13}$$

$$\alpha = \cos^{-1}\{(CA^2 + AB^2 - BD^2)/(2CAAB)\}; \tag{14}$$

$$p = \angle C - 180°. \tag{15}$$

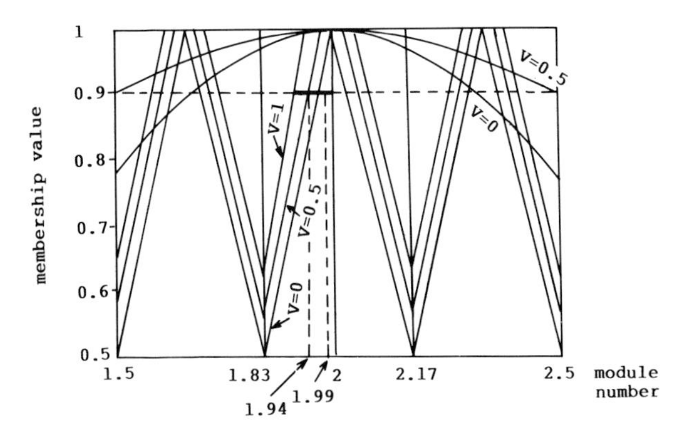


*Figure 4.7 Membership function of horizontal module number '2'.*

## Altitude and roll

The altitude of the robot arm can be obtained by using the pitch value ($p'$) and the distance in CCS according to geometric analysis without procedures of data calculation and data revision.

$$Z = Z' \pm d \sin p'. \tag{16}$$

Since roll is a posture value of rotating the robot's wrist, and this type of movement is not easily perceived by image processing, it is calculated by eye measurement and revised using the $\pi$ type of membership function.

## Calculation of motion data in the macro-motion mode

The algorithm of geometric analysis for the calculation of motion data offers a better criterion to solve the singularity and parameter limitation than that of GIK/CCS because of the geometric analysis using the range value of the WCS. Singularity occurs when the input range value is equal to more than link 2 plus link 3 of the robot, and there occurs an appearance of an immobile robot because of the parameter limitation range that is even within the distance of link 2 plus link 3 of the robot. Therefore, the aim of this study is to eliminate singularity by means of determining the upper and lower bounds even if the human operator makes a mistake in inputting range values that are more than link 2 plus *L34* of the robot. That is,

$$|L2 - L34| \leq \text{Range} \leq L2 + L34. \tag{17}$$

Parameter limitation is solved by calculating the limiting value of the range according to the pitch of the robot. Once a modified range on the standard equal altitude line of the WCS about a given pitch is entered, the parameter limitation value of the GIK/WCS can be easily compared. Therefore, the parameter limitation of the robot is described by the upper and lower bounds of range in which a robot can move for a given pitch and range. That is,

$$\min \text{R} = \{L2^2 + L34^2 - 2L2L34 \cos(90° + E)\}^{1/2}, \tag{18}$$

$$\max \text{R} = L2 + L34, \tag{19}$$

$$\min \text{R} \leq \text{Range} \leq \max \text{R}. \tag{20}$$

The solution of singularity and parameter limitation contributes to the correctness of the robot movements and the,reduction of micro-motion control time. After the singularity and parameter procedure, inherent data of the task object are changed into motion data by the GIK/WCS procedure. The pitch ($\theta_4$) of the robot hand is the angle between the extension line of the link 3 (*L3*) and the length of the robot hand (*L4*) or the end effector (see Figure 4.8).

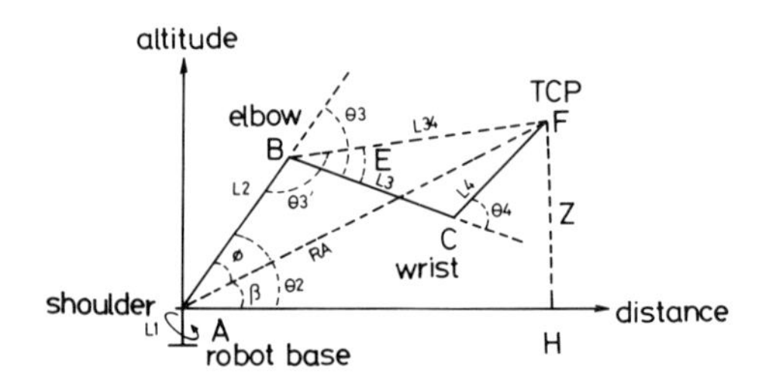


*Figure 4.8 Geometric analysis of the robot by the WCS.*

Though the human operator measures the pitch angle by the horizontal line in three-dimensional (3-D) space, it is difficult to measure the range of the boundary combining the three axes (*X*, *Y*, *Z*) for gaining the joint values within the parameter limitation by this pitch angle. This problem causes interruption of tasks, and is the main reason why the pitch was defined as mentioned above. The relative angle between the extension lines of *L3* and *L4* is used to solve these problems. When this pitch is used, a desired angle, a joint value of shoulder ($\theta_2$) and a joint value of elbow ($\theta_3$) can be obtained from two kinds of triangle relationships.

First, if the pitch of the robot hand is given, the *L34* which is the opposite side of the angle ($180° - |\theta_4|$) is always constant. Therefore the side *L34* and angle *E* can be calculated from the triangle composed of *L3* and *L4*, since the pitch of the robot hand is not related to the movements of other robot joints such as waist joint, shoulder joint and elbow joint:

$$L34 = \{L3^2 + L4^2 - 2L3L4 \cos(180° - |\mathrm{Pitch}|)\}^{1/2}; \tag{21}$$

$$E = \cos^{-1}\{(L3^2 + L34^2 - L4^2)/(2L3L34)\}. \tag{22}$$

*L34* can be obtained by using *L3* and *L4* and acts as a kind of link called the imaginary link *L34*. This reduces one degree of freedom for $\theta_4$.

Second, if the imaginary link *L34* is determined in a situation when the pitch of the robot hand and the range (*RA*) on the WCS (from the shoulder joint to the TCP) are given, it is called the imaginary link *RA* because this range also acts as a link in calculating the robot joint values. The triangle is formed by the shoulder link, *L2*, the imaginary link *L34* and *RA*. From this triangle, the angles $\phi$ and $\theta_3'$, can be determined as follows:

$$\phi = \cos^{-1}\{(L2^2 + RA^2 - L34^2)/(2L2RA)\}; \tag{23}$$

$$\theta_3' = \cos^{-1}\{(L2^2 + L34^2 - RA^2)/(2L2L34)\}. \tag{24}$$

Accordingly, the angles $\theta_2$ and $\theta_3$ needed to move the robot hand can be identified by using the altitude (*Z*), *RA*, $\theta_4$ and the values found from the above two triangles. *Z* and *RA* are the distances from the shoulder joint to the TCP in the WCS. Therefore $\theta_2$ and $\theta_3$ are determined as follows. $\theta_1$ can be obtained by using the horizontal value ($\theta$) of coordinate reading:

$$\theta_1 = \theta \times 30; \tag{25}$$

$$\theta_2 = \beta + \phi, \quad \text{where } \beta = \sin^{-1}(Z/RA); \tag{26}$$

$$\theta_3 = 180° - (\theta_3' - E), \quad \mathrm{pitch} \geq 0'; \tag{27}$$

$$\theta_3 = 180° - (\theta_3' + E), \quad \mathrm{pitch} < 0'. \tag{28}$$

The joint values are calculated by the pitch angle and the position coordinates. The position control of the robot is obtained using the WCS through the algorithm of geometric analysis, that is, the joint values are calculated by the given pitch angle. $\theta_4$ and the roll angle ($\theta_5$) for the posture control of the robot hand are calculated using WCS as follows:

$$\theta_4 = \text{pitch } 13.3 + (\text{roll} - \theta_1)\ 13.3: \quad (29)$$

$$\theta_5 = \text{roll } 13.3 + (\text{roll} - \theta_1)\ 13.3. \quad (30)$$

## *HLC for calculation of motion data in micro-motion mode*

This study attempted to contrast robot motion with human-hand motion for HLC in order to determine micro-motion mode. Robot motion was analysed with the RTM method while human motion was analysed with the Method Time Measurement (MTM) technique. The MGSLM method was used to control position in the micro-motion control-based system.

When the robot grasps an object, the MGSLM method makes the robot hand move along the straight line toward the object in a manner similar to human-hand motion. It performs micro-motion mode more efficiently because the MGSLM method need not consider mutual relationships of axes like UC (unit control) and MUC (micro unit control). The MGSLM method designed in this study can be implemented not only in a straight line but also in the straight line with a slant line according to the CCS (Lee *et al.*, 1988b). Unnecessary robot motion is removed by establishing eight direction vectors which combine the *X*, *Y* and *Z* axes with TCP in order to move the robot in a fashion similar to human motion. The components of these eight vectors are shown in Table 4.1.

*Table 4.1 Components of the eight kinds of vector.*

| | |
|---|---|
| (fore, up, left) | (fore, down, left) |
| (back, up, left) | (back, down, left) |
| (fore, up, right) | (fore, down, right) |
| (back, up, right) | (back, down, right) |

Considering the relationship between the joint values of the eight direction vectors, $\theta_1$ is related only to the left and the right axes, and $\theta_2$, $\theta_3$ to the three cartesian coordinate axes around the TCP. The calculation of joint values related to the vectors is illustrated below. The parameters of other vectors are calculated by the same method changing the necessary variables.

In micro-motion mode, the task is to move $f$ cm forward, $z$ cm upward, and 1 cm leftward along the cartesian coordinate axes around the TCP, (fore $= f$, up $= z$, left $= 1$), there becomes an input. Figure 4.9 shows the position of the robot in the TCP after the robot movements. The TCP describes the situation

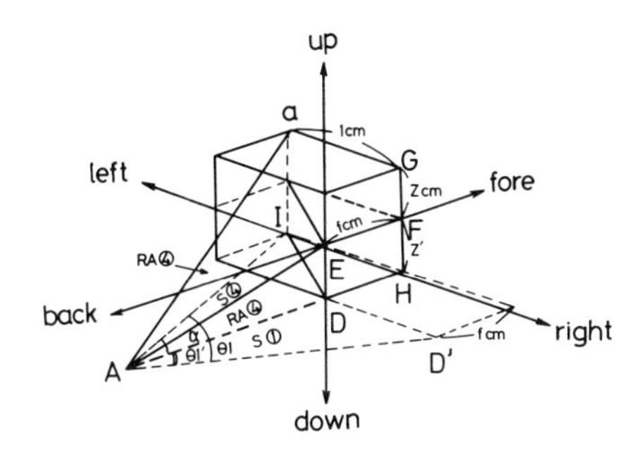


*Figure 4.9* *Position of the robot in the TCP.*

with rotation about $\theta_1{}'$ from criterion of equal azimuth line $AD'$ by the WCS, where the altitude is $Z'$ and the range is $AE$. In reaching the desired point a, the joint value $\theta_1$ can be calculated if the azimuth $\alpha$, the altitude $Ia$, and the range $Aa$ are known. Figure 4.10 shows the side view of Figure 4.9. The imaginary line $a34$, relating point B and point a, comes from Figure 4.9. Here, $p$ is the pitch value and $a_i$ ($i = 1,2,3,4$) is the length of link $i$.

$$a34 = (a3^2 + a4^2 + 2a3a4 \cos p)^{1/2}. \tag{31}$$

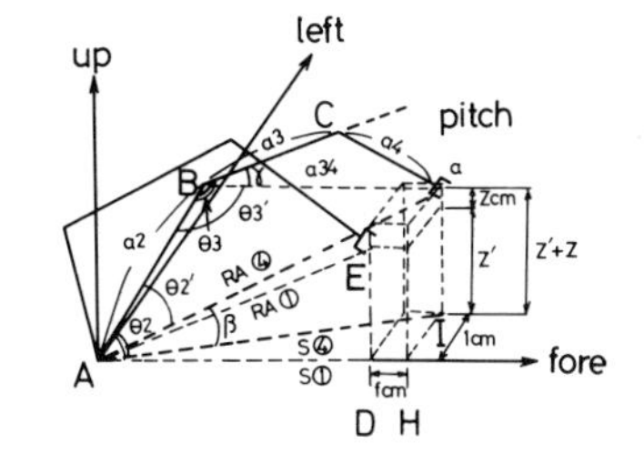


*Figure 4.10* *Side view of the robot in the TCP.*

By calculating $\beta$ and $\gamma$ from the triangles $AIa$ and $Bca$, respectively, $\theta_2$ and $\theta_3$ can also be calculated. These joint values and the variables related to three CCS axes about the eight different vectors are shown in Table 4.2. Consequently, the MGSLM method can control the robot in any direction.

When the task sequence and process are determined for a certain task, the task is analysed with respect to the robot motion with the RTM method, and the task time of the robot is measured. This method analyses ten basic motions into four groups (Nof, 1985). The four groups are as follows:

RTM Group 1: movement element – REACH, MOVE, ORIENT;
Group 2: sensing element – STOP ON ERROR, STOP ON FORCE/TOUCH, VISION;
Group 3: gripper or tool element – GRASP, RELEASE; and
Group 4: delay element – PROCESS TIME DELAY, TIME DELAY.

*Table 4.2 Joint calculation and variables of direction.*

| Specification | Contents |
|---|---|
| Joint calculation | $\theta_1 = \theta_{1'} + \alpha$, |
| | $\alpha = \cos^{-1}\{(AI^2 + AD^2 - ID^2)/2AIAD)\}$. |
| | $\theta_2 = \theta_{2'} + \beta$, |
| | $\theta_{2'} = \cos^{-1}\{(a2^2 + Aa^2 - a34^2)/(2a2Aa)\}$, |
| | $\beta = \tan^{-1}(\|Z' + Z\|/AI)$. |
| | $\theta_3 = \theta_{3'} + \gamma$, |
| | $\theta_{3'} = \cos^{-1}\{(a2^2 + a34^2 - Aa^2)/(2a2a34)\}$, |
| | $\gamma = \cos^{-1}\{(a3^2 + a34^2 - a4^2)/(2a3a34)\}$. |
| Variables of direction | $AH = AD + f$, forward |
| | $AH = AD - f$, backward |
| | $Aa = \{AI^2 + (Z' + Z)^2\}^{1/2}$, upward |
| | $Aa = \{AI^2 + (Z' - Z)^2\}^{1/2}$, downward |
| | $\theta_1 = \theta_{1'} - \alpha$, leftward |
| | $\theta_1 = \theta_{1'} + \alpha$, rightward |

This study combines the RTM method with the MGSLM method. These methods are used to analyse the task sequence and process for a specific task, and to inform the human operator about the process. The human operator inputs inherent data for robot motion using the MGSLM method, while the GIK/WCS converts inherent data into motion data. Therefore, by combining the inherent data of the task object with the motion data, improvement of the robot motion control is achieved.

# Experimental method

## Apparatus

The experimental apparatus for measuring the validity of the proposed approach is shown in Figure 4.11. The Human Interactive Subsystem (HIS) computer which controls the inference engine of the knowledge base, and executes the teaching task, employs an IBM-PC. It receives commands from the human operator, executes the task and provides the feedback of results through the interface. The Task Interactive Subsystem (TIS) computer serves as the drive unit of the ROB-501 robot. It receives motion data from the HIS computer and transforms the movement of joints into the actuator signal. The image processing device is used as an intelligent function for extracting information about the task object from the camera attached to the robot hand. This device uses the PC PLUS board produced by Image Technology Co, which maps image signals from the camera to the image data, and formats the image data stored in the memory on a monitor.

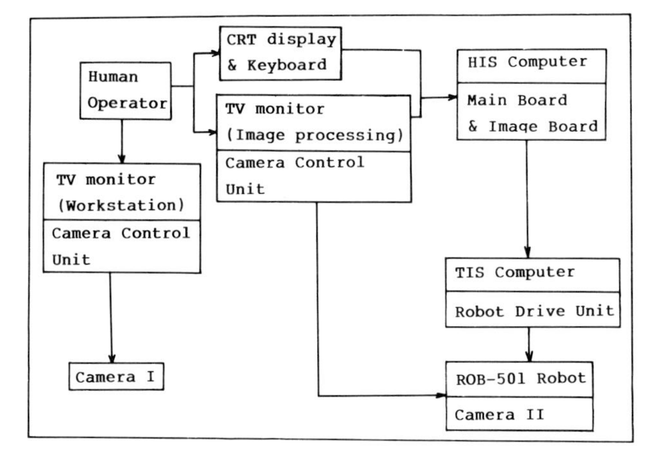


*Figure 4.11 Experimental apparatus.*

## Task

This study involved a simulation of a teaching task in a hostile environment. The main objective was to compare the TES/WCS with the TES/CCS. The TES/WCS system is composed of the integrated image and fuzzy processing system for coordinate reading in calculating inherent data. The GIK/WCS system was used for the solution of singularity and parameter limitation problems for macro-motion. The MGSLM/RTM method was used for micro-motion in calculating the motion data. The performance measures (i.e. dependent measures) were the Root Mean Square (RMS) values of error at coordinate reading and the average teaching time. The experimental task involved lighting a lamp. The sequential procedures consisted of grasping a match on the stand, taking it to a match box, lighting it, moving the light to an alcohol lamp and dropping the match into an ashtray.

## Procedure

The RMS error of the input data was measured to verify the inaccuracy at the coordinate readings. The RMS error was obtained for the input processing of each system for four task objects applied to the lamp lighting task. The average teaching time needed to move the robot to the task object was also measured for each system. This time was composed of the coordinate reading times for the calculation of the inherent data of the task object, added to the macro-motion times and the micro-motion times.

## Results

The results for accuracy comparison are illustrated in Figure 4.12. Figure 4.13 shows the reduction of time, which refers to the average teaching time for the

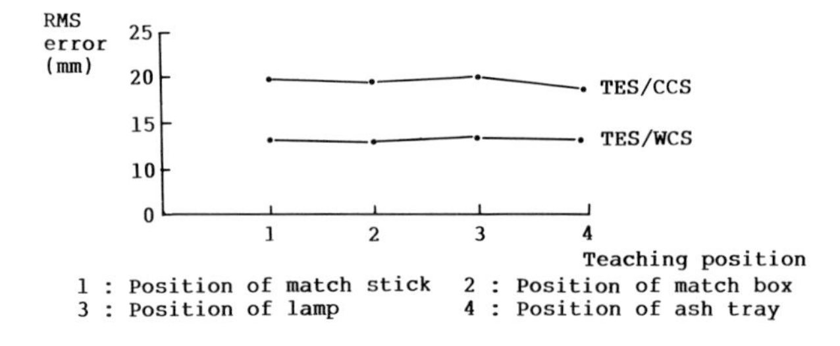


*Figure 4.12 Comparison of the RMS error for the teaching position.*

task order. It was shown that the input processing of the integrated image and fuzzy processing used for coordinate reading in the TES/WCS was more accurate than that of TES/CCS. Displaying equal azimuth lines and equal pitch lines of WCS on the monitor was the criterion for reading the coordinate value of the task object.

This reduction is due to the vagueness about coordinate reading being reduced because the fuzzy processing (for image processing) made the competition group of the inherent data narrower. The result was a more accurate indication of the centre of the task object. As can be seen in Figure 4.13, the coordinate reading time was reduced in comparison with the TES/CCS time, but there was no significant improvement due to the addition of image processing time. The data regarding correctness of robot movements indicated that the task performance times of the human operators were reduced. This was due to the calculation method used for the solution of singularity and parameter limitation which was an algorithm of GIK/WCS. Figure 4.13 also shows that the task-teaching time was reduced in macro- and micro-motion mode.

## Conclusions

In this study, the TES/WCS system was applied to improve robot motion control in an integrated human and robot task environment. The study was conducted

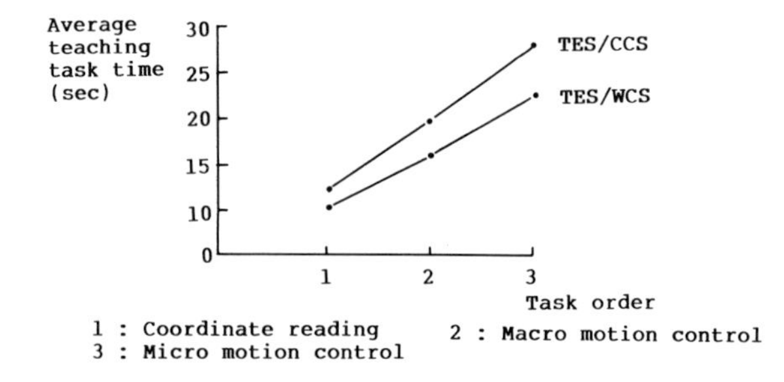


*Figure 4.13 Comparison of the average teaching-task time for the task order.*

using a laboratory simulation. The inaccuracy of coordinate reading was resolved by applying integrated image and fuzzy processing according to the WCS model. This was done in order to obtain inherent data about the position and posture of the task object. The singularity and parameter limitation problems were solved through the proposed geometric algorithm by GIK/WCS to get the motion data of the robot in macro-motion. This led to improvements in the precision of robot movements. The MGSLM/RTM method was used to obtain motion data of the robot in the micro-motion mode. This resulted in removing unnecessary robot motions and decreasing the time of operation. It is expected that these results can also be obtained in real industrial robot operations.

To improve safety and human factors of robots, more research is needed in hardware and software design and integration for sensors, robot language etc. Future robot ergonomics research for motion control should consider other motion analysis methods. For example the RTM method applying Modular Arrangement of Predetermined Time Standards (MODAPTS) should be studied to predict the cycle time of the robot motion required for efficient control.

## *References*


Ballard, D. H. and Brown, C. M., 1982, *Computer Vision* (Englewood Cliffs, N.J.: Prentice-Hall), pp. 119–48.

Croker, K. and Lyman, J., 1983, Research issues in implementing remote presence in teleoperator control. *17th Annual Conference on Manual Control,* **1** (London: MIT Press), pp. 109–26.

Hirai, S. and Sato, T., 1986, Intelligent teleoperation of robots. *Journal of the Robotics Society of Japan,* **14,** 89–92.

Lee, S. Y., Nagamachi, M. and Ito, K., 1988a, A study on a comparison between the ERES/WCS and the ERES/CCS in the advanced teleoperator system. *International Workshop on Artificial Intelligence for Industrial Applications*, 335–40.

Lee, S. Y., Nagamachi, M. and Ito, K., 1989, A study on the ERES/WCS as a part of the advanced teleoperator system. *Transactions of the Society of Instrument and Control Engineers,* **25,** 80–7.

Lee, S. Y., Nagamachi, M., Ito, K. and Oh, C. S., 1988b, A study on a teaching and operating expert system and an error recovery expert system using a world coordinate system in the advanced teleoperator system. In Adams, A. S. and Hall, R. R. (Eds) *Ergonomics International 88: Proceedings of the 10th Congress of the International Ergonomics Association* (London: Taylor & Francis), pp. 645–7.

Lee, S. Y., Nagamachi, M., Ito, K., Oh, C. S. and Lee, C. M., 1988c, A study on a teaching and operating expert system in the advanced teleoperator system. In Burckhardt, C. W. (Ed.) *International Symposium on Industrial Robots* (London: IFS), pp. 441–8.

Nof, S. Y., 1985, Robot ergonomics: optimizing robot work. In Shimon, Y. and Nof, S. Y. (Eds) *Handbook of Industrial Robotics* (New York: John Wiley & Sons), pp. 549–604.

Park, S. K., 1988, 'A study of the manual mode by using the fuzzy sets in an advanced teleoperator system', Master's Thesis, Department of Industrial Engineering, Korea University.

Sheridan, T. B. and Ferrell, W. R., 1967, Supervisory control of remote manipulation. *IEEE Spectrum,* **4,** 81–8.

Takase, K. and Wakamatsu, S., 1984, A concept of intelligent teleoperation system and related technologies. *Journal of the Robotics Society of Japan,* **2,** 62–71.
Tsuji, S. and Ejiri, M., 1984, *Robot Engineering and Its Application* (The Institute of Electronic Communication), pp. 168–71.
Zimmermann, H. J., 1985, *Fuzzy Set Theory and Its Application* (Boston: Kluwer-Nijhoff Publishing), pp. 51–60.

# *Chapter 5*
# *The use of an integrated graphic simulation in robot programming*

**J. Warschat and J. Matthes**

*Fraunhofer Institute für Arbeitswirtschaft und Organisation, Nobelstrasse 12, D-7000 Stuttgart 80, Germany*

**Abstract.** In this chapter, a robot simulation system is presented which effectively supports off-line programming and the layout of workcells. It relieves the user of routine work and helps her/him avoid mistakes which might have expensive consequences. The system consists of five components which are described in detail: interactive programming surface, 3-D (three-dimensional)-graphics modeller, high-level language compiler, IRDATA interpreter, and simulation.

## *Introduction*

Industrial robots are introduced into manufacturing companies for mainly economic reasons, such as the reduction of production costs. Besides, they are expected to increase productivity and improve the quality of product and planning. At present, industrial robots are used mostly in rigidly interlinked production lines where they execute only one or very few tasks. In order to obtain a greater flexibility of these rigid production lines, larger production tasks, such as assembling of whole subassemblies, are being transferred more and more to smaller workcells with only few industrial robots.

The availability of workcells is highly dependent on how fast the program or the product for an industrial robot can be changed. Other important factors are the coordination between robot movements and the periphery of the workcell, and the coordination between robot movements and the production process in the neighbouring workcells. The dominant form of robot programming at this time is the method of on-line programming using teach pendants.

If, for complex assembly tasks, the robot is programmed in this conventional way, it can happen that the respective workcell (or in extreme cases the whole production line) cannot be used during the programming phase, depending on the type of robot control. Therefore off-line programming is expected to avoid expensive downtimes. The use of off-line programming systems with integrated graphic simulation of motions makes it possible to transfer tasks, such as programming industrial robots or optimizing layouts, from the shop floor into work planning. For the production of off-line programs efficient tools must be provided, for instance intelligent or user-friendly interfaces for system handling or a graphics editor for easy layout manipulation of workcells. With these tools critical planning errors can be avoided.

Several simulation systems have been produced already. Very powerful systems, however, have the disadvantage that the costs for hardware and software are very high. Other systems use a conventional Computer-Aided Design (CAD) system for the graphical animation, so that an additional expenditure for the purchase of a CAD system is necessary (McDonnell-Douglas Automation Company, 1981; CATIA Robotic, 1982; Wörn and Stark, 1987). When smaller, low-price computers are used, dynamic characteristics cannot be represented for reasons of CPU time.

These are the reasons why the Fraunhofer Institut für Arbeitswirtschaft und Organisation (IAO) decided to develop a graphics software for medium capacity workstations. This system allows real-time simulations for industrial robots which consider dynamic characteristics. Great emphasis was placed on efficient human–computer interfaces.

## *Developments in robot programming*

### On-line programming methods

Since the 1960s, industrial robots have been programmed on-line, i.e. programming is carried out on the workcell itself. In this so-called Teach-in-Mode, the programmer directs the robot, by means of manual control, to different points on the trajectory and records their spatial coordinates, the positions and orientations of the tool and the corresponding trajectory and tool commands.

For simple tasks and large batch sizes, on-line programming is not only adequate but also justifiable from an economic point of view. For small and medium batch sizes, however, this method is no longer suitable due to the need for a very short set-up time. The reasons why the conventional teach-in method is no longer sufficient for more complex production and assembly structures are as follows:

- The robot cannot be in production during the teach-in process. For complex assembly tasks this can take up to several weeks, i.e. the set-up process keeps the robot occupied for a disproportionately long time. In addition, more time is needed for testing the programmed motions;
- The teaching process can be dangerous for the programmer, as she/he is in close proximity to the robot during parts of the programming phase;
- The robot can only move on the programmed trajectories, i.e. there is no possibility of alternatives to the programmed motions;
- Programs created in the teach-in method cannot be pictured mentally before they are executed by the robot;
- There is no documentation of the programs in the sense of a readable text; and

- The teaching process is long and tedious and thus becomes one of the main sources of programming errors.

A type of on-line programming which removes the above-mentioned disadvantages—at least partially— is the textual programming of motions. Textual program codes are easier for the user to understand because of their language-like syntax and are well documented. Another advantage is that alternative motions can be allowed in the program code. The disadvantage of this method is that the purely textbook program code does not show the real behaviour of the robot when executing the program. These disadvantages stand in the way of a widespread use of robots in industry, and particularly in assembly operations.

## Hybrid programming

Hybrid programming is a first step to avoid the disadvantages that accompany on-line programming (Figure 5.1). Here the programming, or the input of the programming sequence and logic, is typed in as text at a terminal isolated from the actual production process, i.e. off-line. However, the coordinates of the points where the robot has to call during the program must still be put in on the spot, i.e. at the robot, using the teach-in method.

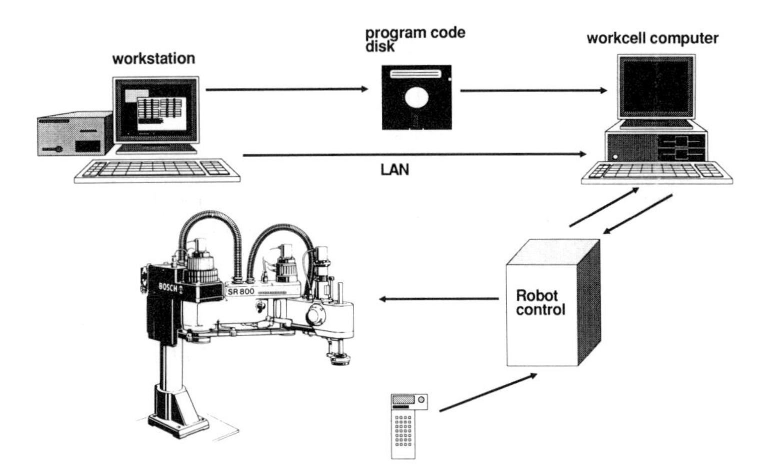


*Figure 5.1 Hybrid programming, a first step to avoid the disadvantages of on-line programming.*

Hybrid programming shortens the set-up time considerably. With this method, the program flow is understandable for the user because the programming is done textually. However, hybrid programming still has the following disadvantages:

- Before the program code is actually transferred to the robot controller, there is no opportunity to check the programmed motions for errors and collisions;

- Testing is still done at the workcell, so that the robot is not available for production during the test phase; and
- The programmer's 3-D mental representations can be overtaxed when programming complex assembly tasks using robots with five or six axes. This problem does not exist with traditional and simple robot tasks like 'pick-and-place' or palletizing.

## Off-line programming in combination with graphically supported simulation systems

The basic function of graphically supported simulation is the visualization of complex processes. The realistic and dynamic form of representation makes it possible to trace the programmed motions of the robot graphically and to realize problems and their consequences at an early stage. If the graphical information is presented well, graphically supported simulation can be a useful aid for decision-making. In future, such simulation systems will be used more frequently as an aid for programmers in production planning.

To use off-line programs, it is important to create a realistic representation of robot and control in a computer model (Figure 5.2). Also, the workcell should be depicted as realistically as possible, including all the peripheral elements, such as worktable, and all objects handled by the robot during the production process.

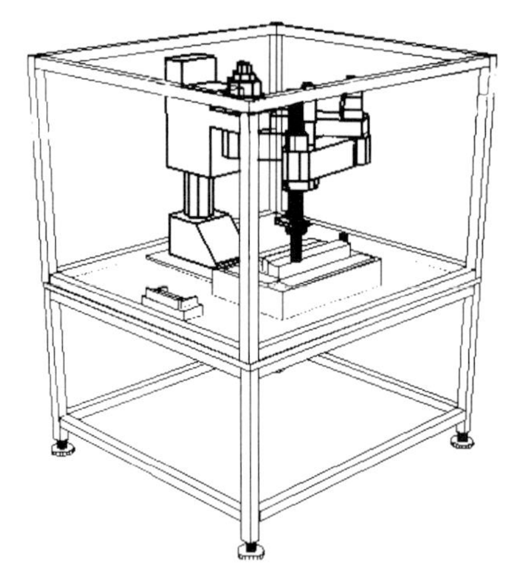

*Figure 5.2 An example of a representation of a workcell in a computer model.*

Simulation systems with combined off-line programming should meet the following demands:

- *Independence from a specific robot*
  Off-line programs are meant to be written only once and should be transferable to robots of different makes but with the same kinematics;
- *Independence from a specific computer*
  The simulation and programming system should be written in a popular and efficient programming language such as FORTRAN, PASCAL, C etc;

- *User-friendliness*
  The programmer is provided with a help system as an aid to decision-making during her/his work on the simulation system;
- *Flexibility*
  The user interface should be flexible enough to allow users with different levels of knowledge to work with the system;
- *Time reduction*
  The times for on-line programming and for testing at the workcell are reduced. This results in shorter downtimes;
- *Modularity*
  The modular structure of the whole system allows functional extensions;
- *Type of software*
  There should be a separation between application-dependent and application-independent software;
- *Modelling*
  Realistic modelling of the simulated environment must be possible; and
- *Independence from CAD systems*
  The system does not need CAD systems for the visualization of robot cells and motions. Thus, there is no additional expenditure for the purchase of a CAD system. A further disadvantage of the use of CAD systems to visualize motions is their static pictures.

A simulation system which fulfils the above demands has the following advantages:

- Detailed layout planning makes it possible to give statements on the choice of robot and tools, on the best possible allocation of the peripheral elements and on the best possible assembly point within the workspace of the respective robot. Thus, the planner can already take optimizing steps before the realization, i.e. without the need to build the robot cell first;
- The possibility of testing different variations of layouts allows the optimization of time and sequences of movements;
- Programming and layout mistakes do not cause damage to the hardware, as virtual collisions can be detected in advance and not through trial and error;
- The robot can still be in production while a new program is being written. This increases effective production times and reduces downtimes; and
- Programming is made safer because the different steps do not have to be determined by an operator and a teach pendant at the robot.

# *GROSS—a simulation and off-line programming system*

## Characteristic features of the system

Graphical Robot Simulation System (GROSS) is a tool for graphical interactive layout planning of workcells and for off-line programming of automatic assembly cells. The following list gives the main features that facilitate the planning of workcells:

- Easy geometric, kinematic and technological modelling of robots, peripheral elements and assembly components. Quick and easy configuration of workcells and simple modification of workcell layouts. These tasks are greatly supported by the graphics editor Interactive Surface for Graphical Design (IAOGRAPH) which was developed at the IAO;

- Comfortable off-line programming of motions. This is done at a comfortable, menu-driven interface. The movements are programmed textually or with the help of a mouse;
- The Industrial Robot DATA code (IRDATA), which is independent of the type of controller, is generated automatically from the high-level language code programmed off-line;
- The control program written off-line is tested by means of the graphically supported simulation of movements in a model of the robot environment. For this purpose, a special simulation module has been developed. The control programs are verified by checking the workspace and checking for collisions; and
- The verified control program is transferred directly from the planning computer to the robot controller via a Local Area Network (LAN).

## Description of the main modules

### *Module for the configuration of objects and workcells*

The general basis for the graphical simulation of movements is a model of the work environment. The robot and its work environment must now be visualized graphically on the monitor (Figure 5.3). This is normally done by means of a commercial CAD system which has several disadvantages. First, there is the additional cost of a CAD system. Second, the use of such a system requires a very long training phase which is not acceptable for a planner of robot cells or a programmer. Finally, the use of a CAD system would be excessive for the tasks described above. Therefore it is necessary to find a straightforward,

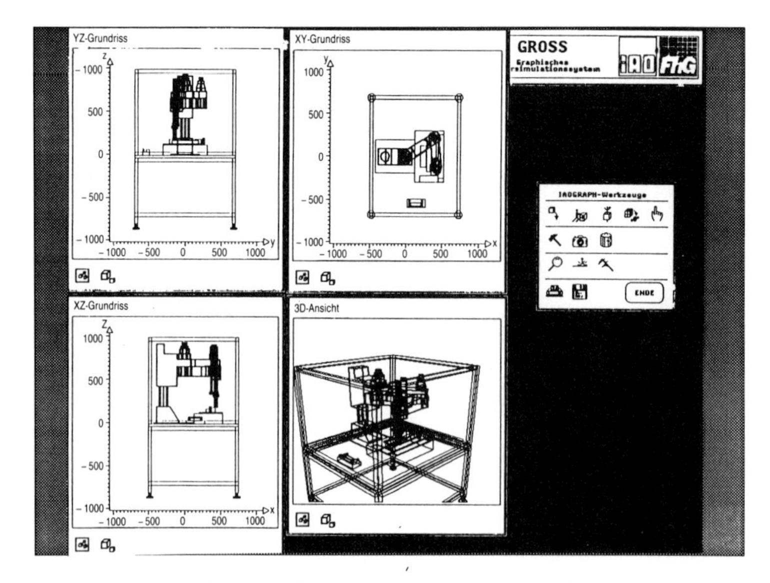


*Figure 5.3 User interface for the configuration of objects and workcells with the help of a graphics editor.*

reasonably priced tool for the configuration of objects and workcells. The graphics editor IAOGRAPH is an efficient tool which meets the above demands. For the modelling of objects and work cells, IAOGRAPH refers to a set of different primitive solids such as cuboids, cylinders and truncated cones (Figures 5.4 and 5.5). The reproduction of objects out of these basic solids can be done as exactly as required.

The geometric complexity of objects directly influences the response time of the system. Therefore, if one takes into account the response times in the

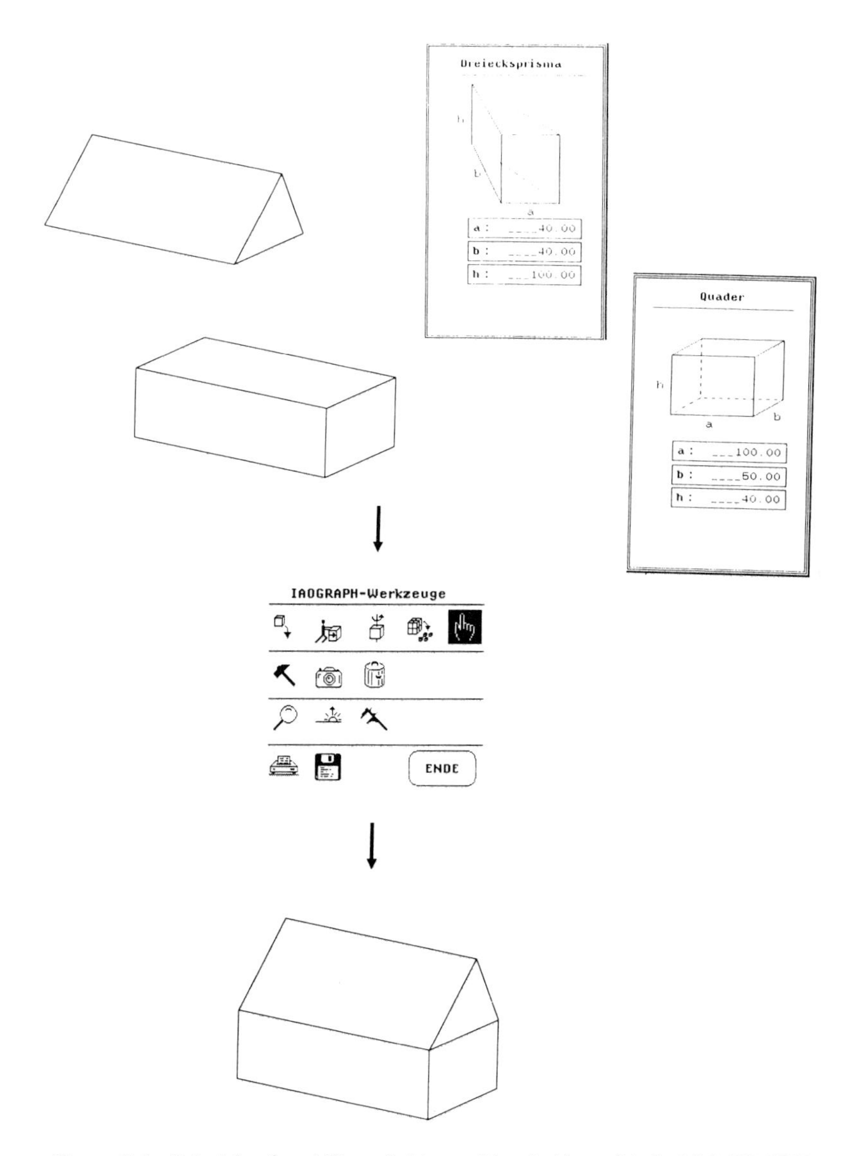


*Figure 5.4 Principle of modelling of objects with primitive solids in IAOGRAPH.*

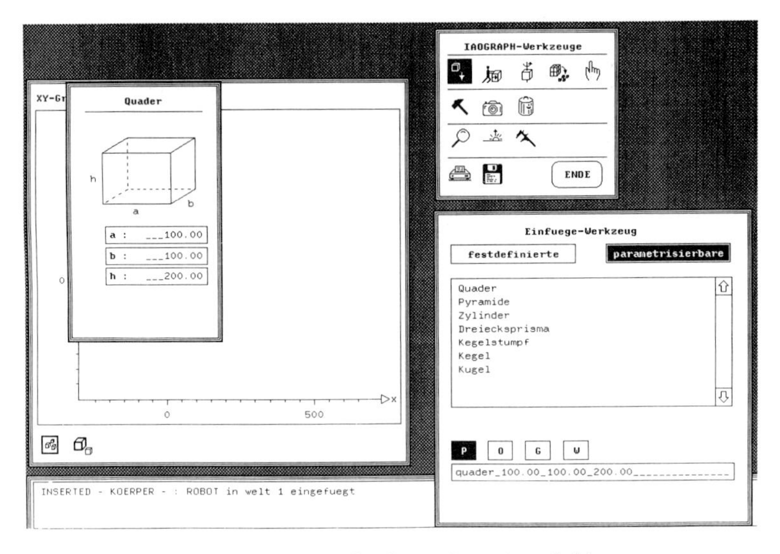


*Figure 5.5 Module for the configuration of objects.*

simulation of motions, the planner will only use approximations for the modelling of objects. Objects which already exist in popular CAD systems like EUCLID, Medusa or AutoCAD, can be transferred into the system IAOGRAPH via various interfaces by conversion of data.

In order to complete the data model of an object, the object is not only described geometrically, but a kinematic and a technological description are given too. Important kinematic information is, for example, the relative positions of the axes in relation to each other and the absolute positions of the axes in the inertial system. Technological factors are, for instance, the type of robot or the positioning accuracy of a robot.

For the configuration of complete work cells, the planner can choose from different classes of objects (Figure 5.6). With the help of the following library of parts: industrial robot; tools; workpiece; workpiece components; suppliers; storage devices; worktables and workpiece carrier, the planner can configure and vary workcells freely. Alternative layout variants of objects or whole groups of objects can be generated easily and quickly by manipulations such as adding and deleting and shifting and twisting (Figure 5.7).

Depending on the robot used and on the workcell layout, it is possible to examine the accessibility of work points such as machines and workpiece locations and to check the corresponding accessibility.

*Module for off-line programming of robot controls*

Conventional textual programming makes routine work necessary where mistakes are inevitable. For this reason, a comfortable, menu-driven, graphical

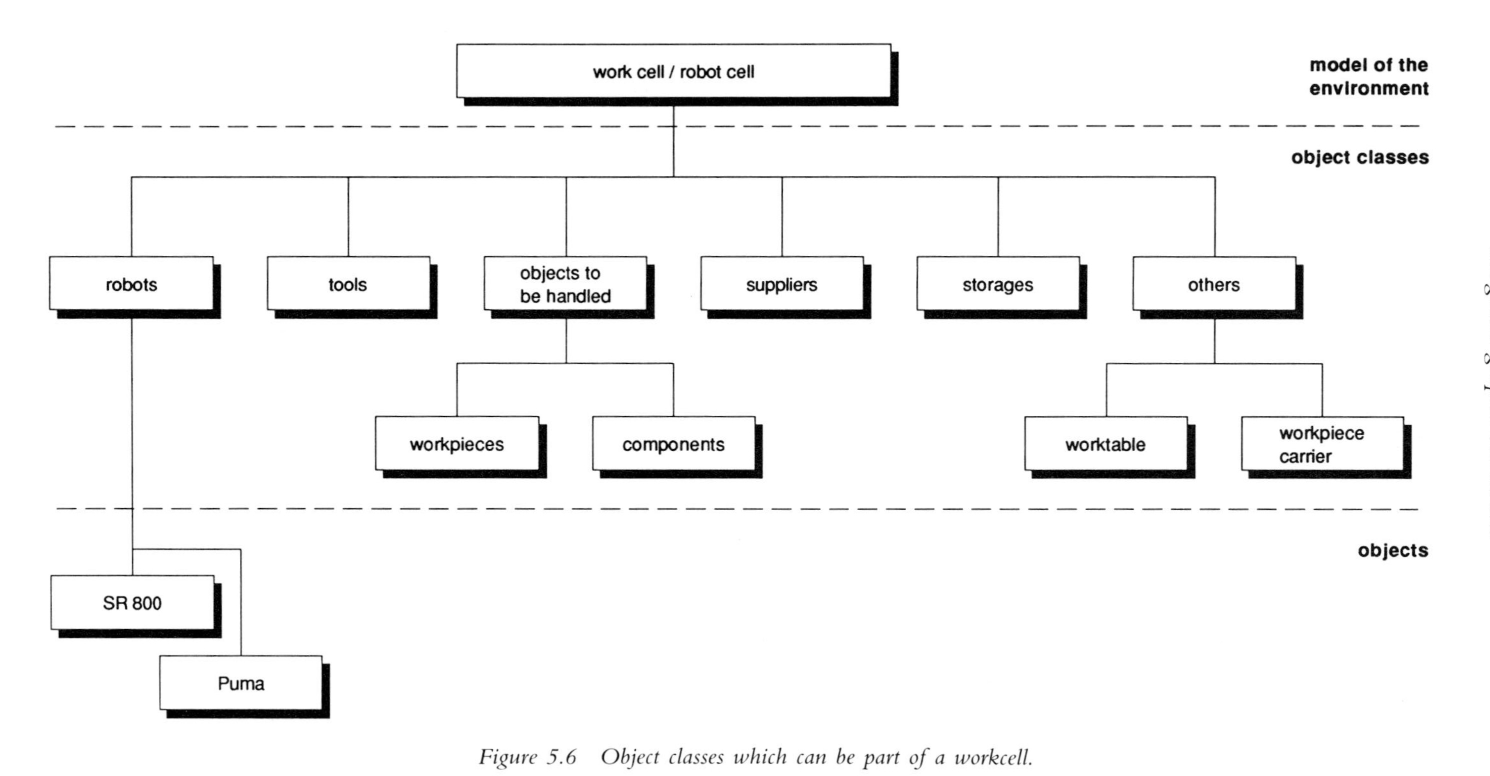


*Figure 5.6 Object classes which can be part of a workcell.*

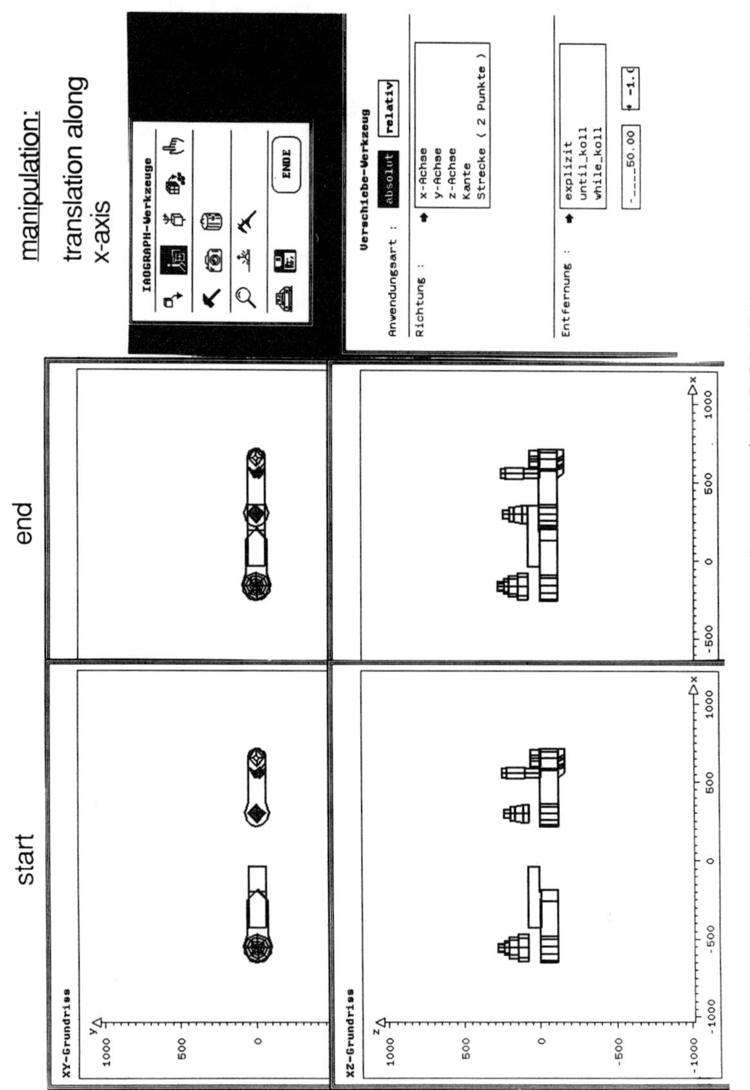

*Figure 5.7* *Manipulation of objects with IAOGRAPH.*

user interface was developed for GROSS. The task of this interface is to relieve the user of routine work in programming and thus to help avoid mistakes. There are two basic modes for programming with the user interface. In the 'learner mode', designed for programming novices, textual programming is done exclusively by activating menus. The interface offers fragments of commands which, when activated, are transferred into the text editor. Here the programmer adds variable names, program names etc. to the fragments of commands. The user-interface design attempts to keep inputs by keyboard to a minimum.

The experienced programmer has the option to enter the whole program code or parts of it into the text editor by keyboard. The user interface was developed for the programming language BewegungsAblauf ProgrammierSprache (BAPS) with its language-specific features which was developed by the Robert Bosch company. Nevertheless, it is possible at any time to generate further specific user interfaces for other programming languages. As mentioned above, the menu-driven user interface supports the programmer in her/his task. As robot programming languages and their syntax are usually modelled on conventional high-level languages, the user interface reflects current programming techniques.

The following paragraphs illustrate the programming process by giving examples of individual commands. For instance, a program name has to be declared at the beginning of programming. The text editor automatically generates a program body (Figure 5.8). Then the programmer must add the name of the program to the program body but does not need to generate the whole program body.

Programs are usually divided into a declaration part, a statement part and, when the programming language BAPS is used, an additional part for the definition of subprograms. For instance, when the declaration part is subdivided further with the function 'declare type of variable', the types of variable permitted in the programming language BAPS are shown (Figure 5.9). When the desired type of variable is determined, the declared variable is taken over into the respective file of variables. During the programming phase this variable is made available to the user in the statement part. Besides, the programmer has access to further lists of this kind which contain, for instance, already declared channels, programming macros, subprograms and GO TO statements.

Menu-supported completion of commands is illustrated by the example of the 'FAHRE' command (Figure 5.10). When specifying motion commands, their characteristics are taken into account automatically, and the user is offered a choice of respective variants. To complete the 'FAHRE' command, the respective variable names or the speed and acceleration values must be added in the text editor. With 'programming macros', the programming interface offers a further function that makes the programming of robot controls more efficient. This function allows the user to record frequently recurring program fragments as macros and to transfer them into the text editor if necessary. In the text editor the selected macro is then completed with the variable name similar to the motion commands.

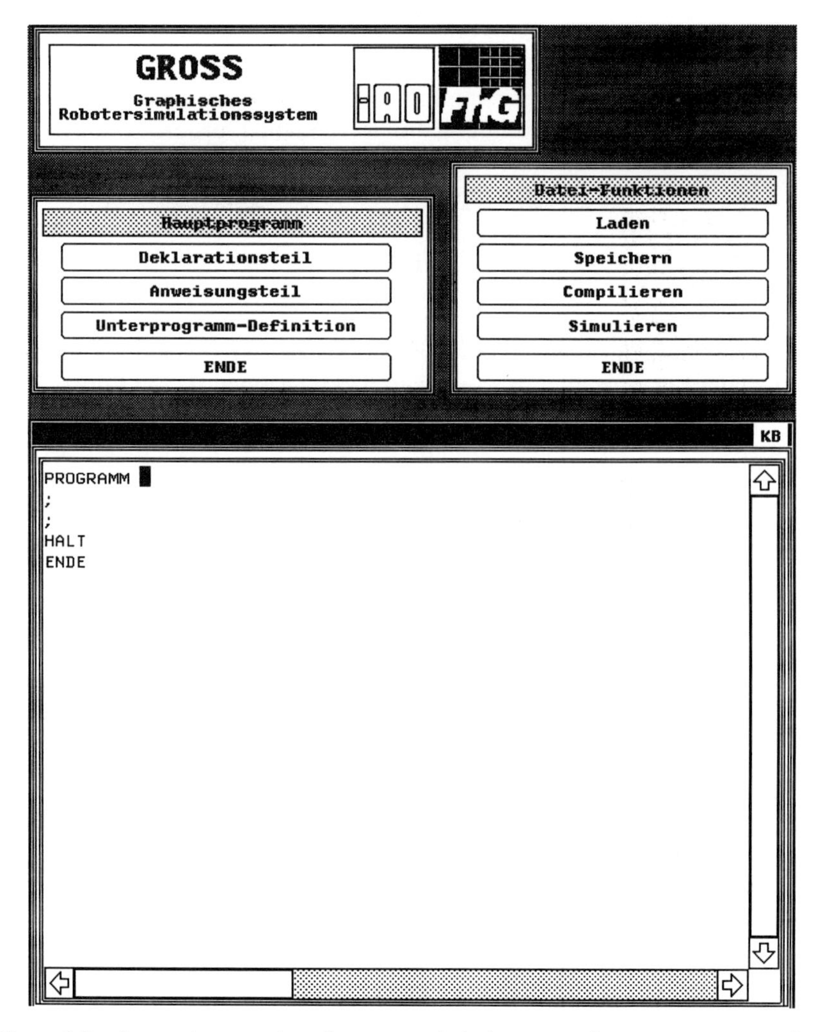


*Figure 5.8 Automatic generation of a program body by means of a programming interface.*

## A module for the graphical simulation of the off-line robot control program

The module for graphical simulation supports the user mainly in three points. The industrial robot moves in 3-D space. Because of its complexity, the user frequently cannot understand it when programming. All the user knows is that the robot moves from point A to point B. It is therefore necessary to validate off-line robot control programs before they are put into operation on the shop floor. For this purpose the programmed motions of the industrial robot are checked for possible collisions with peripheral elements and other objects in the workcell. Also, it is possible to examine whether the work points determined

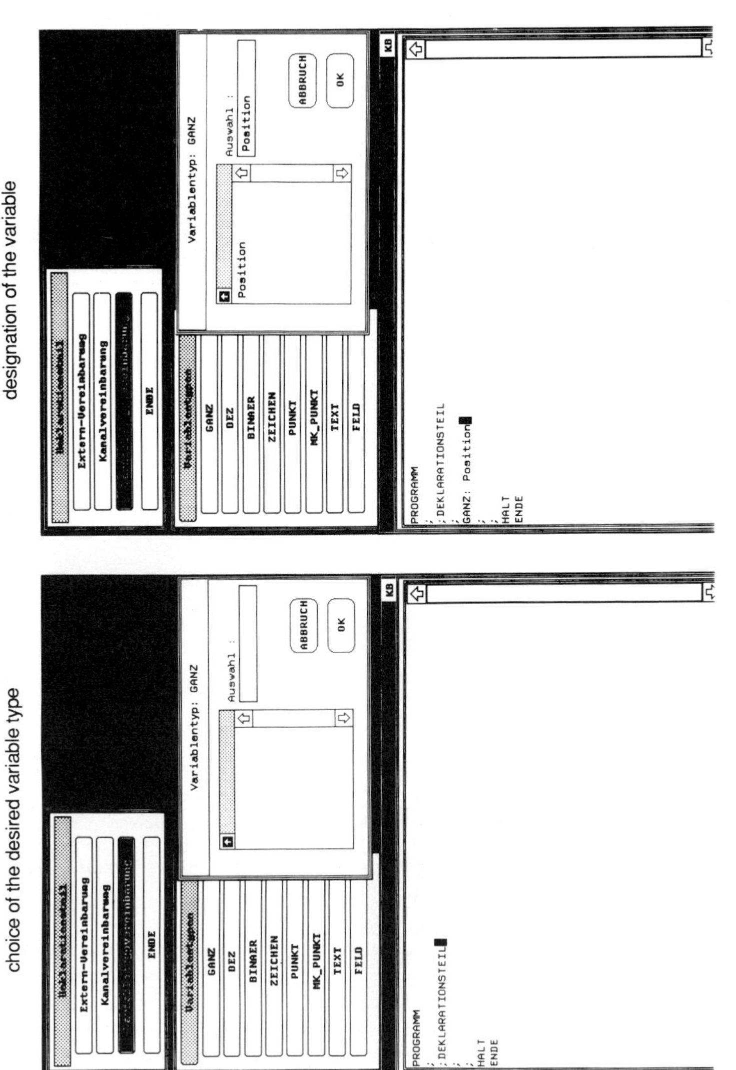


*Figure 5.9 An example from the declaration part: function declaration of variable types.*

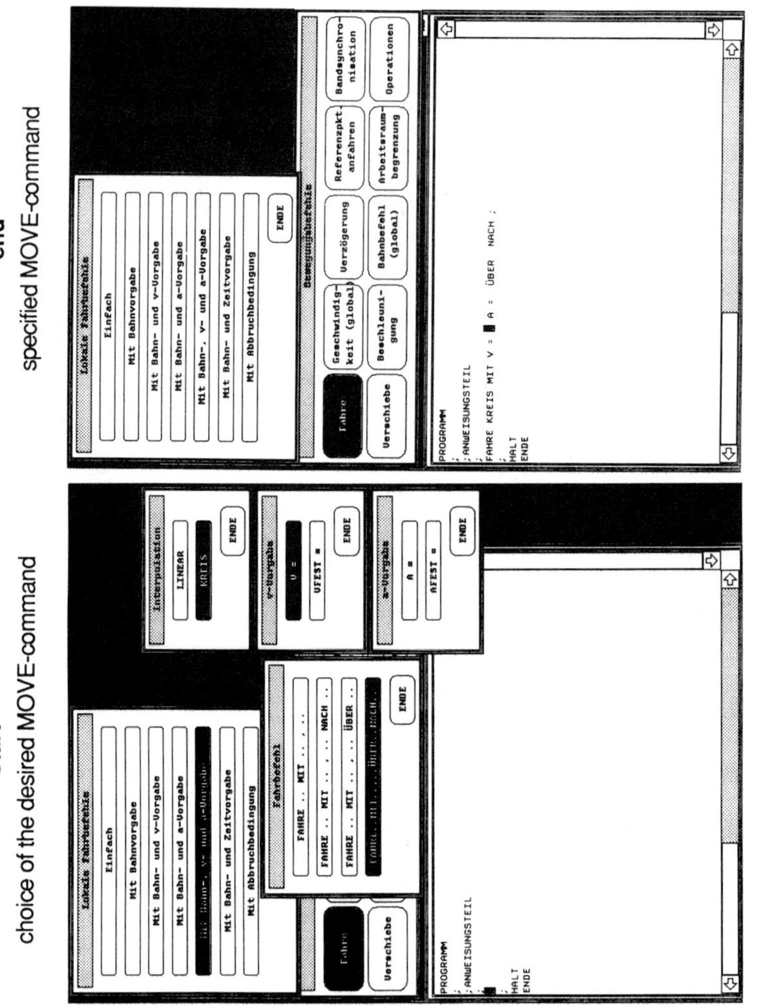


*Figure 5.10* *An example from the statement part: specification and completion of the 'FAHRE' command.*

in the control program are within reach of the robot. Finally, this module offers the possibility of testing alternative parts of the program and thus optimizing motions.

With the help of file functions, the robot control programs written at the programming interface can be edited for graphical simulation. Alternatively, the program can be transferred directly onto the robot control, in case a checking of the control programs through graphical simulation is not desired. For direct transfer of the control program from the computer to the robot control, a communication interface has been installed. Attention must be paid, however, to the fact that almost all robot manufacturers use their own programming language for the programming of their robot controls. These languages differ, sometimes considerably, in syntax, structure and storage capacity. This leads to the following considerable disadvantages for users who work with various different robot controls:

- The programmer has to know the respective programming language of every robot in the company;
- the amount of work needed for the transfer of an already existing program onto a different controller is the same as if the respective program were written anew in a different programming language. This applies even if the robot kinematics are similar; and
- Taught coordinates cannot be transferred from one robot controller onto another.

These disadvantages can be eliminated by using the independent IRDATA format (IRDATA, 1987) as a standardized interface (Figure 5.11). For this purpose, the finished program code is translated by a corresponding language compiler into IRDATA code. This IRDATA code has the advantage that the simulation part does not have to be adapted to every robot programming language. As already mentioned, the simulation system can be extended to other robot programming languages very easily and at any time, using the given programming tools. With

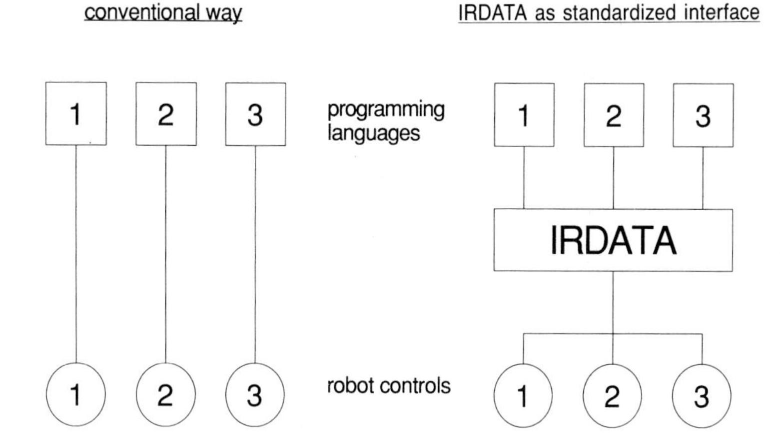


*Figure 5.11 IRDATA, a standardized interface for the transfer of the finished program code to any robot with the same type of kinematics.*

the system GROSS described here, it is also possible to program and simulate robots from different manufacturers, but with the same type of kinematics.

An interpreter as a neutral interface to the simulation module ensures that IRDATA code can be used for simulation. It interprets the IRDATA code independently of the robot language used, and gives the simulation module statements for processing. The simulation module consists basically of three components: simulation of the control; simulation of the kinematics; and graphical simulation. The control simulation takes the statements from the interpreter and, if necessary, starts kinematic algorithms for the calculation of the workspace, the calculation of the final point (if the motion increments of the axes are given), or for the calculation of the motion increment of each axis (if the final point is given). The control and kinematic data records are transferred to the graphics display. Here, visualized sequences of movements are produced for each action. Within each class of equipment (e.g. Selective Compliance Assembly Robot Arm (SCARA) robots) the queued algorithms are independent of individual robots. In case of a transfer onto a different class of equipment, the modules for the calculation of the kinematic chain have to be adapted. The modules' compiler, interpreter and calculation of control and kinematics are executed automatically without any possibility of intervention. During the graphical simulation, however, the user can influence the system directly.

In the data function 'simulation' the user is offered different modes of running the graphical simulation (Figure 5.12). If during the simulation a variable is found in the code whose coordinates have not been defined, the graphical simulation is interrupted. The coordinates can then be put in either explicitly by means of the keyboard, or by activating the corresponding point in the graphic with the mouse.

After the coordinates are determined, the running of the program code is continued automatically. If the programmer notices critical points, such as collisions of the robot with other objects (which the graphic module can point out automatically) or if predefined workpoints are not reached, the programmer can remove these points in the program by corresponding corrections in the program code. This iterative planner–computer dialogue is continued until an optimized program flow is found.

## *Description of the development tool*

### Background

Computer-Aided Engineering (CAE) systems are successful as components of a comprehensive Computer-Integrated Manufacturing (CIM) system and are still growing in importance. Applications in CAE such as interactive modelling, simulation and robot programming are characterized by complex functions and a complex user interface. A user interface is normally built from a graphic system and a collection of simple, interdependent dialogue elements like menus and input fields.

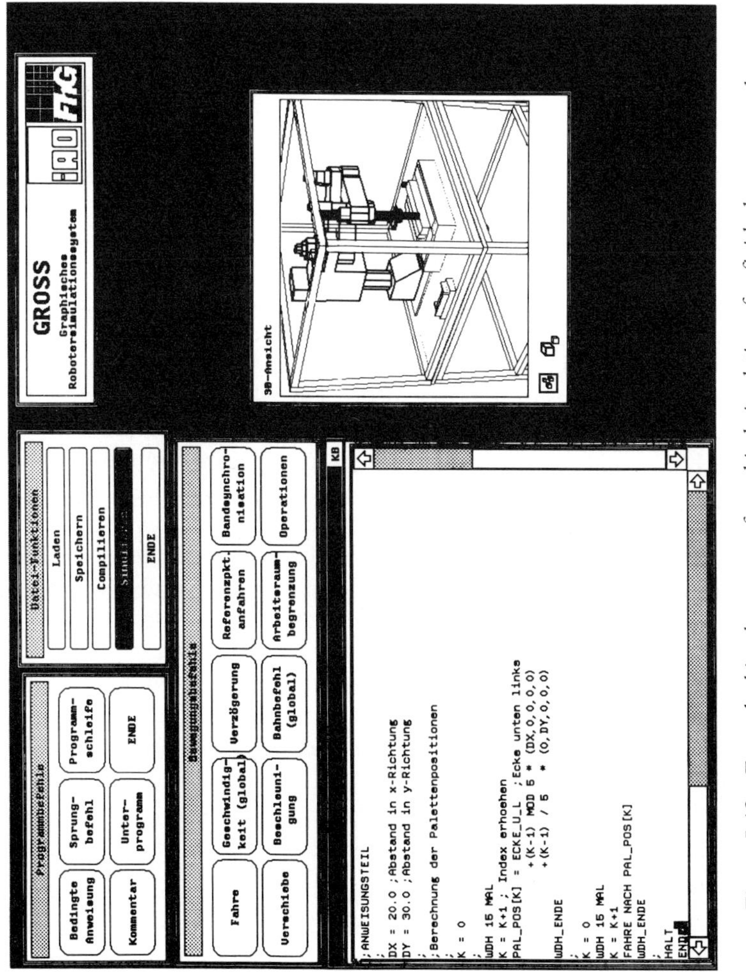

*Figure 5.12* Error checking by means of graphical simulation of a finished program code.

To improve the current system for CAE, the IAO has developed a tool for rapid and efficient construction of window-based user interfaces. This tool, IAOGRAPH, connects 3-D graphics with simple window-based dialogue elements. Thus, 3-D graphic elements found on the screen can be individually manipulated with input devices similar to menus and input fields. When the system was designed, the following features had priority:

- application interdependence;
- direct manipulation;
- possibility for fast prototyping; and
- rapid and simple user-initiated changes.

## User interface

First, a description is given of how the user interface of an application system, constructed through IAOGRAPH, presents itself to the user. In user interfaces it is possible to create several movable windows which are called dialogue blocks in IAOGRAPH. Every dialogue block can contain a number of desired dialogue elements. These elements carry out the communication between user and application in various ways, according to their features.

The dialogue elements are divided into four groups: input/output elements; interaction elements; basic display elements; and 3-D graphic elements. The user performs the input/output and selection of data by means of the input/output elements, for example the input/output fields for strings and numbers of different kinds of menus. Functions of the application program or service functions of the user interface can be called directly with interaction elements such as buttons or icons. Basic display elements, such as text, lines or circles, serve as decoration or explanation of the dialogue blocks. The display 3-D layouts, and the selection of objects within these layouts are realized through 3-D graphic elements.

The control of an application with the user interface is user-initiated. With the help of the mouse or the keyboard, the user can control the application system through every dialogue element which is on the screen. Specific functions can be disabled simply by removing the related dialogue elements or entire dialogue blocks from the screen.

## Construction Language for User Interfaces (CLUI)

The type, appearance, defaults and contents of the individual dialogue blocks and elements of the user interface are defined by the construction language. The appearance of the dialogue block is defined by the position, size and various display characteristics like text fonts and line width. The control of mouse button detection in relation to a dialogue element would be defined by attaching actions to the dialogue element. Such actions could consist of application functions or service functions provided by IAOGRAPH.

Figure 5.13 shows the definition of the dialogue block 'Beispielbaustein', in CLUI and how it appears on the screen. The size, position and name of the dialogue block is defined with the command bo$append_dialog_block.

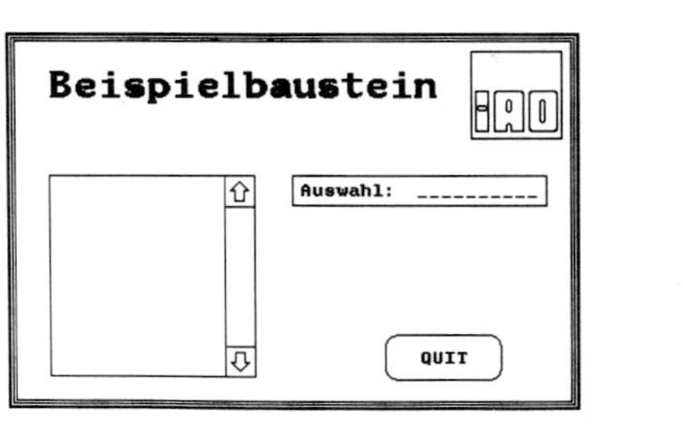


```
/********    Baustein "Testbaustein"    ********/

bo$baustein_anhaengen ( "Testbaustein", 18.00, 16.00, 12.00, 8.00 );

  bo$element_anhaengen ( "Testbaustein", "Überschrift", TEXT, 0.75, 6.75 );
    text$text ( "Beispielbaustein" );
    text$fett ( TRUE );
    text$gross ( TRUE );
    bo$aktion( MAUS_MITTE, bo$doing_nothing );

  bo$element_anhaengen ( "Testbaustein", "Scroll_menü", SCROLL_MENUE, 0.75, 5.00 );
    bo$aktion( MAUS_MITTE, bo$doing_nothing );

  bo$element_anhaengen ( "Testbaustein", "Eingabefeld", E_ZKETTE, 6.00, 5.00 );
    e_zkette$text ( "Auswahl: " );
    bo$aktion( MAUS_MITTE, bo$doing_nothing );
    bo$aktion( CURSOR_OBEN, bo$doing_nothing );
    bo$aktion( CURSOR_UNTEN, bo$doing_nothing );
    bo$aktion( EINGABEANFANG, bo$doing_nothing );
    bo$aktion( EINGABEENDE, bo$doing_nothing );

  bo$element_anhaengen ( "Testbaustein", "Iaoicon", ICON, 9.85, 7.75 );
    bo$read_icon_form ( "Testbaustein", "Iaoicon", "HOGIBOF$ICON:IAO.ICON");
    bo$aktion( MAUS_LINKS, bo$doing_nothing );
    bo$aktion( MAUS_MITTE, bo$doing_nothing );
    bo$aktion( MAUS_RECHTS, bo$doing_nothing );

  bo$element_anhaengen ( "Testbaustein", "Quitbutton", BUTTON, 8.00, 1.50 );
    button$form ( RECHTECK_RUND );
    button$text ( "QUIT" );
    button$groesse ( 2.50, 1.00 );
    bo$aktion( MAUS_LINKS, bo$doing_nothing );
    bo$aktion( MAUS_MITTE, bo$doing_nothing );
    bo$aktion( MAUS_RECHTS, bo$doing_nothing );
```

*Figure 5.13 Description of a dialogue block in CLUI and its appearance on the screen.*

### The graphics editor for user interfaces

The graphics editor for user interfaces helps both the user and the application developer interactively to create and modify user interfaces. The editor is an application system developed with the help of IAOGRAPH. Therefore the editor's user interface offers the same possibilities as other user interfaces developed with the editor. As a result, the editor provides the description of the dialogue blocks and elements as a text file in the language of CLUI. Figure 5.14 shows a screen shot of the graphics editor.

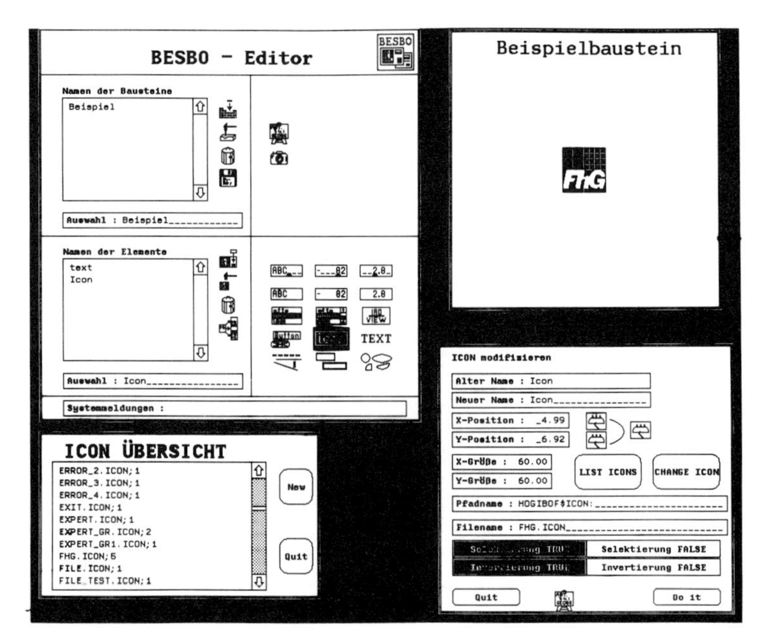


*Figure 5.14 Graphics editor for user interfaces.*

### The creation of an application system with IAOGRAPH

The creation of an application system is done in two steps, which can be carried out independently of each other, depending on the application: developing and testing application functions in the C programming language; and creating the user interface with CLUI and testing it with the help of dummy functions. After the application functions have been completed, they would be tied into an executable program with the IAOGRAPH modules which are filed in the program library. The entire application system can then be tested together with the user interface.

## *Conclusion*

With GROSS, a planning tool has been developed that supports off-line programming of robots and the design of workcells efficiently. The system offers several advantages. It allows realistic modelling of robot cells and the examination of their functionality before they are built. It is further possible to check the robot motions for errors or collisions with peripheral elements and to optimize the motions. The menu-driven user interface relieves the user of routine work and provides support at any time through an effective help system. The independence of the system from other CAD systems for the graphical simulation avoids additional expenditure.

GROSS is being developed further. For instance, a module is planned which helps to find the optimal place of assembly within the workcell. Presently, the IAO is working on the transfer of relevant assembly data, such as assembly sequences and insertion directions from the CAD system MOCAD which was also developed at the IAO (Bullinger *et al.*, 1989). The sequences of movements necessary for assembly are automatically translated into the program code.

## Acknowledgment

We would like to thank Mrs Katrin Schwilk for the translation of this paper into English.

## References

Bullinger, H.-J., Warschat, J. and Richter, R., 1989, Montagegerechter Erzeugnisentwurf auf der Basis objektorientierter Produktmodellbeschreibung. *VDI-Z,* **11,** 67–70.

CATIA Robotic, 1982, *User Manual* (Aachen: Dassault Systems).

McDonnell-Douglas Automation Company, 1981, *Computer Graphics for Robot Off-line Programming.*

IRDATA, 1987, *Allgemeiner Aufbau, Satztypen und Übertragung.* VDI 2863, Blatt 1 (Düsseldorf: VDI-Verlag).

Wörn, H. and Stark, G., 1987, Robot applications supported by CAD simulation. *Robotics and Computer-Integrated Manufacturing,* **3,** 1, 52–62.

# *PART II*
# *SAFETY*

# *Introduction*

The field of robot safety considers a set of human–robot interactions which affect the physical and psychological well-being of the individuals working with robots. The six chapters included in this section offer a review of hazards created by the introduction of robots into human-occupied environments. Hazard reduction strategies and safety solutions are also included. In Chapter 6, the authors present and analyse the robot-related accidents reported in Finland since 1989. They also describe the paradox of robotization, i.e. the fact that industrial robots are often used to alleviate the heavy dangerous tasks, yet they have caused other types of over-exertion injuries.

It has been mentioned that reliability is a prerequisite for safety of robotic systems. In Chapter 7, the author reviews the use of system safety techniques (including Fault and Event Tree Analysis) to examine the safety and reliability of robotic systems. In parallel with the discussions in Chapter 6, this chapter explains how improvements in hardware and software reliability and interfaces can increase safety irrespective of the work layout and human behaviour characteristics.

In Chapter 8, the authors describe the concept of 'six severity level design' (SSLD) for safety operation of manufacturing cells with robots as their main working machines. The proposed robot safety design philosophy integrates guarding techniques with control actions, while considering both the production needs, safety aspects, and interfacing manufacturing process equipment and functions. The practical implementation of the SSLD concept for the automated flexible manufacturing cell is also discussed. It is shown that the proposed system provides for maximum protection to human operators, while causing minimum interruption in the production process.

Following the same line of discussion as the preceding chapters of this section, Chapter 9 begins with a discussion of the variety of hazardous sources that may exist in an increasingly complex advanced production system and the difficulty by which an effective hazard assessment and control can be made. The authors have developed a simulation method to study the human operator actions, focusing on hand motion patterns in removing a faulty product from a robot production line. The proposed simulation is used to specify some design criteria for the human–robot interface in computerized production systems.

During the past 10 years, many authors have emphasized the importance of using robot sensors to increase the safety of robotic work zones. Chapter 10 explains one specific safety design concept through the application of intelligent safety sensors. This sensor system eliminates production process-related inflexibility introduced by fixed barriers and avoids occurrence of unnecessary product defects due to production interruptions. At the same time, intelligent sensors are capable of detecting human presence, thereby improving the level of human safety in a robotic system.

Unlike some highly developed countries, Poland introduced a programme to study issues related to robot safety in the early stages of its advanced production development. Chapter 11 contains a description of a fairly comprehensive programme of robot safety in Poland and the research efforts needed to maintain acceptable safety criteria for their robot implementations.

# *Chapter 6*
# *Industrial robot-related accidents in Finland*

**J. Järvinen[1], W. Karwowski[1], J. Lepistö[2] and M. Rahimi[3]**

[1]*Center for Industrial Ergonomics, University of Louisville, Louisville, Kentucky 40292, USA;*
[2]*Technical Research Centre of Finland, Safety Engineering Laboratory, Tampere, Finland; and*
[3]*University of Southern California, Institute of Safety and Systems Management, Los Angeles, California 90089-0021, USA*

**Abstract.** Industrial robot-related accident data recorded in Finland were analysed. The accident data from two computer data bases maintained by the National Board of Labour Protection in Finland were used. Data base TARE (accident description register) contains accident reports made for insurance companies by the employers of injured workers. Data base TAPS (accident report register) contains accident reports of severe or fatal accidents investigated by occupational safety inspectors from 1977. All reported accidents which occurred between 1985 and 1989 were included in the study. The over-exertion injuries due to manual loading and unloading of the robot systems, and those which occurred during system disturbance control, were common among the studied cases. It was concluded that manual handling of parts and dealing with robot disturbance and malfunction situations appear to be the most dangerous human–robot activities in this study. These tasks should be given more consideration in robot workstation design.

## *Introduction*

Industrial robots eliminate many traditional risks of injury, but also introduce new hazards related to their high speed of motion and unpredictable motion patterns (Karwowski, 1991). During disturbances or robot malfunctions, robot operators may face unexpected situations with limited time to decide the proper action or, in some cases, may need to do the robot's work.

To identify the hazards caused by industrial robots, past accidents need to be carefully analysed. Nicolaisen (1985) states that there is little data available on robot-related accidents, and there may be a high percentage of unrecorded accidents. Accidents involving robot use are difficult to identify in the classification of accidents caused by other machines (Sugimoto and Kawaguchi, 1985; Karwowski *et al.*, 1988).

This report is based on robot-related accident data obtained from two computer data bases, which are maintained by the National Board of Labour Protection in Finland. Data base TARE contains accident reports prepared for insurance companies by the employers of injured workers. Currently, the data base contains the accident descriptions made in 1987 (about 50 000). Data base TAPS contains more than 6000 accident reports of severe or fatal accidents investigated by occupational safety inspectors from 1977. Accidents which occurred between 1985 and 1989 were included in the study.

## *Previous research*

Hazards which can lead to an accident can occur in three modes of robot operation: normal operation, maintenance or teaching (programming). Jiang and Gainer (1987) define the robot accidents as 'contact between the person and a robot either directly or indirectly, leading to a record of the accident'. Carlsson (1985) investigated 36 robot-related accidents which occurred in Sweden in the years 1979–1983. The study was confined to accidents that occurred because of the injured person's contact with a robot while it was operating, or caused by the robot's special functions. Carlsson excluded accidents due to over-exertion when repairing a robot or impacting a stationary robot. However, this type of accident was found to be rare. Most of the accidents (14 out of 36) occurred when the flow of materials had been disrupted and the robot was adjusted manually during normal operation. Almost as many cases (13 out of 36) occurred due to the robot's unexpected movement while programming or repairing. Almost 70% of the accidents involved the operator's finger, hand or head. It was common that the injured person was inside the robot's working area. A survey of robot-related accidents on 190 plants with 4341 robots in use was conducted in Japan (Sugimoto, 1985). A total of 11 accidents and 37 near-accidents due to robots were reported during a 5-year period (1978–1982). Two of the accidents were fatalities. Most of the accidents were due to unexpected movements of robots. Between 1978 and 1987 ten fatal accidents were reported in Japan (Nagamachi, 1988). The accidents occurred during the phases of normal operation of the robot, repair, maintenance or installation work, or during manual adjustment of a workpiece.

In terms of robot accident analysis, Jiang and Gainer (1987) conducted a cause–effect analysis on 32 robot-related accidents reported in the USA, West Germany, Sweden and Japan. They found that line workers were at greatest risk of injury followed by maintenance workers and programmers. Pinch-point accidents (trapping of a part of a human body by the robot) caused 56% of all accidents. Impact accidents (a robot component, tool or workpiece striking the individual) caused 44% of the accidents. Most accidents were caused by poor workplace design or human error. The fatal pinch-point accident, which occurred in 1984 in the USA was also included in the cause–effect analysis. According to Jiang and Gainer (1987), many of the accidents reviewed could have been prevented if attention had been paid to safety throughout all phases of robot implementation: workplace design, robot installation, robot testing and robot operation.

Karwowski *et al.* (1888) compared the safety performance of an appliance manufacturer before and after computer automation of the assembly process. Fewer accidents related to material handling occurred after the automation, partly because of the change from metal to plastic parts. However, some sprains and strains still remained since workers were responsible for loading the product onto the conveyor manually when the robots malfunctioned. Also, tennis elbow injuries and over-exertion type accidents increased.

Jones and Dawson (1985) investigated reliability and safety aspects of 37 robot systems in three companies in the UK. Approximtely 25% of system production time was lost due to disturbances of the normal operation. Some of these disturbances also placed people at risk. Kuivanen *et al.* (1988) investigated the causes of disturbances of automatic production systems and their influence on safety. The studied systems included 26 robot systems among other automated systems. According to the study, in about half of the disturbances workers were exposed to increasing degrees of hazards with accident rates between 1 and 2%.

## *Robots in Finland*

According to the statistics from Suomen Robotiikkayhdistys, the Finnish Robotics Association, there were 427 industrial robots at the end of 1987. This number had increased to 671 by the end of 1989. Almost half of the industrial robots were, and still are, welding robots (Table 6.1). The second largest application area was part handling. According to the Finnish Robotic Association, a device is not classified as a robot if it has less than four degrees of freedom.

*Table 6.1 Robot applications in Finland at the end of 1987.*

| Type of application | Distribution (%) |
|---|---|
| Welding | 48 |
| Part handling | 21 |
| Research and training | 13 |
| Surface treatment | 7 |
| Machining | 7 |
| Assembly | 4 |

## *Robot related accidents in Finland*

Fifteen descriptions of robot-related accidents were found from the TARE data base using the keyword 'robot-'. In addition, three robot-related accident reports, based on occupational safety inspectors' investigations, were found from the TAPS data base. One of them occurred in 1987 and two in 1989.

When interpreting the data the fact that accident descriptions often lack sufficient information must be taken into consideration (Karwowski *et al.*, 1988). Some accidents may not have been recorded in the data base, and some may have had different causal descriptions. Carlsson (1985) pointed to the same problems in his study. All accident descriptions from the present study are presented in the Appendix to this chapter.

Accidents due to operator's over-exertion when loading or unloading the robot workstation or when alleviating the disturbance caused by robot system malfunction were also included. This type of 'non-traditional' robot accident was relatively frequent. In such cases where there is no direct or indirect contact between the operators and the robot, the design of the robot system or the robot workplace has, in one way or another, an effect on the events leading to injury. This type of accident was included to point out that the activities related to serving the robot (loading/unloading) and disturbance handling should be taken into consideration when designing a robot system. In only a third of the cases was the injured person in direct or indirect contact with the robot (cases 2, 8, 10, 13, 17 and 18 in Appendix). Two-thirds of the accidents involved arm and back (Table 6.2). More than half (7 out of 12) of the arm and back injuries occurred due to over-exertion, which was the most common accident type, followed by the pinch-point type of accident, as shown in Table 6.3.

Over-exertion due to lifting or moving a load, or a disorder due to repetitive motions, was involved in seven accidents (Table 6.3). Six of the seven over-exertion type accidents occurred during system loading/unloading, when eliminating

*Table 6.2 Number of accidents by part of body injured.*

| Part of body | No. of accidents |
|---|---|
| Arm from shoulder to wrist | 6 |
| Back, spine | 6 |
| Fingers only | 3 |
| Leg from hip to ankle | 1 |
| Hand (including fingers) | 1 |
| Trunk (not back) | 1 |

*Table 6.3 Number of accidents by accident type.*

| Accident type | No. of accidents |
|---|---|
| Over-exertion or sudden motion | 7 |
| Over-exertion when lifting or moving a load | (5) |
| Disorder due to repetitive motions or inappropriate working postures | (2) |
| Pinch | 6 |
| Fall | 2 |
| Fall, slip or stumble on same level | (1) |
| Fall or slip causing harmful contact while balancing | (1) |
| Struck against object due to own work motion or struck by moving objects | 1 |
| Falling object | 1 |
| Contact with hot substance | 1 |

disturbances or when doing the robot's work while system malfunctioning occurred (see Table 6.4).

The number of accidents by nature of injury is presented in Table 6.5. Twelve of the injuries are classified as sprains or strains and contusions or bruises, all of which are typical manual material handling injuries. Four of the accidents led to a fracture, and three of them were pinch-point type of accidents. The occupations of the injured persons are presented in Table 6.6. Since almost half of the industrial robots in Finland were used in welding applications (Table 6.1), a higher proportion of injured workers worked around welding robots (Table 6.6). Also, repairmen and electricians were involved in three accident cases while working with robots in the course of interruptions and/or malfunctions. Two of the three machinists were pushed against the object by the robot while they were inspecting the part during a work cycle. In the two cases involving supervisors, the supervisor came to help to eliminate the disturbance in the robot system, and injured his back while handling the packages in an awkward position. Most of the accidents occurred in the manufacturing of machines, metal products and vehicles (Table 6.7).

*Table 6.4 Number of over-exertion type accidents by activity of the injured person.*

| Activity | No. of over-exertion accidents |
|---|---|
| Loading and unloading the robot system | 3 |
| Manual material handling during robot malfunction or disturbance elimination | 3 |
| Repair, maintenance | 1 |

*Table 6.5 Number of accidents by nature of injury.*

| Nature of injury | No. of accidents |
|---|---|
| Sprain, strain | 7 |
| Contusion, bruise | 5 |
| Fracure | 4 |
| Cut, laceration | 1 |
| Burn or scald | 1 |

*Table 6.6 Number of accidents by occupation.*

| Occupation | No. of accidents |
|---|---|
| Robot welder | 6 |
| Repairman, electrician | 3 |
| Machinist | 3 |
| Supervisor | 2 |
| Packer, wrapper | 2 |
| Other | 2 |

*Table 6.7 Number of accidents by branch of industry.*

| Industry | No of accidents |
|---|---|
| Machines | 4 |
| Vehicles | 3 |
| Metal products | 2 |
| Food and beverages | 2 |
| Clay and stone | 2 |
| Plastic products | 2 |
| Other | 3 |

## *Conclusions*

Over-exertion injuries due to manual material handling were relatively common in the studied robot-related accidents. This is in direct conflict with an important objective of using robots: to reduce manual material handling and over-exertion injuries. Previous robot accident research is confined to accidents involving injured persons due to direct or indirect contact with robots. However, it seems that robot system disturbances and malfunctions lead to serious manual material handling injuries as well. There appears to exist a new type of injury previously ignored by robot-related accident studies.

We now know that in the case of unexpected disturbances or malfunctions, operators do get involved with the robot work process. Disturbance prevention should be considered at the design stage by paying more attention to the reliability of the system and by designing and teaching safe actions for predictable disturbance situations (see Rahimi and Karwowski (1990) for research issues in this area). Possible disturbance and malfunction situations should be taken into consideration in the physical robot workstation design.

In many instances safety is 'tacked on' at the end of the design phase (Jiang and Gainer, 1987). However, it is important to consider the safety of robot systems and workplaces at the design stage, when inexpensive and creative solutions can be developed. To expand this approach, safety of robots in a production line should also be considered at the design stage. It is recommended that robot safety research should include studies of system disturbances from the material handling point of view.

# *Appendix*

**Accident descriptions (TARE data base)**

*Case 1*

Branch of industry: manufacturing of beverages
Occupation: brewery or beverage worker
Injured part of body: arm from shoulder to wrist
Accident type: over-exertion
Nature of injury: sprain, strain
Description of accident:
Brewery worker's right wrist–elbow zone painful at work—had to move/lift empty beverage pots and pallets, because the robot was broken.

*Case 2*

Branch of industry: food manufacturing
Occupation: packers, wrappers
Injured part of body: arm from shoulder to wrist
Accident type: pinch
Nature of injury: contusion, bruise

Description of accident:
Worker corrected the orientation of a part at the new 'Multivac' deep-drawing machine. For some reason, the robot's gripper fingers moved down when they should have moved to the side. Left arm was pinched between the machine frame and the robot. The machine was on a trial run.

*Case 3*

Branch of industry: manufacturing of metal products
Occupation: welders, flame cutters, etc.
Injured part of body: arm from shoulder to wrist
Accident type: over-exertion
Nature of injury: sprain, strain
Description of accident:
Worker lifted the welding object from the robot table and strained his shoulder.

*Case 4*

Branch of industry: manufacturing of machines
Occupation: welders, flame cutters, etc.
Injured part of body: fingers only
Accident type: struck against object
Nature of injury: contusion, bruise
Description of accident:
When lifting a beam to the table of the robot, worker's grip slipped and the little finger of the left hand got pinched between the beam and the table.

*Case 5*

Branch of industry: manufacturing of machines
Occupation: welders, flame cutters, etc.
Injured part of body: back, spine
Accident type: over-exertion
Nature of injury: sprain, strain
Description of accident:
When lifting a tube of mower frame (weight 7.5 kg) robot welder suddenly hurt his back.

*Case 6*

Branch of industry: manufacturing of machines
Occupation: welders, flame cutters, etc.
Injured part of body: leg from hip to ankle
Accident type: fall, slip or tumble on the same level
Nature of injury: cut, laceration
Description of accident:
When the welder was loading the weld fixture, his foot slipped and his leg hit the end stop (limiter) in the robot's track. Foot swelled.

*Case 7*

Branch of industry: manufacturing of machines
Occupation: welders, flame cutters, etc.

Injured part of body: back, spine
Accident type: over-exertion
Nature of injury: sprain, strain
Description of accident:
Robot welder's back got painful when he took a cover plate from welding robot.

*Case 8*

Branch of industry: manufacturing of other metals
Occupation: smelting works and smelting furnace workers
Injured part of body: hand (including fingers)
Accident type: contact with hot substance
Nature of injury: burn
Description of accident:
Machine operator removed excess material from apportion wing. When throwing the material to a slag cart the skimmer robot threw molten slag to the same cart. The cart was empty, and the slag thrown by the robot spattered from the cart to the hand of the operator. The operator was wearing short leather gloves.

*Case 9*

Branch of industry: manufacturing of vehicles
Occupation: machine setters, machinists and tool makers
Injured part of body: back, spine
Accident type: fall
Nature of injury: contusion, bruise
Description of accident:
When loading the jig of the robot the worker was taking the sideplate (weight 30.6 kg) of centre beam from truck pallet by using a hand magnet. The magnet slid and came off the plate, and he staggered and struck his back on the jig of the robot table.

*Case 10*

Branch of industry: manufacturing of vehicles
Occupation: electricians
Injured part of body: arm from shoulder to wrist
Accident type: pinch
Nature of injury: contusion, bruise
Description of accident:
Worker's right elbow was pinched between two parts of the robot.

*Case 11*

Branch of industry: manufacturing of metal products
Occupation: welders, flame cutters, etc.
Injured part of body: fingers only
Accident type: falling object
Nature of injury: fracture
Description of accident:
The injured employee worked with the robot and struck the part with hammer; the part fell down on his left little finger.

*Case 12*

Branch of industry: manufacturing of electrotechnical products
Occupation: maintenance worker
Injured part of body: arm from shoulder to wrist
Accident type: over-exertion
Nature of injury: sprain, strain
Description of accident:
Charger man's right shoulder got painful when moving battery cells to a robot station.

*Case 13*

Branch of industry: manufacturing of plastic products
Occupation: plastic product workers
Injured part of body: fingers only
Accident type: pinch
Nature of injury: fracture
Description of accident:
Plastic worker's right hand forefinger was pinched between two parts of the pneumatic plastic plate trimming automate (robot) of vacuum forming machine.

*Case 14*

Branch of industry: manufacturing of clay and stone products
Occupation: technical, scientific, juridical, humanistic and artistic work
Injured part of body: back, spine
Accident type: over-exertion
Nature of injury: sprain, strain
Description of accident:
The robot station jammed and shift's supervisor went to help the workers. When he threw plate plackages down from the conveyor belt he strained his back.

*Case 15*

Branch of industry: manufacturing of clay and stone products
Occupation: packers, wrappers
Injured part of body: back, spine
Accident type: over-exertion
Nature of injury: sprain, strain
Description of accident:
When the robot station jammed the operation supervisor of the packing section had to throw packages to the floor and he strained his back.

## Accident descriptions (TAPS data base)

*Case 16 (1987)*

Branch of industry: manufacturing of plastic products
Occupation: machine and engine repairmen and other maintenance workers
Injured part of body: arm from shoulder to wrist

Accident type: pinch
Nature of injury: fracture
Report:
Injured was supervising the plastic injection moulding machine. The machine was set for automatic run. The worker put his arm inside the machine through the opening in the safety gate. The guiding pin of the die pierced his arm.

The injection moulding machine was used to manufacture soap boxes, which the industrial robot picked up from the machine. The robot placed the pieces of soap into these boxes and then closed the boxes. The plexiglass (500 $\times$ 500 $mm^2$) of the safety gate was removed and the robot was programmed to pick up the soap boxes out of the injection moulding machine. The development work of this set of machines was not yet complete, and the system was not ready for full production. The injection moulding machine had had malfunctions at the beginning of the shift, and it had been used manually. The injured had removed faulty boxes a few times, until he switched to automatic run. When he again removed a piece of plastic from between the dies, he did not stop the machine, but reached his left hand into the machine through the opening in the safety gate.

The guiding pin of the die pierced the arm and the arm was pinched between the dies. The dies opened up automatically, and the arm was released. The injured was hospitalized. His radius was fractured, and the arm has been operated on three times.

The injured had completed a two-year process technology course in vocational school, and he had about 5 years' experience in corresponding work. There were no witnesses to the accident.

Causes of the accident:
The dies of the injection moulding machine closed and it was possible to reach the danger zone through the opening in the safety gate; the machine was not stopped before removing a piece of plastic from the die; and the design work for linking the injection moulding machine and the industrial robot was unfinished.

Injection moulding machine Battefeld 300/100 HK, importer HBS-Kone Oy. Industrial robot: Jarmek.

Prevention of similar accidents:
The structure of the machine combination should be made to meet the requirement of safety regulations (reaching to the danger zones of either machine must be prevented); and technicians should be given instructions about how to maintain the machines safely.

*Case 17 (1989)*

Branch of industry: manufacturing of cars
Occupation: machine setters, machinists and tool makers
Injured part of body: trunk (not back)
Accident type: pinch
Nature of injury: fracture
Report:
Injured worked in a brake drum cell. Automatically functioning loading robot (loader) pushed the injured against a brake drum. Some ribs were fractured. The operation of the loader had not been stopped.

The injured worked in a so-called *drum cell*, where car brake drums are machined. The work sequence included washing, turning with Computer-Numerical Control (CNC) vertical lathe and drilling in CNC machine centre. The part was moved by automatically functioning portal loader (loading robot).

The loader moved the parts from vertical lathe to intermediate storage. The injured went to measure the brake drum which had just arrived from the lathe. Before this, he had stopped the loader by turning the percentage control to the position which he thought was a zero position.

While performing measurements the loader moved slowly and noiselessly. The gripper portion of the loader moved down on the injured person's shoulders and pushed him against the brake drum being measured. When the resisting force increased sufficiently the loader stopped. Alarmed by the cries for help, two technicians went to stop the robot using the stop button. However, the injured was released only after the engineer moved the gripper portion. The injured was transferred to hospital by ambulance. The accident resulted in rib fractures.

The injured had finished vocational school, and had worked for two years in his job.

Causes of accident:
The injured did not stop the operation of the loader using the stop button, but used the percentage control, which he apparently failed to turn all the way to the zero position. It was found in the investigation that the percentage control may easily remain in, for example, 3% position when it is turned carelessly; and there were no safety arrangements to prevent access to the danger zone during the operation of the loader, and there were no arrangements to stop the loader when the worker entered the danger zone.

Loading robot Fibro, manufactured in 1986, was equipped with gripper which can grip the outer periphery of the brake drum.

Prevention of similar accidents:
The investigation pointed out that measuring the part at this point of the workstation is not necessary, and that access to the danger zone can be prevented by installing a fixed wall in front of it. The wall should be dimensioned in a way that the pinch point between the upper edge of the wall and the loader cannot be reached; and if it will be necessary regularly or repeatedly to enter the danger zone because of measurements or other reasons, the stopping of the loader or preventing its entry to the danger zone must be guaranteed. Equipping the gate with a limit switch, light cell arrangements, etc., should be considered.

*Case 18 (1989)*

Branch of industry: manufacturing of lifting and moving equipment
Occupation: machine setters, machinists and tool makers
Injured part of body: back
Accident type: pinch
Nature of injury: contusion, bruise
Report:
The injured worked in a machining cell. The loading robot pushed him against the machine. The accident resulted in a minor back injury. Access to the robot's work area was not prevented.

The injured worked in a machining cell which consisted of external and internal cylindrical grinding machines which the robot loaded and unloaded, and a round table from which the robot picked up the unfinished parts and onto which it placed the finished parts. The robot served both grinding machines, otherwise the work done in each machine was independent from the other machine.

After the grinding cycle of the external grinding machine, the worker went to measure the part which was still fixed in the grinding machine. The robot pushed him against the grinding machine. The injured escaped and caused about 300 N lateral force to the robot arm. The cotter pin in the arm broke and the robot stopped. The accident resulted in minor upper back injury.

The injured could be considered as an experienced, skilled worker. After training at the machine-shop school of the factory, he had worked for nine years with numerically controlled machines and had spent four weeks in this job.

Causes of accident:
There was free access to the robot's work area through the gap between the grinding machine and the control cabinet; and the injured, who should have known the operating principles of the system, probably made a judgement error, or he assumed that he had enough time to measure the part before the robot picked it up from the grinding machine.

Industrial robot Mantec-Fanuk, model M1, age about 5 years, cylindrical grinding machine Shaudt, age about 5 years.

Prevention of similar accidents:
Access to the robot's work area has to be prevented during automatic run. The limit switches, light cells or other safety devices should take care of stopping the movements of the robot if the danger zone is entered; if the robot has to be operated while in the danger zone, the operation must be controlled with the 'dead man's control' or the speed or forces are restricted (momentarily max 40 N, continuously max 150 N). The manual drive control device must have an emergency stop, controls other than manual drive must be used while in the dnager zone and the other controls must automatically be switched off while the manual drive control is used.

# *References*


Carlsson, J., 1985, Robot accidents in Sweden. In Bonney, M. C. and Yong, Y. F. (Eds) *Robot Safety* (Bedford: IFS Publications and New York: Springer), pp. 49–64.

Jiang, B. C. and Gainer, C. A., 1987, A cause-and-effect analysis of robot accidents. *Journal of Occupational Accidents,* **9,** 27–45.

Jones, R. and Dawson, S., 1985, People and robots: their safety and reliability. In Bonney, M. C. and Yong, Y. F. (Eds) *Robot Safety* (Bedford: IFS Publications Ltd and New York: Springer), pp. 65–81.

Karwowski, W., 1991, Human-robot interaction: An overview of perceptual aspects of working with industrial robots. In Kumashiro, M. and Megaw, E. D. (Eds) *Towards Human work: Solutions to Problems in Occupational Health and Safety* (London: Taylor & Francis), pp. 68–74.

Karwowski, W., Rahimi, M. and Mihaly, T., 1988, Effects of computerized automation and robotics on safety performance of a manufacturing plant. *Journal of Occupational Accidents,* **10,** 217–33.

Kuivanen, R., Lepistö, J. and Tiusanan, R., 1988, Disturbances in flexible manufacturing. In *Preprints of the 3rd IFAC/IFIP/IEA/IFORS Conference on Man-Machine Systems*, vol. II (Oulu, Finland: IFAC), pp. 472–75.

Nagamachi, M. 1988, Ten fatal accidents due to robots in Japan. In Karwowski, W., Parsaei, H. R. and Wilhelm, M. R. (Eds) *Ergonomics of Hybrid Automated Systems I* (Amsterdam: Elsevier), pp. 391–6.

Nicolaisen, P., 1985, Occupational safety and industrial robots. In Bonney, M. C. and Yong, Y. F. (Eds) *Robot Safety* (Bedford: IFS Publications and New York: Springer), pp. 33–48.

Rahimi, M. and Karwowski, W., 1990b, A research paradigm in human–robot interactions. *International Journal of Industrial Ergonomics,* **5,** 1, 59–71.

Sugimoto, N., 1985, Systematic robot-related accidents and standardization of safety measures. In Bonney, M. C. and Yong, Y. F. (Eds) *Robot Safety* (Bedford: IFS Publications and New York: Springer), pp. 23–9.

Sugimoto, N. and Kawaguchi, K., 1985, Fault-tree analysis of hazards created by robots. In Bonney, M. C. and Yong, Y. F. (Eds) *Robot Safety* (Bedford: IFS Publications and New York: Springer), pp. 83–98.

# *Chapter 7*
# *Robot safety and reliability*

**K. Khodabandehloo**

*Department of Mechanical Engineering, University of Bristol, Queen's Building, University Walk, Bristol BS8 1TR, UK*

**Abstract.** Over the past decade robot safety received considerable attention within the domain of factory automation. More recently, however, safety in the use of robotics outside factories or processing plants has become a matter of great international concern. Domestic robots and those intended to assist nurses and surgeons in hospitals are examples of cases where safety and reliability are considered critical. This chapter examines the safe performance of robot systems which depends on many factors, including the integrity of the robot's hardware and software, the way it communicates with sensory and other production equipment, the reliable function of the safety features present and the way the robot interacts with its environment. The use of systematic techniques such as Fault and Event Tree analysis to examine the safety and reliability of a given robotic system is presented. Considerable knowledge is needed before the application of such analysis techniques can be translated into safety specifications or indeed 'fail-safe' design features of robotic systems. The skill and the understanding required for the formulation of such specifications is demonstrated here based on a number of case studies.

## *Introduction and background*

Recent developments and the new applications of microelectronics and computer technologies have helped the evolution of programmable manufacturing equipment. Industrial robots normally classed as programmable machines utilize the enabling computer technologies for their operation. However, the anatomical configuration of industrial robots differs significantly from that of conventional machine tools as a robot's active work volume is considerably greater than the physical construction of the arm allows. The work volumes for a selection of robot arm configurations are shown in Figure 7.1, illustrating the extent of physical reach in each case (Hartley, 1983). It is important to note that the large work volume of a robot is desirable and in part represents an important feature of its flexibility in its industrial use. Although it is possible to restrict the reach of the robot by mechanical means as a safety measure, this option limits the use of the robot to specific installations and reduces a flexible machine to dedicated hardware.

Human–robot interaction must be considered in order to identify the hazards as well as the risks associated with the use of robots in every application. For

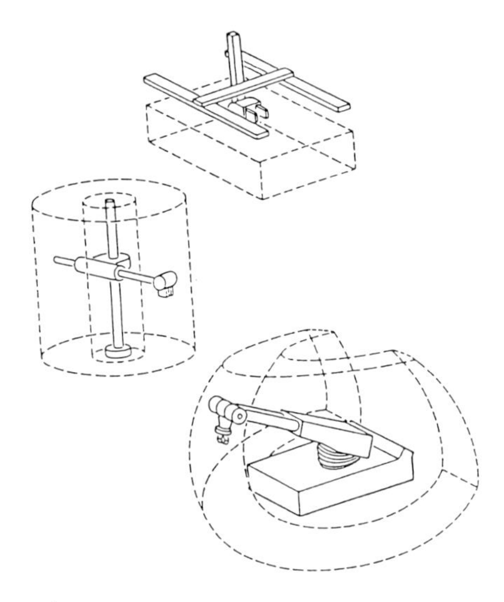

*Figure 7.1 Work volume (envelope) of typical robots.*

instance, an operator required to program a robot will need to be in close contact with the robot during programming but when the robot is carrying out a specific task automatically the level of contact is reduced. In this case any safety assessment will need to consider the operator interactions with the cell, which may exclude the situation where the robot is operating without close human supervision. In contrast, a robot assisting a surgeon in an operation will need to receive instructions from the surgeon and then carry out a specific task which may involve close contact with a patient. The robot interactions with both the surgeon (operator) and the patient must be examined from the safety viewpoint. Furthermore, the safety of those who come into contact with the robot also depends on the reliability of the system during operation, i.e. the system's ability to tolerate failures becomes a crucial factor. The more advanced applications not only require a good physical work volume or reach from the robotic arm, but the range of tasks for the robot is becoming more and more varied. Many applications require two or more robot arms, the use of sensors and artificial intelligence (Khodabandehloo and Rennell, 1988). Safety and reliability remain the most important issues in robotics due to the increased sophistication in the use of control hardware and software and the nature of applications such as surgery, where close interaction between people and the robotic system is a functional requirement. There is no doubt that autonomous operation of robots can be achieved and this will in the long term reduce the interactions needed between programmers or operators and robots. In so far as safety is concerned, there is a clear distinction between the situations where human–robot interaction is necessary as part of the routine operation of a robot in a factory, and the interactions of an advanced robot with a human, say a patient in surgery, whilst it is performing its intended task.

## The coupling between safety and reliability

The sight of a robot producing erratic movement due to failures in its control system can seem rather comic as long as no one gets hurt. But people have been involved in accidents under such circumstances and several deaths have been reported over the past few years. The assessment of safety in a robotic system requires a thorough analysis of the potential hazards presented by the system during its operation. Identification of the hazards requires the consideration of both system design and operation. The specification of a robot system design cannot be divorced from its intended use, however the safe operation of a robotic system relies on a clear definition of safe working procedures and the provision of appropriate safety features to support the practice of such procedures by people operating the robot. Accidents result from both equipment malfunction and bad operating practices. The responsibility of the equipment designers and manu-acurers must be to minimize the possibility of failure and to design out features that lead to unsafe operator practices (Jones and Khodabandehloo, 1985).

Failure in a robotic system can be considered from two related, but different, perspectives: reliability and safety integrity. Reliability analysis in its broadest form considers all failures that produce a system output different from that which was desired or specified under a given set of conditions. It is relevant to note that no distinction is made between such failures having a safe or unsafe outcome. The safety integrity of a system, however, takes account of the coupling between safety and reliability by considering the failures that have an unsafe outcome (Khodabandehloo *et al.*, 1985). A system can therefore be very unreliable, but possess a high safety integrity: that is to say, the system may malfunction regularly, but the possibility of a breakdown having an unsafe outcome can be very low.

## Safety and reliability analysis techniques

The established techniques of Failure Mode and Effect Analysis (FMEA), Fault Tree Analysis (FTA), and Event Tree Analysis (ETA) have been in use for many years (MacCormick, 1981). FMEA can be used to examine all possible component failure modes and to identify their first order and final effects on the system. FTA and ETA may be applied at various levels for examining the errors and failures in the system software and hardware. FTA is a top-down technique for assessing the way in which several failures can cause a single outcome or a system failure (termed the top event). Standard logic gates (AND, OR etc.) are used to combine several failures taking into account their logical interactions. ETA is a 'forward' technique which may be used to examine the propagation of an initiating event (or failure) with the presence of a number of other events, failures, faults or conditions. Analyses have been performed to evaluate a number of robotic systems. Case studies have been included which consider the major factors in both design and operation of industrial robotic systems.

Throughout the following analyses a distinction is drawn between a hazard that is realized with a person present, and those with the machinery operating alone. In the former case, there is the possibility of an accident resulting in injury to a person, but in the latter, damage or harm can only happen to the machinery. Damaged equipment may prove to be a costly mishap, but it is obviously not as important as considerations of personnel safety. The event trees also refer to certain events as 'lucky escapes'. When a hazard is present and no means of preventing it have been implemented, the fact that the hazard is not realized is merely fortuitous. Although no actual accident occurs, there is effectively nothing preventing it in these cases. The analysis also shows how equipment behaviour and working practices can combine to produce potentially dangerous events.

## *Case study A: study of a stand-alone robot*

An hydraulic pick-and-place robot has been investigated for reliability using FTA and FMEA. The robot system considered consists of five joints, each driven and controlled by an hydraulic servo-mechanism as shown in Figure 7.2. Pressure is generated by a conventional motor and pump assembly. Hydraulic fluid is pumped from the reservoir through a filter and a check valve, which ensures flow in one direction. Pressure is kept below the maximum allowable value by means of an unloading valve, usually factory preset. An accumulator is used to assist the pump in supplying additional fluid under a high flow demand. The motion of each hydraulic actuator, which is controlled by a servo-valve, is transmitted to the appropriate limb of the robot either directly or by some mechanical transmission (rods, chains, gears, etc.). Each limb is coupled to a position transducer, in this case an absolute optical encoder, giving the joint angle codes. A multiplexer allows scanning of each code in turn. In playback mode the code for each joint is compared with a corresponding code from the memory giving the joint angle for a previously taught point. A digital difference is converted into an analogue signal according to a prespecified motion control algorithm. After amplification the signal is applied to the appropriate servo-valve. In teach mode the motion of the arm is controlled by the operator using a teach pendant. When a specific point is to be recorded, the codes from the joint encoders are entered into the memory sequentially.

An end-effector, usually a two jaw gripper or a weld gun, is used. For the purposes of this example, a pneumatically operated gripper was considered. Apart from the teach pendant the operator interface consists of a control panel and an emergency stop facility.

### FMEA for the robot

FMEA tables have been prepared for all relevant components of the hydraulic robot. Table 7.1 shows a sample of these tables for the hydraulic parts. Each

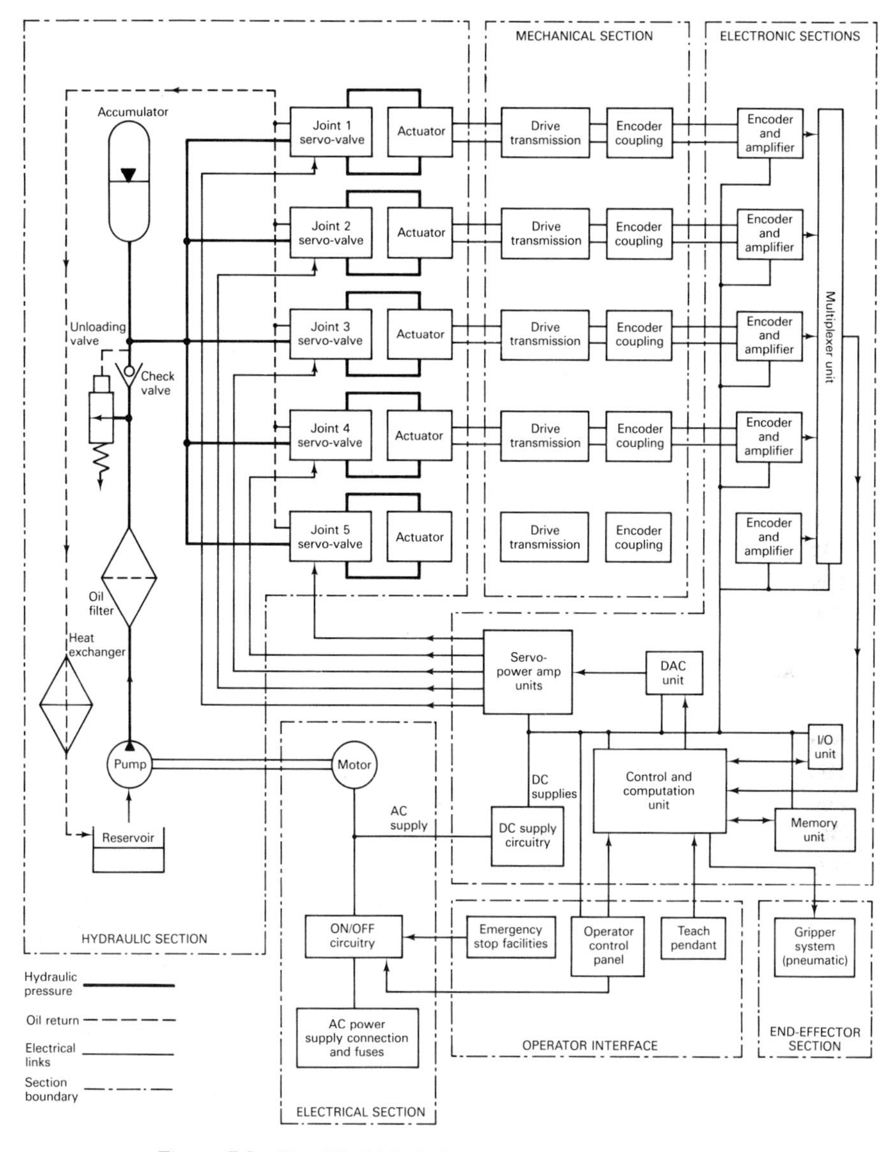


*Figure 7.2 Simplified block diagram for an hydraulic robot.*

component and its major failure modes are considered individually and their effects on other components as well as on the whole system are identified. Classification of the failures in accordance with their possible consequences is made in four categories as given in Table 7.1. Compensating provisions that could reduce the probability of occurrence of a certain outcome are also tabulated. Similar tables for pneumatic, mechanical, electrical and electronic sections have been produced, but are not presented here. Major consequences that reflect upon

*Table 7.1 Sample failure mode and effect analysis.*

| No. | Component | Failure mode | Effect on | | *Class | | | | Failure rate | Detection | Compensating provisions and remarks |
|---|---|---|---|---|---|---|---|---|---|---|---|
| | | | Other components | Whole system | 1 | 2 | 3 | 4 | (Faults/$10^6$h | method | |
| 1 | Pump | Vane wheel breakage | Pump casing may be ruptured | Pressure loss. No operation | | X | | | 604.5 (Op. time) | Inspection | Rupture will cause rapid loss of hydraulic fluid. This could lead to a fire. Pump must be replaced. |
| | | Leakage from casing | Gradual loss of hydraulic fluid | Air may enter the system | | X | | | | Inspection | Air in the hydraulic system increases the wear of hydraulic drive components. Jerking in the robot movements may result. |
| | | Vane wheel jammed | Pump motor will be overloaded | No hydraulic pressure | | X | | | | Rotation monitoring | Overheating of the motor will occur. Current limiting is needed. |
| | | Pump wear | Fewer unload-valve operations | High fluid flow demand may not be met | X | | | | | Efficiency testing | Wear in the pump causing a reduction in efficiency will slow the movements of the robot, thus slowing down the work. |
| 2 | Filter | Blockage | Excessive pump/filter pressure | No fluid flow to the system | | X | | | 107.6 (Op. time) | Pressure sensing | Pressure increase as a result of blockage will cause filter to rupture. By monitoring the pressure difference across the filter this can be identified. |
| | | Rupture | Hydraulic fluid contamination | Rate of wear increases | | | X | | | Inspection | Pressure monitoring across the filter can indicate filter failure. Contamination in the system can lead to servo-valve failure and increased wear in moving hydraulic components. |
| 3 | Unload valve | Failure to unload | Pressure rises above component's rated value | Excessive pressure in the system | | | X | | 31.3 (Op. time) | Pressure sensing | By monitoring the pressure the failure can be detected and system automatically shut down. Failure of other components such as servo-valves/piping may also result. |
| | | Continuous return to reservoir | Slow actuator operation | System pressure low. Slow robot movements | | X | | | | Pressure sensing | Robot movements will be slow. Fault can be detected by pressure sensing. |

*Table 7.1 (cotinued)*

| No. | Component | Failure mode | Effect on: Other components | Effect on: Whole system | *Class 1 | 2 | 3 | 4 | Failure rate (Faults/$10^6$h | Detection method | Compensating provisions and remarks |
|---|---|---|---|---|---|---|---|---|---|---|---|
| 4 | Check valve | Failure to open | Pressure rise before unload valve | No fluid flow. Piping rupture may result | | | X | | 10.7 (Op. time) | Fluid flow sensing | No robot operation can be expected. Pressure increases between the unload valve and the pump which can cause rupture in the piping. |
| | | Failure to close | Reverse flow to the pump | Fluctuations in pressure occur | X | | | | | Fluid flow sensing | By sensing the fluid flow in the reverse direction the fault can be identified. |
| 5 | Pressure gauge | Incorrect reading | | | X | | | | 11.7 (Op. time) | Inspection | Calibration is needed during routine maintenance. Failure can lead to incorrect human action. |
| 6 | Relief valves | Failure to open | Damage to drive transmission can result | | | X | | | 8.7 (Op. time) | Force sensing | Valves protecting the transmission system can be backed up by a force/torque sensor that causes shut down in case of a collision. |
| | | Failure to close | Undesirable actuator motion will result | Drop in driving power for the relevant joint | | | | X | | Fluid flow sensing | Relief valve failure in open mode will cause uncontrollable actuator movement and in some cases the appropriate joint can become free to move. |
| 7 | Accumulator bladder | Ruptured | No accumulator operation | Gas may leak into the system | | X | | | n/a | Pressure sensing | When system is depressurized gas can enter the system affecting the stiffness of the arm. |
| 8 | Dump valve | Leakage | Loss of fluid | Slight drop in pressure | | X | | | 7.1 (Op. time) | Inspection | Sufficient oil leakage can lead to a fire. Slight movements may result. |
| 9 | Servo-valves | Null position drift/offset | Slight actuator movements | Robot creeps in one direction | | X | | | 158.0 (Op. time) | Inspection | Servo-valve adjustment is required. Repeatability is degraded. |
| | | Failure in closed mode | No actuator movement for the relevant joint | Robot joint will not drive | | | | X | | Inspection | Servo-valve failure in the closed mode will not allow movement of the corresponding joint. This can cause a collision between the robot and other equipment. |

*Table 7.1 (continued)*

| No. | Component | Failure mode | Effect on: Other components | Effect on: Whole system | *Class 1 | *Class 2 | *Class 3 | *Class 4 | Failure rate (Faults/$10^6$h | Detection method | Compensating provisions and remarks |
|---|---|---|---|---|---|---|---|---|---|---|---|
| | | Failure in open mode | Unwanted drive of actuator | Joint will move until end-stop is reached | | | | X | | Inspection | Uncovenanted movement will occur and the arm could collide with other objects or machinery. The extent of the damage can be reduced by tactile or other type of sensors (vision?) for stopping the arm. Note that some form of pressure-dumping mechanism is needed if the arm is to be stopped fast, otherwise emergency stop action will be delayed due to stored pressure in the system accumulator. |
| 10 | Actuators | Housing leakage | Gradual loss of fluid | Robot arm creeps | | X | | | 56.0 | Inspection | Creep of the appropriate joint will occur with worse effects under high loads. |
| | | Housing or seal rupture | Loss of fluid | Unwanted robot movement | | | | X | | Inspection | Rupture can cause the corresponding joint to move freely. |
| 11 | Piping and seals | Leakage | Gradual loss of fluid | Can cause fall of pressure | | X | | | 115.0 (Op. time) | Inspection | Leakage can cause creep of robot arm. |
| | | Rupture | Loss of fluid | Unwanted robot movement can result | | | | X | | Inspection | Rupture of piping (and seals) between servo-valves and actuators can cause the corresponding joint to become free or move unexpectedly. Collision between the robot arm and other equipment is inevitable. |
| 12 | Reservoir | Leakage: total fluid loss | Air is pumped into the system | No system opeation | | | X | | n/a | Fluid level sensing | This can lead to a fire. All joints will become free if loss of fluid is sudden and goes unnoticed. |
| | | Leakage: slow fluid loss | Air can enter the system | Robot arm may creep | | X | | | | Fluid level monitoring | Sufficient quantity of oil spill can lead to a fire. Air bubbles in the system can cause jerks in the arm movements. Wear in hydraulic components will be increased. |

*Table 7.1 (continued)*

| No. | Component | Failure mode | Effect on | | *Class | | | | Failure rate (Faults/$10^6$h | Detection method | Compensating provisions and remarks |
|---|---|---|---|---|---|---|---|---|---|---|---|
| | | | Other components | Whole system | 1 | 2 | 3 | 4 | | | |
| 13 | Heat exchanger | Oil leakage | | Loss of fluid | | X | | | 52.1 (Op. time) | Inspection | This can lead to a fire. |
| | | Insufficient cooling | Oil temperature increases | Rise in system temperature | | | X | | | Tempeature sensing | Too high a temperature can cause further damage to oil cooler and possibly rupture. Disintegration of fluid can also occur. |

Op. time = operation time.
*CLASSES: 1, Safe—negligible effect on the system; 2, Marginal—failure will degrade system but will not cause system damage or injury to personnel; 3, Critical—failure will degrade system performance, could cause system damage and/or possible injury to personnel; 4, Catastrophic—failure will cause severe system damage and/or possible injury or death.

the safety of equipment, workpiece and personnel are identified. These include:

1. undesirable robot movement in playback mode;
2. undesirable robot movement in teach mode;
3. arm runaway when switching on;
4. no emergency stop action when demanded; and
5. arm 'creep' or degradation of repeatability.

These consequences become the top event for FTA.

## FTA for the robot

For each of the above top events a fault tree can be constructed, tracing the outcome back to all possible random component failures or their modes of failure. The trees may be modified to include systematic faults as well as software and human errors. However quantification of the latter sources of 'failures' is much more difficult. Figure 7.3 shows the hierarchy of the combination events that contribute to the top event of undesirable arm movement in playback (or automatic) mode. Each event is partitioned into other combinations of events further down the tree until a 'basic' event which can be assigned an independent probability is reached (e.g. a component failure mode). Where appropriate the branch of the tree is terminated by an event for which the required failure rate data is available. Figure 7.4 expands branch 3 which relates to joint 1 drive system failure. This joint produces a rotary motion and its mode of action is shown in Figure 7.5. The movement is produced by two hydraulic actuators connected to a single servo-valve. A rack and ring gear system is used to transmit the motion to the limb of the robot arm. A joint is considered as failed if the failure of a single component causes the corresponding limb to move freely or not to move at all. Other failures within the drive system that can cause this top event are covered by other branches of the tree in a similar fashion.

As mentioned before, each of the 'basic' events can be represented by a Boolean variable and when assigned a failure probability the two can be combined to give a probability model for the top event. In this particular case the model is simple, because all the variables are combined by an OR gate. The probability of the outcome over a specific period of time is approximately given by the sum of failure probabilities of individual basic events over the same time period, which is also the upper bound in this case. Failure data for a large selection of components are available commercially. The Systems Reliability Service (SRS) data base (Systems Reliability Service, 1984) and the MIL HDBK-217D (MIL 217D, 1982) are two sources that may be used in such study. The SRS data base provides failure rate data for a large number of components. Apart from its very comprehensive coverage it has the advantage that it can give the failure rate of a specific mode of failure for a particular component. In addition one could extract information about occurrence of a certain outcome under a given set of environmental conditions.

The assessment of top events using FTA points to problem areas and can prove useful in improving the reliability of a system, thus reducing the possibility of

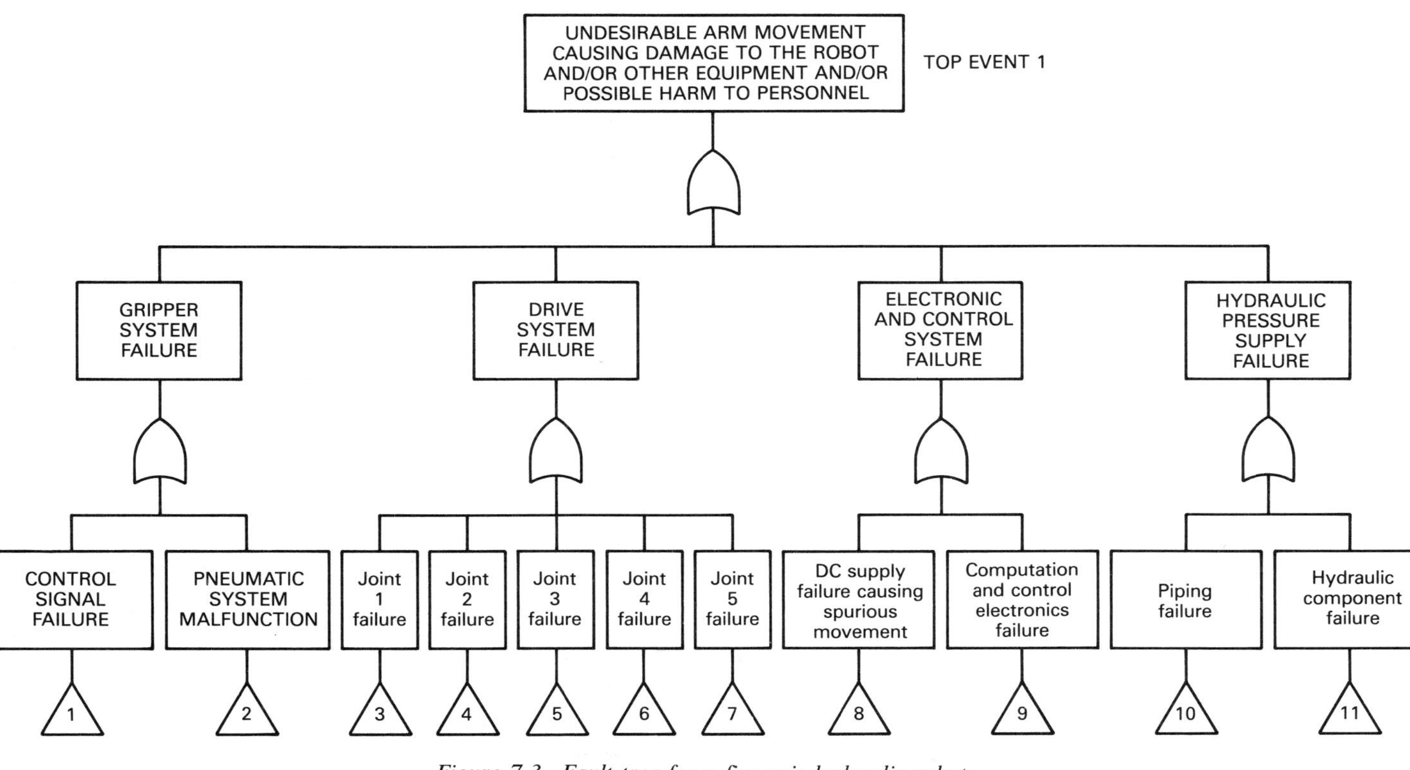


*Figure 7.3 Fault tree for a five-axis hydraulic robot.*

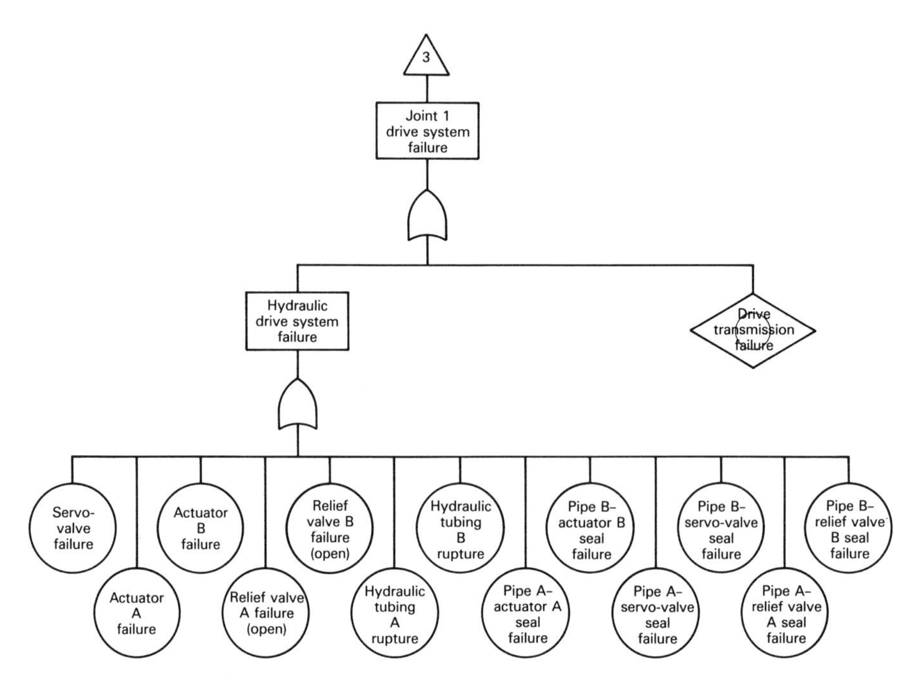


*Figure 7.4 Fault tree branch (3) for rotary motion drive system shown in Figure 7.5.*

accidents resulting from hardware failures. The evaluation of the top-event failure rate of the fault tree in Figure 7.4 can be an indicator of the safety integrity of the robot. An example of how this may be achieved for another similar robot is given in Case Study C.

## *Case study B: operator–robot interaction in a welding cell*

A robotic cell designed to perform an arc-welding task is used here to highlight the main safety issues in the operation of the cell by an operator (Jones and Khodabandehloo, 1985). Particular attention is paid to those events that necessitate operator–robot interactions. The cell is considered in the presence and the absence of safeguards. ETA has been applied as a systematic means by which safe/unsafe modes of the system and the operator practices can be understood.

A number of implications for training arise from this analysis. Naturally, the involvement of personnel varies with the task that they are given and so their training needs are different. Firms also vary in their work organization. In some, production supervisors might play a major role in problem identification and solution, whilst maintenance personnel might take on this role elsewhere. However, certain points can be raised without being specific about the type of personnel involved.

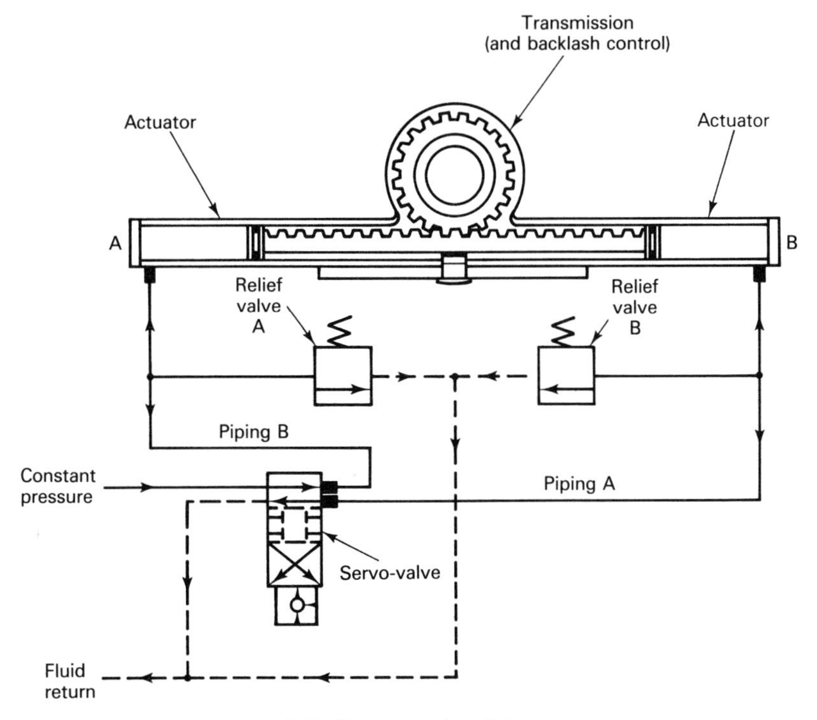


*Figure 7.5 Rotary motion drive system.*

## System description

Figure 7.6 shows the basic layout of the system with only the inherent safeguards built into the equipment. Extra safeguards may be added to enhance this basic system. The turntable has two positions with two headstocks. The robot is a conventional five- or six-axis electrically-powered robot and includes a typical control system. The weld unit, including the gun adapted to fit onto the robot, is a conventional piece of modern equipment commonly used by manual welders. The robot, the turntable and the weld unit have their own controls which are interlocked and provide the necessary communications for the welding operation. There are two main interlocks for the operator: a start button, which initiates welding operation; and an emergency stop button that puts each piece of equipment into a state which prevents any motion or action of the equipment.

The robot controller is assumed to be the system supervisor, although the operation of the system is initiated by a start button. The person loads and clamps the parts to be welded on one side of the turntable. When all the parts have been placed onto the turntable for welding, the start button is pressed. This sends a signal to the robot controls which in turn send a signal to the turntable causing it to rotate horizontally until the component to be welded is in position. The robot then starts the welding operation, switching the weld gun on or off as necessary through the weld-unit control. The weld parameters are previously

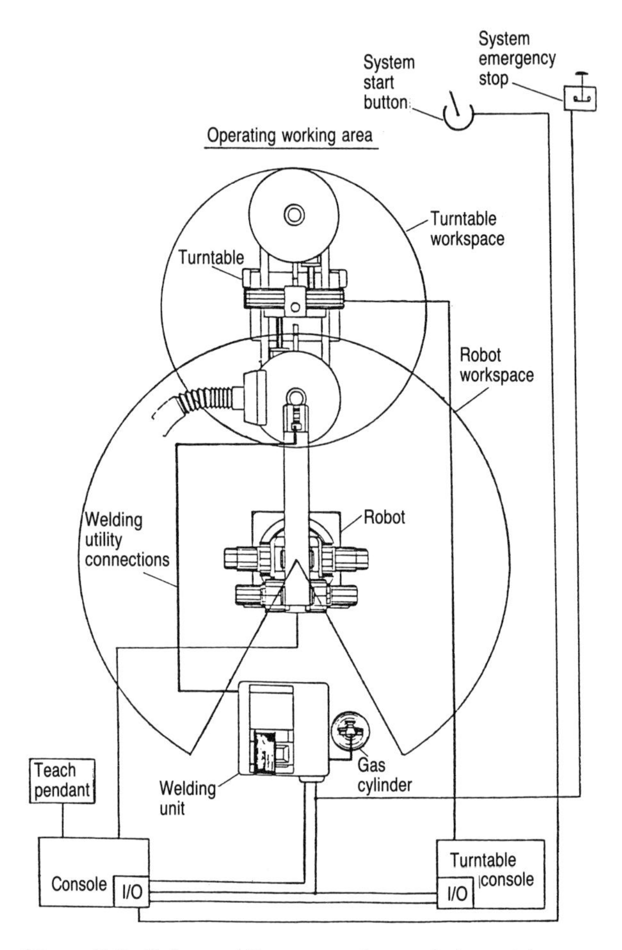


*Figure 7.6 Robot welding system layout (schematic).*

set so that the flow of $CO_2$ and water is regulated for correct welding. During the welding process vertical rotation of the headstock is required, allowing access to several other weld positions on the component. The request for vertical rotation at the teach stage comes from the robot and the turntable acts by unlocking, moving and re-locking the headstock. A signal from the turntable is sent to the robot after the completion of each vertical rotation so that the robot continues welding. When the final welding sequence is completed the robot returns to its rest position, away from the turntable, and requests the headstock to return to its home position. The turntable in turn replies to the robot when it is locked at the home position. (The turntable must lock itself at each position so as to prevent its motion during welding.) Meanwhile the person removes any previously welded parts and loads a new set of unwelded pieces.

Once the welding operation is completed by the robot, it moves to its rest position and then returns to the start of the procedure awaiting the signal to begin the operation again. The turntable and the weld gun will also remain in their final states ready to initiate another sequence.

ETA has been performed for this system (excluding any safeguards). Three initiating events have been chosen because of their frequent occurrence and the hazardous steps they involve.

For the manual arc welder, the welding process produces certain hazards. These derive from the noxious fumes, the bright arc flash and the risk of electrocution. When automation is substituted, these hazards may be considerably reduced for the operator but the new equipment produces an entirely different set of hazards. This equipment has the ability to make large, sudden and powerful movement. The robot or turntable can move suddenly whilst carrying out their programmed movements and catch a person unawares. It is also possible for the equipment to move as a result of a component failure or a fault in the software. With the welding torch on the end of the robot arm, the hazards of arc welding can occur in conjunction with the unexpected movement of the robot arm. This produces the extra hazards of possible electric shock and burns. The consequence of these various hazards can be as severe as a fatality (see Table 7.2).

*Table 7.2 Some hazards of the robot system and their consequence.*

| | | |
|---|---|---|
| Welding equipment | Gas fumes<br>High voltage | Dangerous concentrations of fumes, electric shock, burns (could result in a fatality) |
| | Arc-flash | 'Arc-eye' and conjunctivitis |
| Robot | Sudden movement | Trapping, crushing, person pushed into machinery (could result in a fatality)<br>Collision with the turntable (could result in damage) |
| Turntable | Sudden movement | Trapping, crushing, person pushed into machinery (could result in a fatality)<br>Collision with the robot (could result in damage) |

## Weld equipment failure: event tree I

The event tree in Figure 7.7 examines the initiating event of weld-equipment failure during operation. One such event would be caused by blockage of the weld gun causing a restriction on the $CO_2$ arc shield gas. The first four events after this initiation event relate to the actions of the person involved in attending to the problem. The person may enter the working area (event 1) to perform the necessary adjustments/cleaning/minor repair work with the robot in its rest position (i.e. away from the workpiece and the turntable). The robot at this stage is waiting for the start button to be depressed.

The person should remove power from the robot, the weld gun and the turntable before entering the working area of the equipment. Access to the

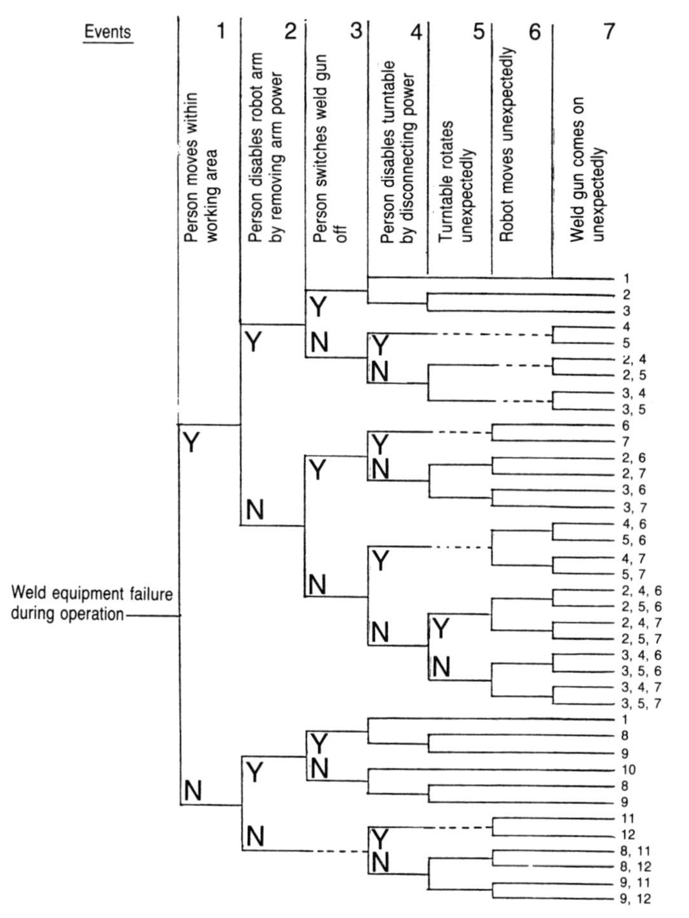


*Outcomes*

1. Intrinsically safe
2. Hazard from turntable motion
3. Hazard from turntable motion not realized — 'lucky escape'
4. Hazard from weld gun (e.g. weld flash, high voltage)
5. Hazard from weld gun not realized — 'lucky escape'
6. Hazard from robot motion
7. Hazard from robot motion not realized
8. Potential damage due to turntable motion
9. Potential damage due to turntable motion not realized
10. No hazard likely to occur for equipment

*Figure 7.7 Event tree I: weld equipment failure.*

control consoles is assumed possible without the need for anyone to enter the equipment work areas. No machine movement is assumed to occur or to be needed during the disabling of the equipment. Events 2, 3 and 4 allow the examination of the possibilities that exist when the robot, the weld gun and/or the turntable are not disabled.

Events 5, 6 and 7 relate to the unexpected movement of the turntable and the robot, and the activation of the weld gun. It is assumed that no hazards are created by unwarranted robot or turntable movements while the equipment is being disabled.

Only two outcomes from these possible events are intrinsically safe, both of which demand that the equipment be disabled (branches leading to outcome 1). Naturally if the equipment is disabled then unexpected movement cannot occur. One of the outcomes has produced no likely hazard for the equipment (outcome 10). This was because the robot and the turntable were disabled and no access was necessary. The weld gun produces no hazard in this situation as it is incapable of independent movement.

The greatest hazards are present when the robot, the turntable and the weld gun are not disabled. It is under these conditions that the robot and/or the turntable movements can occur, and the weld gun could become activated. The underlying reasons behind unexpected movements are to a large extent related to malfunctions within the system elements.

## Person enters working area: event tree II

The event tree in Figure 7.8 considers the initiating event of a person entering the working area for adjustments and programming. This differs from the previous event in that the robot could be required to move and so would not necessarily be disabled. As such unexpected movement of the robot arm can present greater hazards during programming and adjustment than with the recovery from weld-equipment failure, should major adjustments to the weld gun or the turntable fixtures be made, then the robot will need some reprogramming.

The first event after the initiating event relates to the activation of the emergency stop (event 1). If the system emergency stop is pressed at any time, then the whole system is inactive and thus there are no hazardous outcomes in this tree. This assumes that failures that cause the robot or the turntable to move and that have no power to them (e.g. brake or lock failure) do not occur. If the system emergency stop has not been activated, then there are several hazards present, unless the turntable and the weld unit have been switched off (events 2 and 4 respectively) and the teach pendant emergency stop is pressed (event 6).

If the turntable and the weld gun are not switched off then the hazards from unexpected turntable movement (event 3) and unexpected action of the weld gun (event 5) remain. Unexpected movements of the robot, both large and small at slow and fast speeds (events 7 and 8) are possible during the programming mode of a robot either because of a lack of specific provision or failures causing such events. To perform tasks such as programming in a safe way, the system is required to have a number of safety features. The restriction of large arm movement by the robot and its speed are examples. With complex tasks such as arc welding, however, one is required to operate the robot at normal speed

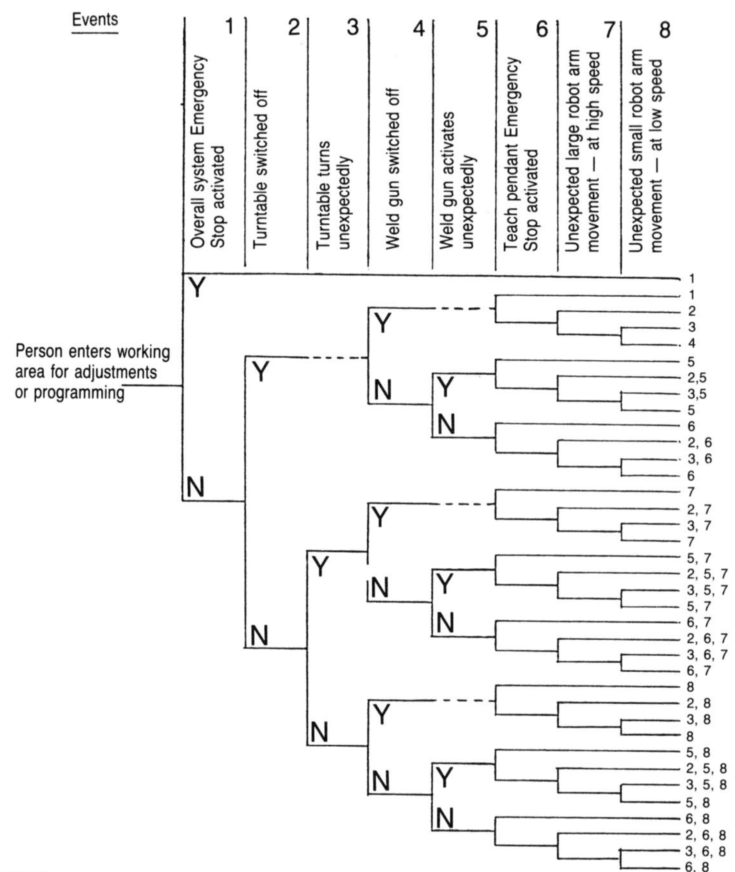


*Outcomes*
1. System inactive/put into safe mode
2. High level of hazard due to robot collision with equipment or person
3. Hazard due to robot causing equipment damage and possible harm to person
4. Task performed satisfactorily
5. Hazard from weld gun (e.g. weld flash, high voltage)
6. Potential hazard from weld gun — 'lucky escape'
7. Hazard from turntable motion

*Figure 7.8 Event tree II: person enters working area.*

and at the same time optimize a number of parameters, wire-feed rate being one of them. This requires close contact with the moving robot arm which cannot be restricted easily. Outcome 4, 'task performed satisfactorily', is the final result when the procedure for correct teaching is followed with restricted robot speed, and attention is paid to the programming required.

The critical steps that should be followed while teaching a robot in this situation are: (1) to remove power from the weld gun and the turntable; and (2) to follow correct programming procedures for the robot. Some aspects of safety related to programming will be presented later in case study C.

The outcomes, as shown by these event trees, present hazards ranging from a serious hazard of robot collision due to large unexpected movement at high speeds, to potential hazards from unexpected turntable motion and from unexpected activation of the weld gun. The greatest hazards arise when no machinery is switched off and a large unexpected robot motion occurs: no emergency stop is activated. This brings up three points: the possibility of such occurrences should be minimized by good design followed by an analysis of the possible effects of failures; working practices should ensure the immobilization of equipment; and safety interlock facilities should disable the machinery in a safe way (i.e. their use should not present other hazards).

## Consideration of the system with safeguards

Introducing safeguards can reduce the hazards, but could also alter the way in which work is carried out. Such aspects of the production design are examined here, and their effectiveness in preventing hazards is assessed. The same robot system as shown in Figure 7.6 is considered, but with a number of different safeguarding options.

As shown in Figure 7.9, the safeguards include fencing around the robot which encloses the robot and turntable workspaces with only two access points: an access gate and a loading area safeguarded by a light guard. The light guard provides protection in the event of intentional (but unauthorized) and possible unintentional intrusion of personnel into the enclosed area, but at the same time allows the worker to load and unload the turntable. The gate allows access to the enclosed area for the appropriate tasks of programming or adjustments. The gate and the light guard are interlocked to the equipment controls in such a way that any intrusion will remove power from the system. However, for operations such as teaching or repair and adjustments, an override mechanism is proposed which allows the gate to open once activated.

## System operation with the safety system

The only major differences here are related to the way in which the operator performs specific tasks. Before programming or attempting to perform adjustments or repair work, the operator may need to enter the workspaces of the turntable or the robot. The action of the operator here is dependent on the way the system access gate operates. Some of the possible options are:

1. Opening of the access gate is controlled by a locking device and an override mechanism which is key operated. Once the key is used access to the equipment workspaces can be gained with power to the equipment either enabled or disabled. Once the override key has been used and the gate is opened to allow access, the case is identical to that of the systems without the guards (Figure 7.6).
2. Similar to option 1, the access gate is prevented from opening by a locking device. But the override key disables the locking mechanism, only allowing selection of teach mode (i.e. prevents selection of playback with the gate lock open). The locking

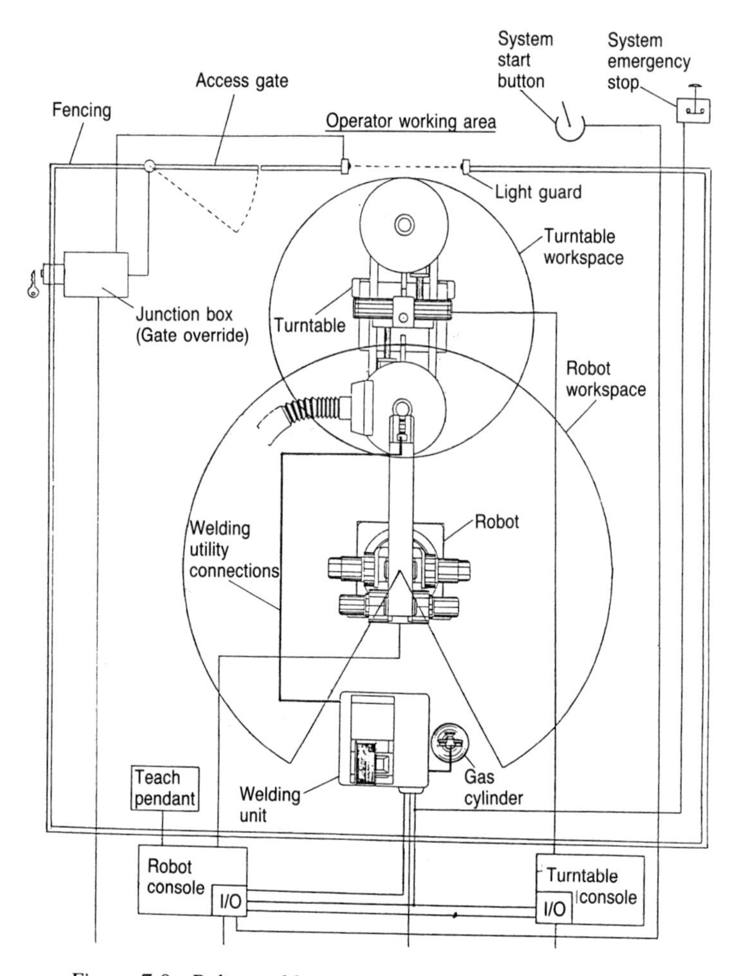


*Figure 7.9 Robot welding system layout including safeguards.*

mechanism must be such that the override key can only allow selection of playback (or automatic) mode with the gate locked shut.

3. The access gate is again prevented from opening by a locking device. The override key in this situation would not only disable the lock but would also produce a routine disabling of the equipment. This would mean that the equipment is brought to a halt and all machine operation or movement is disabled. In addition only teach mode is selectable on the robot if the equipment is enabled.

## Implications of the safety system for event tree I

Assuming the weld-equipment failure has occurred and the operator is required to perform some adjustments, then for the three options mentioned above the following implications arise.

In option 1 access can only be gained by an authorized person holding the override key. Any attempt to enter the cell through the turntable access gap, protected by the light guard, while the turntable is in motion will put the system into an emergency stop mode. (Note that the turntable physically blocks the access gap when it is in position for loading.) The use of the override key will allow full access to the system with the system in its normal state. In this situation event tree I (Figure 7.7) fully presents the outcomes, and the safeguarding measures may be seen to have had very little effect except that of preventing unauthorized access.

In option 2, again no one can enter the working area without the use of the override key. However, the use of the override key is assumed to prevent play-back operation. This reduces the hazards from the robot as speed is assumed to be restricted. The person must, nevertheless, disable the robot before carrying out any repair work.

In option 3 the use of the override key will produce a different result. It will halt the operation and automatically disable the robot and other equipment, thus removing the hazards from their unexpected behaviour. Events 2, 3 and 4 (Figure 7.7) are therefore irrelevant and the path to be followed leads to outcome 1.

## Implications of the safety system for event tree II

In this case, the person involved requires access to perform adjustments or programming with the equipment workspaces. As may be seen in Figure 7.8, putting the system into emergency stop (event 1) would lead to a 'safe' state. If the override key is used the following alternatives arise as the access gate lock is disabled.

In option 1 the negative path after the initiating event for event tree I is still valid and leads to the hazards stated. This is because the situation is identical with that of the system without the safety barriers.

In option 2, however, the robot is prevented from performing playback operations. The large high speed movements that could occur are prevented (except in failure situations) as program execution is not possible. The hazard created by outcome 2 is thus reduced.

In option 3 all machinery is disabled and thus any decision by the user to resume power to the equipment will be a conscious one. This eliminates the possibility of omission of manual steps leading to the disabling of all the equipment. The hazards that could exist would then depend on whether or not good working practices are adopted; i.e. no equipment is enabled unnecessarily and is disabled when no longer needed.

## Some implications of the ETA in case study B

A number of aspects of robot system operation are discussed in the light of this assessment. Perhaps the clearest set of implications of this work is that for equipment design. The three options for the access gate present problems for the interlocking of the safety measures with the production equipment. For certain designs of robots, option 1 may be the only one possible. In this situation the choice of disabling the equipment is solely a personal one. Putting the robot into teach mode (option 2 or 3) will be difficult to achieve with the internal design of a number of robots. Since both of these options are beneficial, their implementation suggests implications for standards for robot designs (MacCormick, 1981). Besides being able to select teach mode or playback by remote means, there are distinct advantages in having status information from the turntable and the weld gun transmitted to the robot. This would allow the robot to halt should the weld gun or the turntable fail or operate in an unexpected way. Naturally this requires a high level of sophistication for these pieces of equipment.

It is also of benefit to make safeguarding as routine and straightforward as possible. Easy recovery would be a great advantage. In option 2, this would mean that the use of the override key would not result in a lengthy program initialization procedure. It would be easier to engage playback mode and start the program fairly rapidly from outside the fencing. With option 3 the override key would automatically disable motion, but re-enabling should not require re-initialization. The advantage of this is that the operation would be easy and errors less likely. There would also be little motivation to speed up operation by circumventing the system in some way. Systems that have laborious access procedures tend to be altered (by-passed) in an informal manner. This eases the actions required but at the expense of lowering the safety level.

## Failures in the system

Throughout this exposition of the technique of ETA, an assumption has been made about the types of failure in the safety system. These have been assumed to result in equipment becoming inoperative and an emergency stop resulting i.e. a safe mode failure. A FMEA of the system could discover otherwise, but complete safety integrity is certainly an ideal towards which the technology should aim. For all failures that result in hazards, a rigorous attempt should be made to reduce the possibility of their occurrences. This can be studied by FMEA and FTA. Equipment with a higher reliability should be introduced and if necessary a redesign of the system carried out to decrease the probability of failures. There is a need for analysis of this kind (FMEA, ETA, FTA) at the design phase, which also results in long term cost savings.

A 'fail-safe' safety system would only result in a safe condition if the emergency stop produced no undesirable movement by any of the equipment in the system. For example, if the robot design is such that once power is removed the arm can move, then damage is likely because of the possibility of collision. Having brakes on all joints would overcome this, but the possibility of collision depends on the integrity of the braking mechanism for each joint. Design changes following an analysis should aim to reduce the probability of such failures.

## *Case study C: analysis of robot system hardware and software*

A study of a typical robot cell employing the PUMA 560 robot is presented here. The robot is used to perform a loading and gauging task within an integrated cell to be presented later. A description of this robot is given in the PUMA manual (Unimation (Europe) Ltd, 1980). It should be noted that the PUMA robot is used as an example here and all robots considered to be in the same class as the PUMA would, if considered in such a study, lead to similar results. Figure 7.10 shows the hardware configuration of the PUMA robot in schematic form.

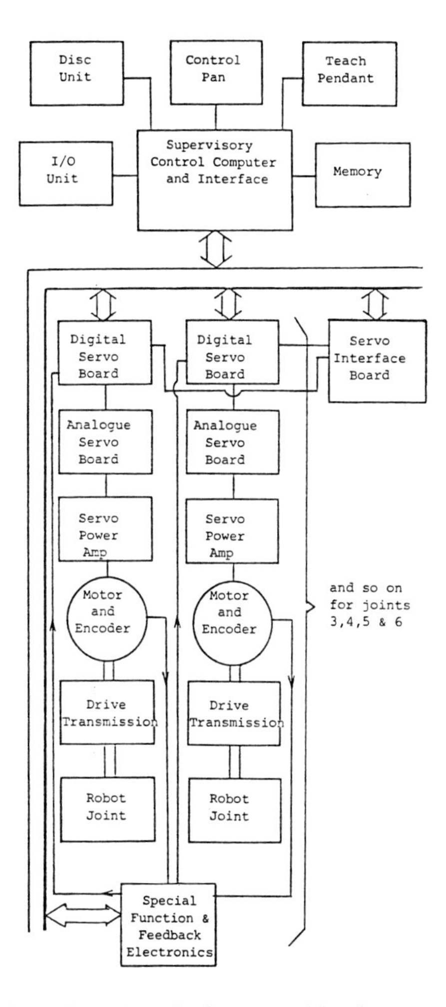


*Figure 7.10 Schematics of robot control hardware configuration.*

## The PUMA robot cell configuration

A PUMA robot is arranged in a cell with the configuration shown in Figure 7.11. It is programmed to transfer parts from the pallet to a Computer-Numerical Control (CNC) lathe where it will be machined. It is then transferred to a turn-around station for the robot to re-hold it with the corrrect orientation before loading it onto the gauging station. The part is released when loaded on the gauging station and its diameter is measured. If the machining tolerances are acceptable after the gauging process, the part is returned to the pallet via the turn-around station. Unacceptable parts are rejected and dropped into a bin. The flow diagram of Figure 7.12 shows the sequence of operation of each device with the robot acting as the system supervisor.

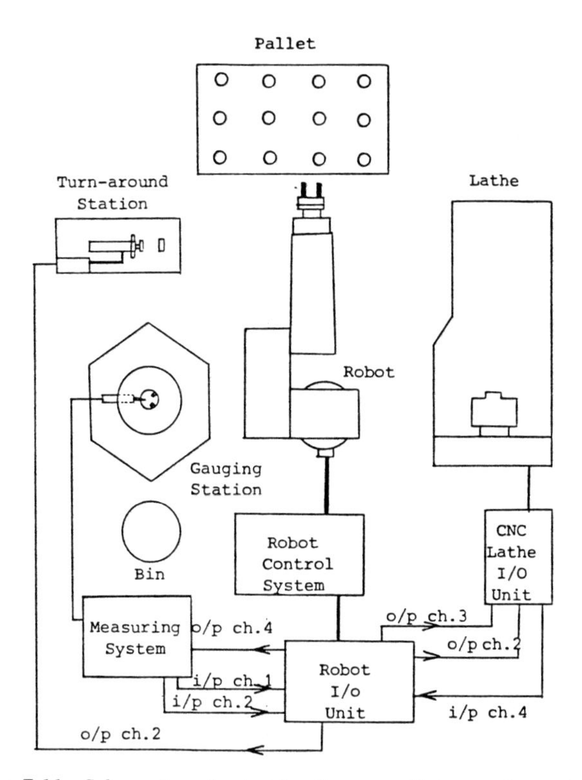


*Figure 7.11 Schematics of robot loading, machining and gauging cell.*

## Analysis of software/hardware interaction

The fault tree in Figure 7.13 considers the top event of 'undesirable robot movement causing likely damage or harm'. There are two separate causes of this which could be interrelated but are examined in isolation here: robot control hardware failures, and robot application software related failures. The embedded software that includes the VAL operating system as well as the firmware that

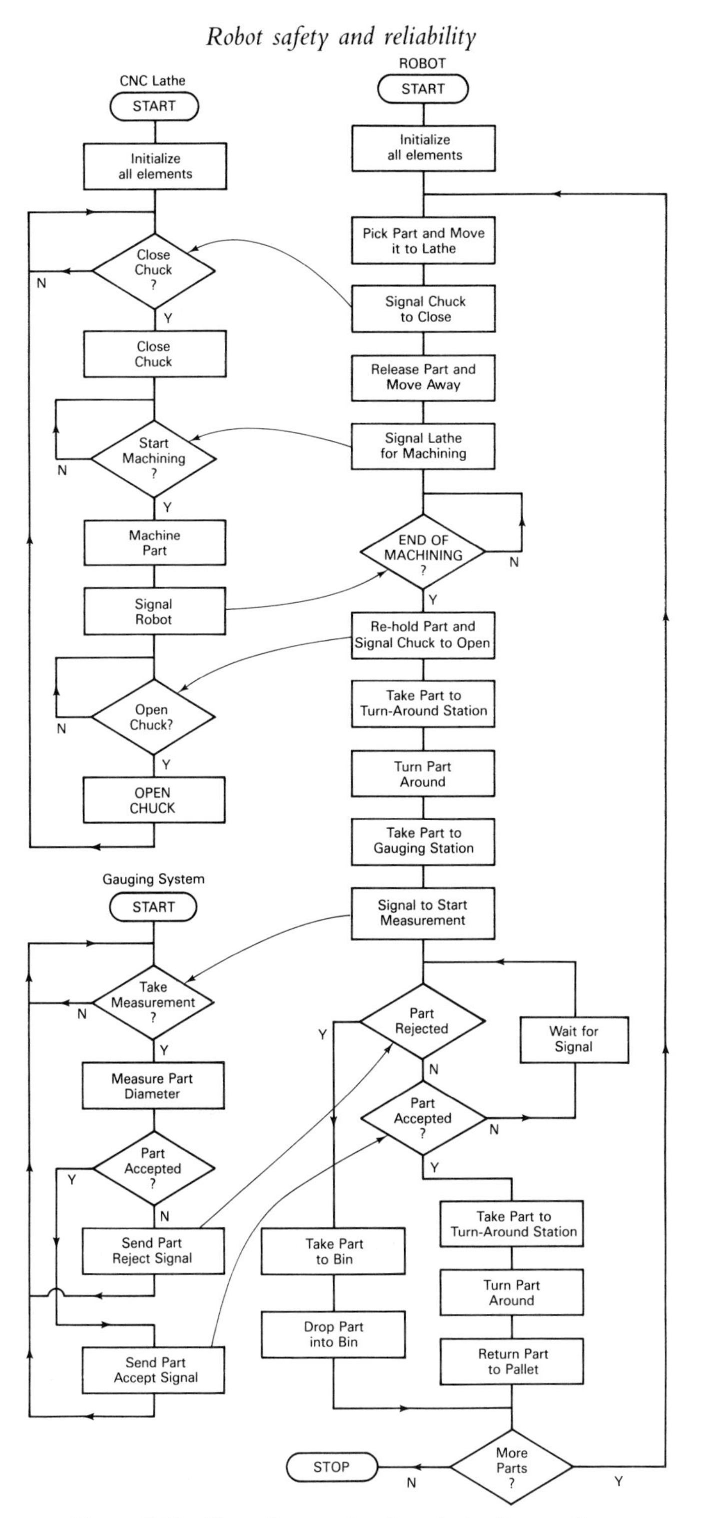


*Figure 7.12 Flow diagram for the robot cell operation.*

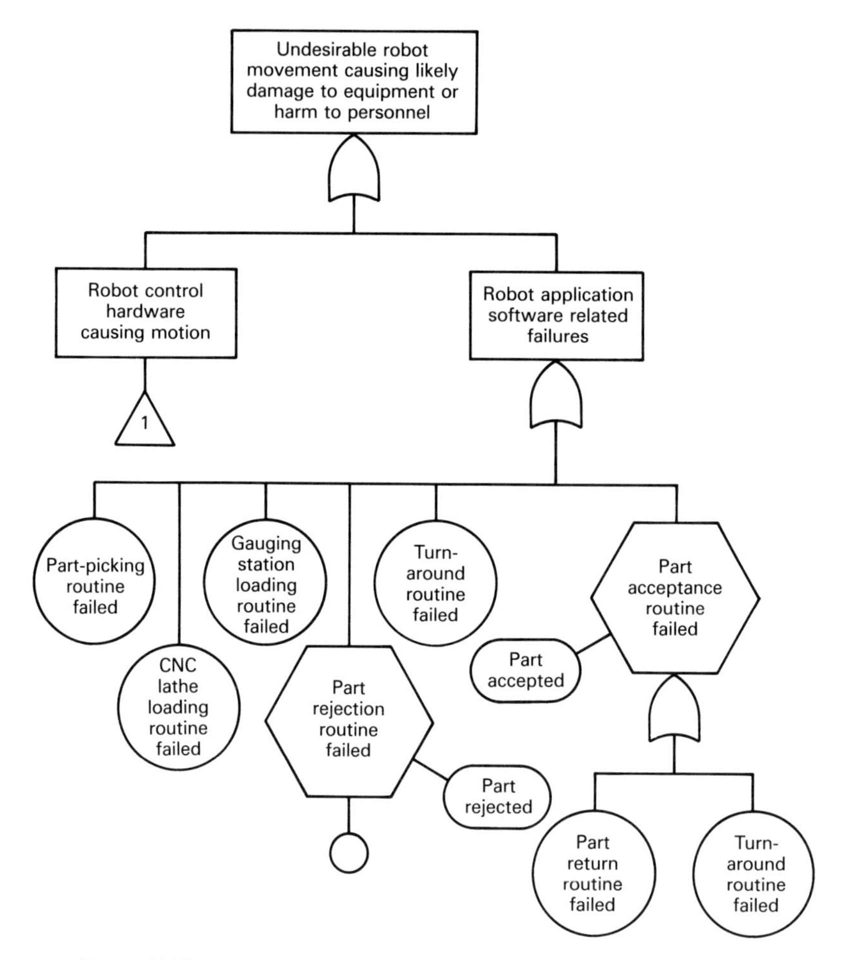


*Figure 7.13 Fault tree for the top event of undesirable robot motion.*

controls the movements of each joint are assumed fully operational. This is valid in so far as the application software for performing the task contains no errors and has been executed satisfactorily for one complete cycle via all possible routes under all likely conditions. To illustrate how the robot cell can be analysed for integrity two specific subprograms, respresentative of a number of program modules for the robot cell, are chosen.

*The part picking routine* requires the robot to move to a pallet position, where a part is present, grab the part and remove it from the pallet. The robot then moves to the intermediate position ready to load the part onto the CNC lathe (Figure 7.11). The subprogram that would perform the part picking task is as follows.

1. APPRO PART, 50
2. OPENI

3. MOVES PART
4. CLOSEI
5. DELAY 0.5
6. DEPARTS 100
7. MOVE INTEM

The event tree shown in Figure 7.14 examines the possible errors that could cause the top event of the fault tree in Figure 7.13. The initiating event considered is 'part picking routine started'. The likely parametric errors in the approach instruction (step 1) are considered with the consequences stated. The next event examines the states of the gripper. If the gripper fails to open (step 2) the movement that follows (step 3) will result in gripper–part collision. With the gripper open, the correct execution of the MOVES instruction will be a safe one unless the location specified is incorrect. Step 4 closes the gripper to grip the part. A delay of 0.5 s is included to ensure that the gripper pneumatic drive system has sufficient time to act. The possible errors that could have safety implications are examined in the event tree with the consequences stated for steps 4 and 5. If the delay is too short the robot may move (i.e. depart from the part location) before the part is gripped. This would cause the part to be held in a different position which could result in a collision during the operations that follow. The DEPARTS instruction produces a motion in a straight line, retracting the part from the pallet hole. If the distance specified is too small, the part will remain partially inserted in the hole leading to a part–pallet collision when the next step (MOVE INTEM) is executed. The safe way of defining the

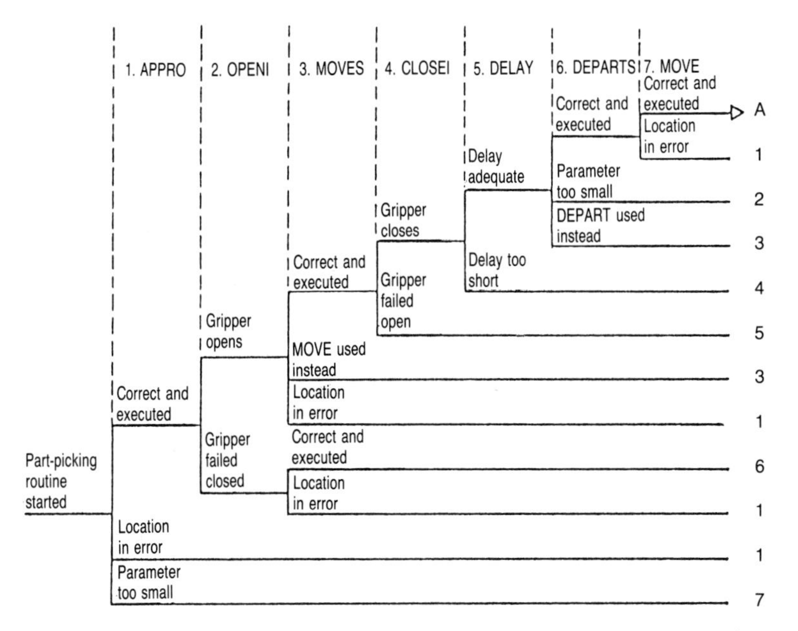


*Figure 7.14 Event tree for the part-picking routine (see list of outcomes).*

depart distance is to have it slightly greater than the length of the part. A distance of 100 mm was adequate in this case, with the consideration of the physical constraints of the cell.

*The lathe loading routine* follows after the part picking routine with the robot at the intermediate location (INTEM). To load the part onto the lathe, the position named CHUCK is used. This is the required location where the part will be held for maching. The programming steps corresponding to the operation flow diagram of Figure 7.12 are as follows.

```
 8. APPRO CHUCK, 100
 9. MOVES CHUCK
10. SIGNAL 1                      to close the chuck
11. OPENI                         to release part
12. DELAY 0.5
13. DEPARTS 100
14. MOVE INTEM
15. SIGNAL 3                      to start lathe operation
16. DELAY 0.5
17. SIGNAL -3                     to deactivate signal to lathe
18. 100 IFSIG - 4 THEN 100        to check if lathe is finished
19. APPRO CHUCK, 100
20. MOVES CHUCK
21. CLOSESI                       to grasp part
22. SIGNAL -1                     to open chuck
23. DELAY 0.5
24. DEPARTS 100
25. MOVE INTEM
```

The event tree in Figure 7.15 is constructed for examining the possible outcomes of errors or failures in this piece of software—steps 8 and 9 produce the required motion for placing the part in the chuck. The chuck must be open before step 9 is executed (it is assumed that an initialization routine opens the chuck at the start of the operation). As the state between events 8 and 9 in the tree indicates (Figure 7.15a), a collision involving the part, the gripper, and the chuck will result if the chuck jaws are closed. This could be caused by an incorrect signal or failure in the chuck actuation system. Step 10 enables a signal for closing the chuck assuming the previous steps are successfully executed. The consequences of errors for step 10 are shown in the event tree. Having placed the part into the chuck and successfully released it (step 11) the robot should move away before machining can start. Steps 13 and 14 move the arm safely to the intermediate point. The consequences of errors in these two instructions are presented by the appropriate branches of the tree.

Output channel 3 of the robot is used to signal the CNC lathe to start machining. It is imperative that this signal is active for only a short period of time to allow its recognition by the CNC lathe (steps 15, 16 and 17). Omission of step 17 or failure to set channel 3 back to its inactive state can have severe

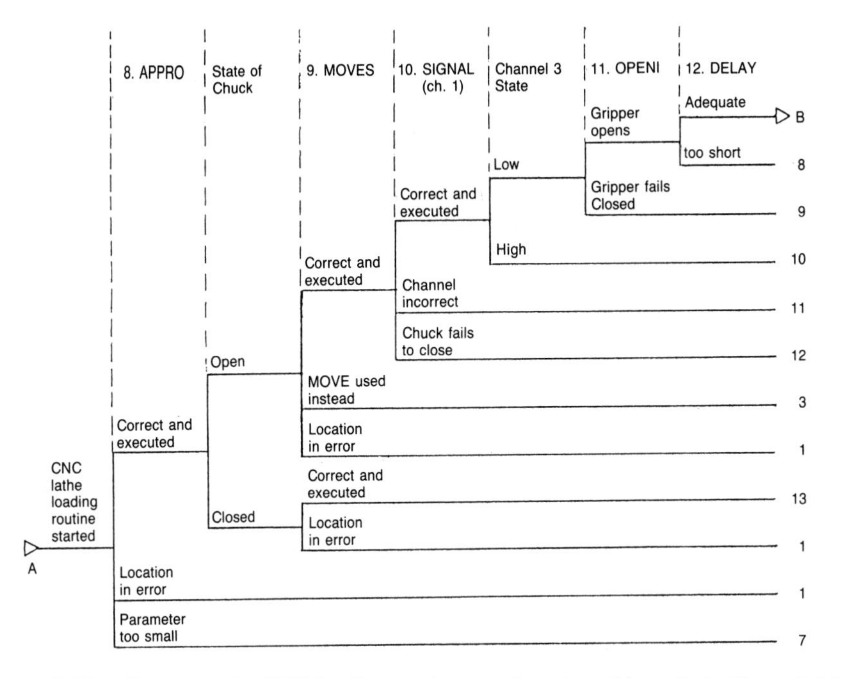


*Figure 7.15a Event tree for CNC loading routine, continuation of branch A, Figure 7.14.*

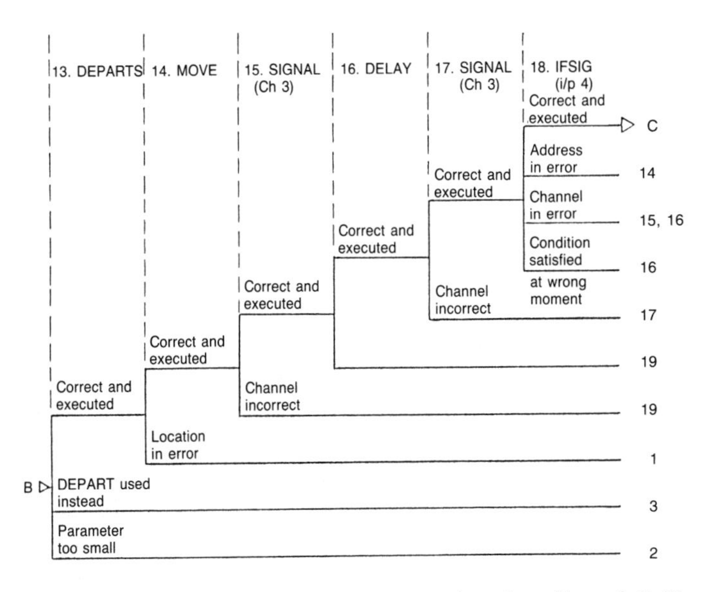


*Figure 7.15b Event tree for CNC loading routine, continuation of branch B, Figure 7.15a.*

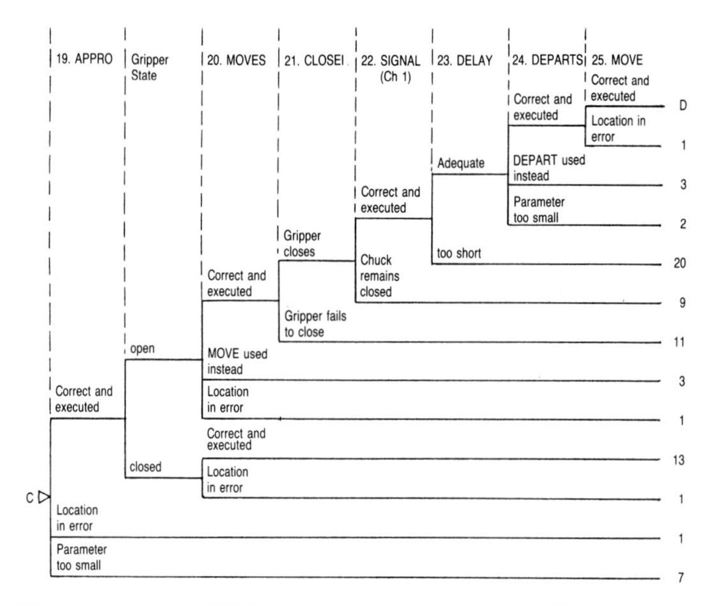


*Figure 7.15c Event tree for CNC loading routine, continuation of branch C, Figure 7.15b.*

consequences. Failures that cause channel 3 to remain active will result in CNC cutting tool–robot collision as well as part–gripper damage because machining will start as soon as the chuck is closed. (It has been assumed that the chuck must be closed before machining is possible.) This occurs immediately after step 10, and the event related to the channel 3 state in the event tree examines the outcome.

When machining is complete (see Figure 7.11) a signal from the CNC lathe via the robot's input channel 4 informs it to remove the part from the chuck. Step 18 enables the robot to accept this signal and act accordingly. A failure causing this signal to be activated at the wrong time can lead to unsafe consequences, including robot–CNC lathe collision. The event tree (Figure 7.15) presents the possibilities. Steps 19 to 23 move the robot to the chuck location to remove the part. The possibilities of errors in these instructions and their consequences are given by the appropriate branches of the event tree (Figure 7.15c). It is important to hold the part, i.e. close the gripper, before opening the chuck by deactivating the signal on channel 1 (step 22). If a fault occurs causing the chuck to remain closed then damage will occur when the robot tries to remove the part. Step 23 produces the required delay for pneumatic actuators before the part is removed at step 24. The robot once again returns to the intermediate location before the turn-around operation is initiated. The remaining parts of the program are not presented here, nevertheless the routines described represent the majority of the safety-related problems characteristic of this kind of software.

Branch 1 of the fault tree in Figure 7.13 relates to the failures within the robot control hardware. The way numerical assessment for such a tree can be performed in terms of component and submodule failure rates is also presented.

It is not possible fully to examine the exact way in which failures can cause undesirable robot movement. This is mainly due to the lack of detailed design description made available by the manufacturers. The fault tree branch of Figure 7.16 gives the major sources of failures that can cause undesirable motion. This fault tree applies to any six-axis electric robot with a similar control configuration as shown in Figure 7.10. Two states are examined. The first considers the failures that can cause robot movement with the arm power on, and the second examines this with the arm power off. When power to the electric motors of the robot is disabled, the brakes are automatically activated on three major joints (1, 2 and 3) to stop motion due to the stored energy in the arm. When servo-power is on, the movements of the robot are controlled by the control system. In this situation a failure in any one joint or in the control system supervisory computer hardware can cause aberrant arm movement. This is indicated by the OR combination of all the failures within the module. Brake failures can cause problems if the failure activates the braking system whilst servo-power is on. A gripper failure is also a contributory factor to undesirable movement of the robot as illustrated by the ETA of the software. Rigid and flexible couplings play an important role and their failure can result in a joint becoming free to move. Further breakdown of these basic faults has not been attempted and it has been assumed that any failure in the modules causes the top event. This assumption will produce an over-estimation in the failure probability to be calculated since not all the failures will affect the motion of the arm. This is acceptable for the following reasons:

1. the correct motion of the robot arm requires the simultaneous operation of all joints. A failure in any one joint can cause erratic motion in that joint or cause it not to move at all. Either case will result in an overall aberrant behaviour of the robot arm; and
2. the majority of the electronics hardware consists of large- and medium-scale integrated circuits. The failure modes of such circuits are impossible to identify under all conditions of use. Such circuits usually perform a critical function and any failure irrespective of the mode will inevitably cause a system failure. This is particularly the case for microprocessor and control computer hardware.

Furthermore, a detailed understanding of the embedded software for the programmable devices is needed to consider the conditions under which failure modes influence the operation.

## Numerical analysis of robot failure probability

Failures occurring in the electronic, mechanical, or electromechanical components can be a major cause of undesirable movements. Failures could occur because of systematic design errors or random failures. Systematic faults can be detected during the design process by formal computer analysis techniques and testing under the conditions specified for the unit. There remain random failures that

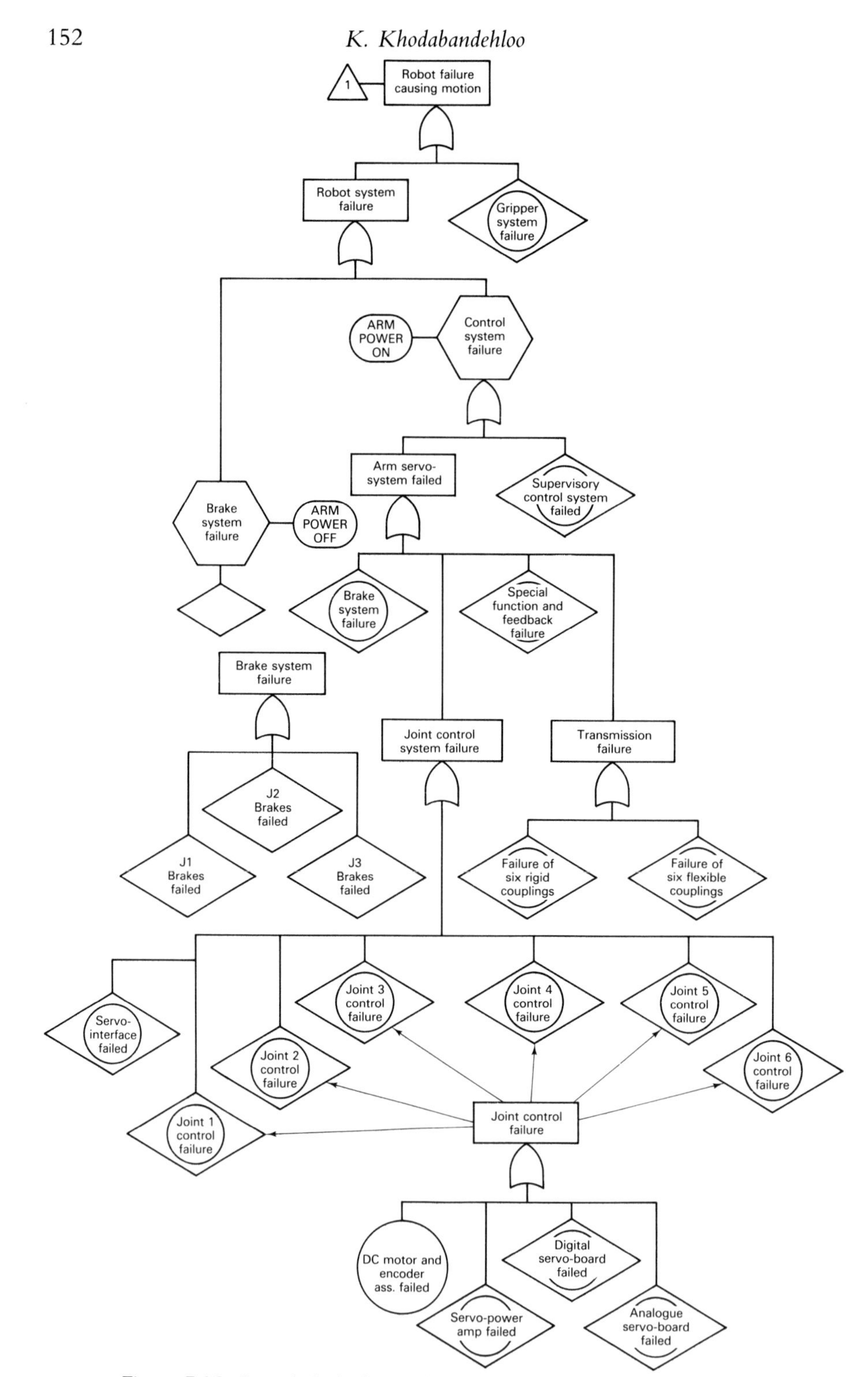


*Figure 7.16 Branch 1: fault tree for the robot failures causing motion.*

could occur during normal operation. The MIL 217D handbook (MIL 217D, 1982), the SRS Data Bank (SRS Data Bank, 1984) and the NPRD-2 handbook (NPRD-2, 1981) provide the failure rates used for predicting the likely probability of undesirable movement over a particular period of time.

Table 7.3 presents the failure rate estimates of typical modules that influence the motion of the arm. The failure probability of branch 1 may be estimated using the failure rates given in Table 7.3 and the relationship

$$P_j = 1 - \exp(-f_j t) \tag{1}$$

for each element, together with

$$P_{\mathrm{TOT}} = \pi_{j=1}^{n} (1 - P_j)$$

$$= 1 - (1 - p_1)(1 - p_2)(1 - p_3)\ldots(1 - p_n), \tag{2}$$

where $P_j$ is the failure probability of the $j$th module with failure rate $f_j$ over time period $t$ and $P_{\mathrm{TOT}}$ the failure probability for branch 1.

*Table 7.3 Failure rates for the robot modules.*

| Item | Number | Failure rate (Faults/$10^6$ h) | | | Source |
|---|---|---|---|---|---|
| | | 10°C | 30°C | 50°C | |
| Digital servo-card | 6 | 33.0 | 62.6 | 154.5 | A |
| Analogue servo-card | 6 | 12.6 | 21.4 | 64.9 | A |
| Servo-amplifier | 6 | 13.9 | 17.9 | 25.0 | A |
| Servo-interface | 1 | 23.4 | 27.1 | 38.2 | A |
| Special function and feedback electronics | 1 | 20.1 | 27.8 | 42.8 | A |
| DC motor and tacho | 6 | 20.0 | 20.0 | 20.0 | B |
| Brakes (motor) | 3 | 4.3 | 4.3 | 4.3 | C |
| Flexible coupling | 6 | 17.0 | 17.0 | 17.0 | B |
| Rigid coupling | 6 | 1.2 | 1.2 | 1.2 | B |
| Gripper system | 1 | 114.0 | 114.0 | 114.0 | A |
| Supervisor processor interface and memory | 1 | 487.0 | 487.0 | 487.0 | D |

*Sources:* A, MIL 217D (1982) MIL 217D models used for the components of each board; B, Systems Reliability Service (1984), SRS Data Bank; C, NPRD-2 (1981); D, Daniels (1979).

Table 7.4 gives the total failure probabilities over 1000 h at three different case temperatures for the electronic modules. A computer program has been used for predicting the failure probabilities. This suggests that a failure probability of about 78% is to be expected over 1000 h of operation at 30°C. The operating case temperature of electronic components can make a considerable difference.

*Table7.4 Failure probaility of branch 1.*

| Case temperature (°C) | 10 | 30 | 50 |
|---|---|---|---|
| Failure probability | 0.712 (71.2%) | 0.779 (77.9%) | 0.908 (90.8%) |

It should be emphasized that the elements in this robot are representative of almost every industrial robot of this type and complexity, and that the failure rates used are for typical modules.

In view of the financial investment in software and hardware, it is cost effective to have an overall microprocessor-based monitoring unit. The only disadvantage is that this technology is software based and there are no defined rules that indicate this is universally acceptable. However, using the techniques illustrated one could prevent errors in software development if the programs are of a manageable size. Testing the overall design for operation by the introduction of actual faults could also be a means of further proving the integrity of any safety software and hardware. There remains, however, one important point to be considered which relates to the level of improvement achieved. Assuming the microprocessor-based monitoring system has a failure rate equal to that of a system similar in complexity to the digital servo-card (Table 7.3), both the robot control system and the monitoring system must fail for the robot to produce an undesirable motion. This reduces the previous failure probability to

$$P_N = P_{TOT} \cdot (1 - \exp(-62.6E-6 \times 1000)) \quad (3)$$

$$= 0.047 \text{ (or } 4.7\%)$$

where $P_N$ is the new failure probability at 30°C, which clearly demonstrates improved safety integrity for the hardware.

## *Future research in advanced robotics*

The analyses performed so far have considered hardware, software and robot–operator interactions. It is possible to consider refinements in the studies to achieve a more detailed assessment of safety integrity. Several assumptions have been made in the process of quantifying the probability of failures causing undesirable robot motion. As far as safety is concerned it is clear that not all movements caused by failures result in a collision or an accident involving a person. Thus a number of refinements in the analysis may be achieved.

Attention must be paid to the failures that can cause injury or death. Although the assessments presented so far have to a large extent demonstrated the qualitative process by which this may be done, a numerical evaluation of the risk to people is a useful refinement. For the purposes of achieving a measured improvement, the risk factor described numerically can be used to compare various designs.

The numerical assessments, particularly where a failure probability is defined, cannot fully describe the frequency of a possible outcome. In the case of a robot system interacting with a number of machines, the layout of the cell is a major factor. Figure 7.17 shows a typical robotic cell with a six-axis point-to-point robot loading and unloading parts in a machining cell consisting of a milling machine and a lathe. To consider the probability of failure in the robot cell causing an undesirable motion, resulting in a collision or trapping of a person, one must study the sequence of moves by the robot and the phases during which a failure can cause an accident. With reference to Figure 7.17, this can be explained by noting that the failure of the 'boom' joint producing extension of the arm will not cause an accident if the robot is travelling between points A1 and I2, I4 and I5, and the same in reverse motion. The reason for this is clearly the fact that the extension of the 'boom' joint avoids the machine tools within the work volume of the robot whilst it is travelling between these locations. The same may be applied to other joints as the robot travels through the various programmed parts. It should be stated that a joint runaway will be detected and the sequence stopped by any conventional robot once it completes a given move, but this is not the case when the robot is travelling between two programmed points. Further analysis of this aspect of safety is possible using Fault and Event Tree techniques not presented here.

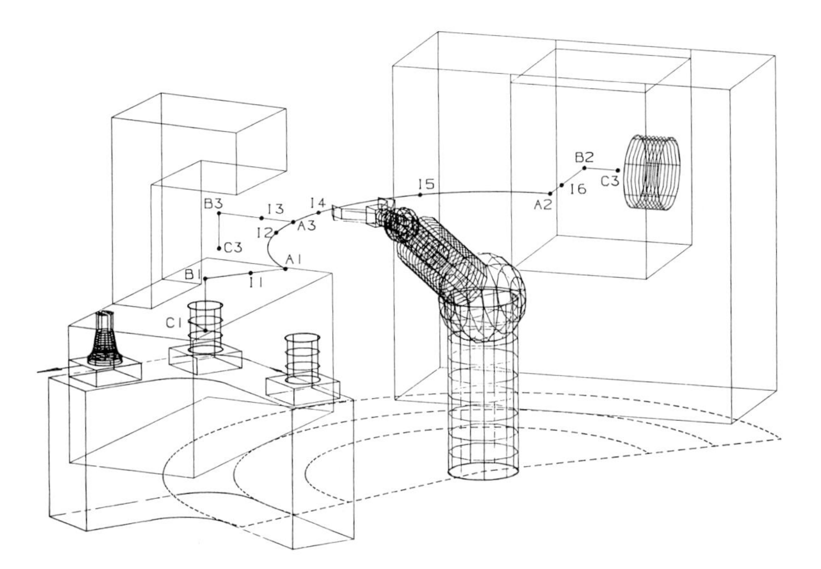


*Figure 7.17 A typical robotic machining cell.*

Another aspect of robot safety relates to the physical anatomy of the arm. A study of the risk of collision for the configuration of the cell layout may be further refined by the examination of the coupling between the joints of a robot. Without being specific about the kinematic characteristics of robots, it is possible to demonstrate this aspect of robot safety by a brief examination of two proposed robots for an advanced application such as brain surgery.

The robot in Figure 7.18 is a conventional articulate arm which may be used to position and manipulate appropriate surgical instruments in a supervised biopsy operation. It is assumed that the surgeon defines the robot movements and that the robot performs each move automatically.

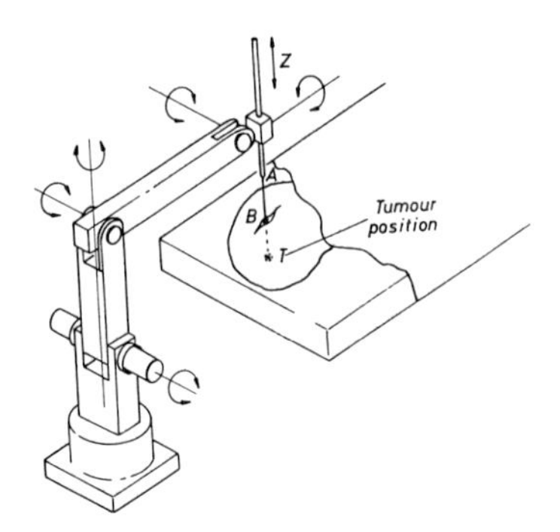


*Figure 7.18 An articulate robot arm for brain operation (schematic).*

Consider the situation where the robot is required to insert a biopsy probe in the head. The important sequence of moves for the robot may be defined in the following way:

1. position the tool outside the head at point A (Figure 7.18);
2. move to position B (outside the head) in a straight line;
3. obtain confirmation from the surgeon to continue; and
4. move to position T in a straight line and stop.

It may be noted that motion of the arm of the robot requires several joints to be moved simultaneously in order to achieve the straight-line trajectory required. The robot configuration is such that several joints must move in order to manipulate the tool into position A (Figure 7.18). A failure in any one joint may cause a collision of the arm with the patient. It is of course appropriate to lock the robot at this position and use a linear axis to drive the probe into the head. Nevertheless the patient is still exposed to the hazards from undesirable robot motion due to failures which may also include brake failures. Indeed the failure of the linear axis inserting the probe is critical, but this may be considered equally as critical with the surgeon or any other robot arm configuration. The relevant issue is whether a safer configuration of the robot can be defined. Figure 7.18 shows the schematic of a tool positioner (Khodabandehloo, 1990) that is inherently safer for this operation. The main reason for this is related to the fact that the robot arm is physically outside its work volume and the only movement that brings the tool in contact with the patient is the $Z$ motion, produced by a linear axis, the same as before. It can be shown that a runaway on any one of the joints, once the tool is outside the head, will not bring the robot arm or the tool into collision with the patient. This is a considerable design improvement for safety.

The anatomical configuration of a robot can be designed to minimize the risk of an accident despite possible failures in the system. It is clear that with the increasing application complexity facing robotics, with reports of robots planned for use in surgery, domestic and other critical applications, human safety must be ensured. Systematic analyses of robotic systems must be an essential activity in the design, production, installation and operation of any robotic system. FMEA, FTA and ETA provide the means for such analyses to be performed thoroughly.

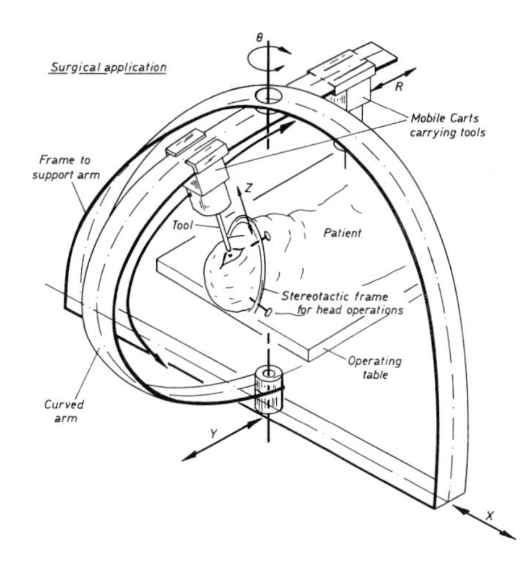


*Figure 7.19 Tool position and configuration to assist in brain surgery.*

# Discussion and conclusions

The predicted number of robots for use in the home has been estimated to be over 5 000 000 in the next decade (Cato, 1989). Although such figures can only be regarded as estimates, there is no doubt that the use of robots worldwide will increase dramatically as technology develops. The nature of applications will also become more sophisticated with a significantly greater level of routine robot–human interaction. The studies here have considered a number of cases, focusing on the safety and reliability issues of robotic systems. In particular, robot safety integrity has been examined with consideration of hardware, software and robot–operator interactions.

Software in a robot can easily be influenced by external hardware faults with severe consequences. Inclusion of safety features in both hardware and software can reduce the likelihood of failures producing aberrant behaviour. This often demands specific software requirements of the robot programming language and should therefore be a major factor in deciding what robot to use for a particular application. Assessment of the integrity of the software as part of its development procedure has an important role to play in ensuring overall safety and must be undertaken formally.

Robot hardware must also perform satisfactorily. Any hardware fault that affects the motion of the robot arm has safety implications for the equipment and the personnel regardless of the state of software. Fault tree analysis can be applied as a numerical assessment in addition to identification of critical failure interactions. This is demonstrated in terms of erratic movement, which is quantified as the probability of failure over a particular period of time (case studies A and C). For the example of a typical six-axis electronic robot, we have estimated a failure probability of about 0.78 per 1000 h at 30°C. It should be noted that similar analysis can be performed for any robot, and in fact the numerical assessments apply equally to any other electric robot with the same level of complexity.

Every undesirable motion will not necessarily lead to a collision or an accident. It is the environmental interactions of the arm and its position relative to the rest of the equipment, in addition to the distances involved that govern this interactive behaviour. Methods for reducing such occurrences due to robot hardware failures are presented, suggesting that it is impossible to eliminate undesirable movement. Nevertheless, preventative design measures can be taken by using high quality components, improving the environmental conditions of electronic components, and adding monitoring electronics to reduce the likely occurrence of erratic robot movement in any mode of operation.

ETA has been applied in assessing the safety aspects of a robotic cell in case study B. Specific consideration has been given to a welding cell. The event trees constructed for this cell have identified a number of outcomes that put personnel as well as equipment at risk. Removing people completely from such a cell, which is frequently suggested in the literature, is still far from being viable. Since, in the near future, it will not be possible to remove people altogether from the production process, a number of hazards will remain. An analytical method of assessing such risks is presented in anticipation of removing or mitigating them by effective training and safeguards. Three safeguarding options have been considered in case study B with the following conclusions:

a simple safety interlock mechanism provides no more of a safeguard than preventing unauthorized access—a minor improvement upon a system without safety barriers (option 1);

a more complex safety interlock mechanism would demand additional interfacing facilities from the equipment within the cell and this is not yet available (option 2); and

an intelligent safety system that carries out routine disabling would require a continuous two way communication between the production equipment and the safety system controls. A more complex interfacing facility would be needed in this case, and specific provisions would have to be made for such facilities in future designs of automation and safeguarding equipment (option 3).

Several critical steps rely upon the correct action on the part of the person involved. As previously stated, the choice of safety system does not affect these greatly. In event trees I and II, the equipment can be disabled by the safety system in options 2 and 3 but the selection of lower speed and the appropriate programs are still tasks which operators must perform. One obvious hazard occurs when the robot is moved in the wrong direction simply because a wrong button is pressed. Correct procedures must be established and then taught for all the tasks needed for the continued operation of the robot system. This is particularly important for such things as programming, powering up and turning off the system or maintaining any piece of equipment. To do this, a complete record of the contents of each program should be kept on every machine so that alterations to programs are visible to all and the correct ones are selected.

Along with this, a sound working knowledge of the layout of the teach pendant should be established. This is especially important if the functions of some of the buttons change for different types of motion. Since variations in the design of robots have an important role in how hazards are controlled, complete training is necessary for each type of robot to be used.

For option 1 the override mechanism allows free access to authorized personnel in any operational mode (automatic playback or teach). The onus for maintaining safety during periods of access rests solely on those who carry out the work. Prevention of high-risk events can only be achieved by correct disabling and selection of operating modes. In option 2 the override mechanism would automatically prevent playback mode. Nevertheless equipment is still powered on and so if maintenance were to be carried out, the correct disabling procedures would still have to be adopted. In option 3 all equipment motion and action is disabled by the use of the override key. In this option, a positive decision has to be taken to power up any piece of equipment and increase the hazard level.

In the last two options, playback mode is not possible with the override mechanism activated. However, for a complex task such as arc welding, there is a need to observe the operation of the robot clearly at playback speed. What is more, the robot program will have to be run through a few times to carry out the fine tuning of the weld parameters. To do so a worker will need to be within the cell workspace while the system is fully automatic. This is obviously a high-risk activity, but often a necessary one. One solution to this problem could be the introduction of a permit-to-work system for such periods.

FMEA, FTA and ETA are useful techniques for the assessment of safety and reliability in a robot system. However, in the application of such techniques a number of drawbacks must be borne in mind. The choice of failure or events in each case is critical. With a wrong or inadequate choice the analysis may overlook hazardous consequences and the decisions taken will not necessarily safeguard the system against these omitted faults or failure events. It should be stressed that a great deal of care is needed at this stage of the assessment. Detailed knowledge of the operation of the system, design of equipment and the interlocks are necessary for a high level of analysis. An identified hazardous outcome might be caused by other events not considered by the analysis and awareness of such possibilities is necessary to make the analysis complete. Perhaps most important of all is that the techniques as applied here emphasize the hazards that may occur and not their relative frequency. The techniques are thus used for maximizing the safety integrity and minimizing the hazards in some cases. Furthermore, it would not be an easy task to assign probabilities to a number of events in the trees. Human actions are frequently considered and their quantification is notoriously difficult.

Despite the drawbacks, the analysis provides some very useful information. It gives a framework within which knowledge of equipment design can be clarified and decisions can be made on appropriate safety measures. It also aids in the development of an awareness of the hazards present in a robot system and hence provides a basis for the selection of the necessary contents of safety training.

System assessment during the design and development phase is the best method of achieving safe and effective production. Advances in vision, sensory equipment and artificial intelligence, and their integration with robotics and automation systems, will be the major growth areas of technology. Analysis of the kind demonstrated can play a substantial role in ensuring inherent safety integrity in such systems.

## References

Daniels, B. K., 1979, Reliability and Protection Against Failure in Computer Systems. NCSR R17, The National Centre of Systems Reliability, UKAEA.

Hartley, J., 1983, *Robots at Work: a Practical Guide for Engineers and Managers.* (IFS (Publications) Ltd).

Jones, R. H. and Khodabandehloo, K., 1985, The safe use of robots. Robots and Safety—HSE/Industry View, Institution of Mechanical Engineers Conference.

Kato, I. C. H., 1989, Towards the age of minor robots. Proceedings of the 20th Century Symposium on Industrial Robots, October 4–6, Tokyo, Japan.

Khodabandehloo, K., 1990, Arrangements for supporting a mobile article. British Patent Application No. 900258.5.

Khodabandehloo, K. and Rennell, I., 1988, Intelligent robots guided by vision. In *Computing: the Next Generation* (Chichester: Ellis Horwood).

Khodabandehloo, K., Sayles, R. S. and Husband, T. M., 1985, *Safety Integrity Assessment of Robot Systems, Safety of Computer Control Systems* (Oxford: Pergamon Press).

MacCormick, N. J., 1981, *Reliability and Risk Analysis: Methods and Nuclear Power Applications* (Academic Press).

MIL-217D, 1982, *Reliability Prediction of Electronic Equipment.* US Military Handbook.

NPRD-2, 1981, Nonelectric Parts Reliability Data. Reliability Analysis Center, Rome Air Development Centre.

Systems Reliability Service, 1984, *SRS Data Bank.*

Unimation (Europe) Ltd, 1980, *PUMA Technical Manual.*

# Chapter 8
# Six severity level design for robotic cell safety

**B. C. Jiang and O. S. H. Cheng**

*Industrial Engineering Department,*
*Auburn University, Alabama 36849–5346, USA*

**Abstract.** A robot usually works with other machines in a manufacturing cell. When considering safety, attention should be given not only to the individual machines and robots, but also to their interactions due to processing needs. This paper presents the Six Severity Level Design (SSLD) concept for safely designing a manufacturing cell. This design philosophy integrates guarding techniques with control actions, considers both production needs and safety concerns and interfaces machine functions with process requirements. This philosophy has been implemented in an unmanned manufacturing cell that makes a family of parts. The hardware and software, as well as their interactions with the manufacturing process and control actions are described. The designed system has the advantage of providing maximum protection to the operators while causing minimum interruption in production. Future research needs in the areas of sensor reliability, system integration and training are also discussed.

## Introduction

A manufacturing cell is a collection of machines and robots that performs a specific function, such as manufacturing a family of parts, and is the basic structure for a cellular manufacturing system (Black, 1983; Black, 1988; DeGarmo *et al.*, 1988). Recently, there has been a great emphasis on how to design and control a manufacturing cell, but safety considerations have often been overlooked. The safety guarding in such a system can be classified into two levels—the individual machine level and the system level. At the individual machine level, safeguarding devices are used to prevent contact with dangerously moving parts (OSHA, 1982), to prevent human errors and to protect humans from ejected objects. At the system level, consideration is given to the interactions between the machines and the manufacturing process performed in the cell. Robots are relatively new machines designed for automating a manufacturing process, serving as a material-handling device or as a process-assisted (e.g. painting, welding, and drilling) device. It may be questioned if a robot can be treated as traditional machinery, but a review shows that most of the guarding principles for traditional machines are still applicable to industrial robots (Liou *et al.*, 1989). For robot guarding, there are many guidelines and standards available in the literature (JISHA, 1983; Bonney and Yong, 1985; National Safety

Council, 1985; RIA, 1987; Verein Deutscher Ingenieure, 1987). Table 8.1 shows the definitions, descriptions and application examples of various guarding techniques (NIOSH, 1975; MTTA, 1982; OSHA, 1982; Strubhar, 1984; Robinson, 1985; Etherton, 1988).

Although a robot usually has to work with other machines, less attention has been paid to system level safety. Collins (1986) mentioned the importance of space requirements for designing a robotic workstation to avoid pinch points. Linger (1984) proposed the concept of the 'production adapted' safety system, i.e. a 'safety system based on high knowledge about the process'. He also presented a safety system that interfaces sensors, a control system and a power/brake system to provide maximum protection to human operators. A safety system for automated production to maintain the highest level of safety with lowest loss of production has also been proposed by Kilmer (1985), who suggested a safety sensor system to detect three levels of safety by region: perimeter penetration detection, intruder detection within the workstation and intruder detection very near the robot. None of these authors have, however, considered the interactions between a robot and the machines during the manufacturing process.

## *Six severity level design (SSLD)*

The following principles were used when this design philosophy was formed:

1. The design should be based on the knowledge of the manufacturing process (Linger, 1984);
2. Safety measures should be provided for system failure, human mistake, and misuse of the safety devices (Linger, 1984; Barnett and Switalski, 1988);
3. The control action should depend on the severity of the risk;
4. The guarding technique should consider both individual machines and the interactions among the machines;
5. Warning/alarming devices should be used first for personnel awareness; and
6. Guarding devices should be provided to isolate the hazard sources from the personnel. Control actions should be taken according to the level of severity of the danger.

Table 8.2 shows the SSLD. The columns of warning signs/signals show what type of warning devices can be used to call appropriate attention to intruding personnel and surrounding personnel. Warning signs and signals (both auditory and visual) are often used in a workplace as safety measures. The signs and signals must be simple and effective. Size, colour and location are the major considerations in warning sign design. Table 8.3 summarizes the principles for appropriately selecting a warning sign or signal (Bailey, 1982; Eastman Kodak Company, 1983; Salvendy, 1987).

The column of safeguarding devices in Table 8.2 shows the types of device that can be used to isolate personnel from hazardous sources and to communicate with the system control device as necessary. The purpose of such safeguarding devices is to prevent contact between the source of the hazard and other objects. Safeguarding devices can be divided into three general categories with 12 individual guard types (Jiang *et al.*, 1990), as described in Table 8.1.

The robot control action column in Table 8.2 shows the type of control action that can be taken in response to different levels of severity. Robot control actions should be taken if there is a potential hazard to operators, to reduce the robot speed or to stop the robot. These actions should be activated only if the guarding devices and the warning signs and signals cannot prevent the hazard source and the personnel from coming dangerously close. To consider both safety and production, an appropriate robot control action should be taken according to the severity of the danger.

### Severity Level 1—staying within the perimeter of a general work area

In a manufacturing cell situation, the personnel is on the factory floor but still outside the cell. They should be aware of the cell operation. The major safety concern is to attract the operator's attention and to prevent him or her from approaching the cell area. At this level, the use of warning signs is sufficient and no robot control action must be taken. The size of sign must be large enough, its colour must be bright and different from the machine (or background) colours, and its location must be where people will see it when they pass.

### Severity Level 2—approaching the cell area

As a person approaches the manufacturing cell area, although direct contact between the hazard source and the person is unlikely, a perimeter identification device (e.g. fencing with interlocks) should be installed to minimize unauthorized entry. A yellow warning light can also be installed to indicate that there are machines running nearby. No robot control action need be taken because the person is still at a distance from the robot working area.

### Severity Level 3—entering the cell area

This is the level where the interaction between personnel and robot begins to occur. The personnel should be brought to maximum alertness, and the robot should take steps to reduce the risk level. As a person advances into the cell area, pressure mats can serve as a buffer between the person and the machines. The activated pressure mats can trigger a flashing yellow warning signal to attract the attention of personnel. Robot speed can also be reduced to reduce the energy released.

### Severity Level 4—approaching the working area

The working area is defined as the robot working envelope plus the hazardous area for each machine. Photoelectric sensors (either a single sensor or a light curtain) can be used as the guarding devices. The warning light should become red to indicate the severity of the dangerous situation. A repetitive buzzer can

*Table 8.1 Summary of guarding techniques.*

| Methods | Description | Advantages | Disadvantages |
|---|---|---|---|
| 1. Enclosures | | | |
| Fixed barrier | isolation devices; barrier to enclose the hazardous part in an immovable shield | simple; provides maximum protection; requires minimum maintenance | interferes with visibility; requires other means of protection for maintenance personnel |
| Adjustable barrier | barrier that is adjustable depending on the size of the stock | flexible—can be adjusted to suit varying sizes of stock | does not provide maximum protection; interferes with visibility; requires adjustment and frequent maintenance |
| Self-adjusting barrier | barrier that has an opening that is only large enough for the different sizes of workpieces | flexible—can be adjusted by machines automatically to suit varying sizes of stock | does not provide maximum protection; interferes with visibility; requires frequent maintenance and adjustment |
| 2. Interlocks | | | |
| Barrier | switching devices; protection barrier interlocked with machine controller | removable; removing of the barrier stops the machine operation | high initial cost; requires careful adjustment maintenance |
| Gates | | | |
| Manual | gate that can be opened manually | can be opened for maintenance purposes; opening of the gate stops the machine operation | high initial cost |
| Automatic | gate opened automatically by the machine cycle | frees the operator from the need to open and close gate | high initial cost; may catch operator's hands by automatically closing the gate |
| Sensing systems | | | |
| Photoelectric | consists of light sources and controls that are interlocked with the machine controller | high visibility | high initial cost; does not protect operators against mechanical failure |

*Table 8.1 (continued)*

| Methods | Description | Advantages | Disadvantages |
|---|---|---|---|
| Radio-frequency (RF) (capacitance) sensing | consists of RF source, and antenna to detect the intrusion into the capacitance field | high visibility | high initial cost; does not protect operators against mechanical failure; requires regular calibration and maintenance |
| 3. Other type | | | |
| Stroke confiner | physical stop to limit the motor of a machine | low cost; achieves worker protection without the addition of mechanical or electrical guards | limited stock range |
| Pressure sensing system | force/torque-sensitive device used to prevent the machine from exerting excessive force/torque | protects both the operator and the machine | high initial cost; allows accidents to occur even when the machine stops |
| Two-handed controls | use two control buttons necessary to activate the machine | prevents the operator from accidentally activating the machine | limited to certain types of machines |
| Warning devices | signs/signals to alert the person of dangerous situation | easy to install; low cost | less effective as a single method of safeguarding |

*Table 8.2 Six Severity Level Design.*

| Severity level | Safeguarding device | Warning | | Robot control action |
|---|---|---|---|---|
| | | Visual | Auditory | |
| 1. Staying within the perimeter of a general work area | — | Signs | — | None |
| 2. Approaching the cell area | Fencing (rope) with gate | Yellow | — | None |
| 3. Entering the cell area | Pressure mats | Yellow and flashing | — | Reduced speed (5) |
| 4. Approaching the working area | Light curtains | Red | repetitive buzzer | Reduced speed (1) |
| 5. Entering the potential danger zone | Pressure mats and light curtain | Red and flashing | continuous buzzer | Stop (no reset) |
| 6. Entering immediate danger zone | Photo-sensors | Red and maximum frequency flashing | continuous buzzer | Stop (reset) |

*Table 8.3 Principles of using warning sign or signal.*

A. Use visual signs when:
- the message is complex, long, or will be referred to later
- the message deals with location in space
- the message does not call for immediate action
- the receiving location is too noisy for auditory signals
- the person receiving can remain in one place
- the person receiving is overburdened by the auditory system

B. Use the auditory signals when:
- the message is simple, short, or will not be referred to later
- the message deals with events in time
- the message calls for immediate action
- the person receiving has to move around for the job
- the person receiving is overburdened by the visual system
- the receiving location is too bright or too dark when adaptation integrity is necessary
- the displayed information occurs randomly and must immediately capture the attention of the operator

C. When using visual signs:
- consider printing style, size, and location
- use stereotype or standard symbols
- select appropriate colours—
    1. red is used to alert that the system is inoperative
    2. flashing red is used to denote an emergency condition that requires immediate action
    3. yellow is used to indicate a marginal situation in which caution is necessary or unexpected delay may be encountered
    4. green is used to indicate all conditions are satisfactory
    5. white is used to indicate transient or alternative conditions
    6. blue is used to indicate an advisory situation
- ensure luminance and sharp contrast with the background
- ensure that they are shaded or out of direct sunlight
- ensure that they are not subject to colour detection confusion

D. Auditory signals should:
- have a signal level 15 db above the masked threshold (defined as the level required for 75% correct detection) for 100% detectability and for a rapid response
- have a signal level less than 30 db above the masked threshold to minimize generated to the operator
- have a single pitch between 150 and 1000 Hz
- have a pulse spacing compressed to increase urgency
- cease only afer the operator responds appropriately
- be consistent with others already in use in the plant
- not be confused with the noise generated from the machining process

*Source:* Bailey (1982), Eastman Kodak Company (1983), Salvendy (1987).

be used not only as an additional warning to the personnel involved, but also to attract the attention of other personnel in the same general area. The measures used at this level are to ensure that maximum attention has been drawn from the involved personnel, and to stop the intrusion with help from others if necessary. The robot's speed should be reduced to a minimum. The control actions do not completely stop production but minimize the energy released from the robot.

### Severity Level 5 – entering the potential danger zone

This zone is defined as the area that the robot will be moving into and the area of the machine it will be working with for its next step. Both pressure mats and light curtain are activated at this level. Figure 8.1 shows the four zones under control. The system software keeps checking which zone the robot is working in (called immediate danger zone) and which zone the robot will move to (called potential danger zone). A red flashing light and a continuous buzzer can be used as the warning devices to indicate that the personnel must leave the potential danger zone before the manufacturing process can continue. The robot should freeze its motion if there is an intrusion in the Level 5 area; however, the control should be such that the process can continue at any time once the personnel leaves that area. At this level, all the power is still on, and the software is still in control.

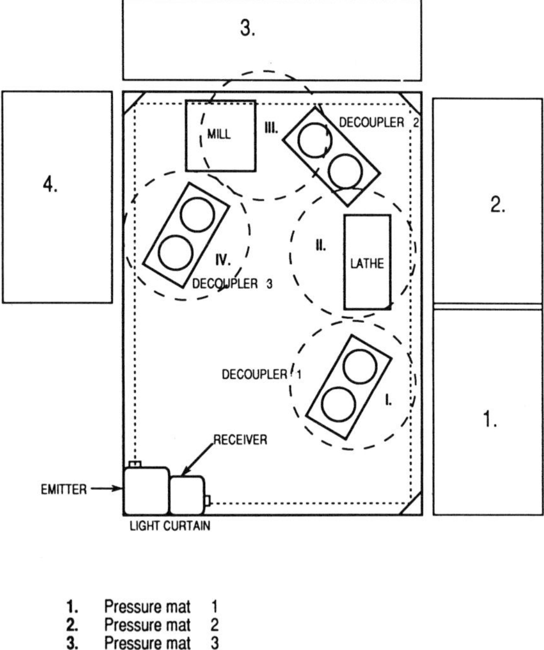


1. Pressure mat 1
2. Pressure mat 2
3. Pressure mat 3
4. Pressure mat 4

I. The area surrounding decoupler 1
II. The area surrounding lathe machine
III. The area surrounding decoupler 2 and milling machine
IV. The area surrounding decoupler 3

*Figure 8.1* Four zones under control.

### Severity Level 6 – entering immediate danger zone

When a person is in the immediate danger zone, all possible actions should be taken to prevent contact between robot and personnel. In addition to the zone detection as described for Level 5, proximity sensors can be affixed to the robot arm as redundant sensing devices to check if there is an object near the arm. Warning signs and signals should operate at maximum capacity (e.g. fast red flashes and continuous buzzing) to indicate the immediate danger. The robot should be stopped immediately as the emergency stop button is activated. All the power is turned off in the system, and the software is no longer operational. A restart procedure must be followed before the system can resume its operation.

## *SSLD for a manufacturing cell*

The SSLD has been implemented in the design of a manufacturing cell. This section describes the cell formation, the manufacturing process in place and the implementation of the SSLD.

### Cell structure

In this research, the safety system is applied to an unmanned manufacturing cell that is based on the principles of Integrated Manufacturing Production Systems (IMPS). The IMPS is based on the linked cell manufacturing system and combines the Just-In-Time (JIT) production philosophy and the Kanban system (DeGarmo *et al.*, 1988). JIT production is set up to 'produce and deliver finished goods just in time to be sold, subassemblies just in time to be assembled into finished goods, fabricated parts just in time to go into subassemblies, and purchase materials just in time to be transformed into fabricated parts (Schonberger, 1982)'. For a linked-cell manufacturing system, a good scheduling or production control method is used to implement JIT production. Conventional central production planning schedules all manufacturing processes simultaneously and 'pushes' the materials into the manufacturing system step by step (Monden, 1983). If the actual requirements are different from the original plan, however, it is very difficult to restructure the whole plan in terms of timing. Therefore, this top-down scheduling method fails to meet JIT's quantity and time requirements. At Toyota, a technique of looking at the production flow conversely has been introduced. The technique uses Kanban to 'pull' the parts in reverse order to the final assembly cell from the preceding manufacturing cells. In other words, the system is a 'pull' system in which production activities are initiated only when parts are pulled, or requested, in order to minimize the inventory level. This is the IMPS to which the robot-safety system has been applied here. This unmanned cell is composed of one robot, one Computer Numerical Control (CNC) lathe, one CNC mill and three decouplers (Black and Schroer, 1986). They are arranged in a circular layout with the robot in the centre (Figure 8.2).

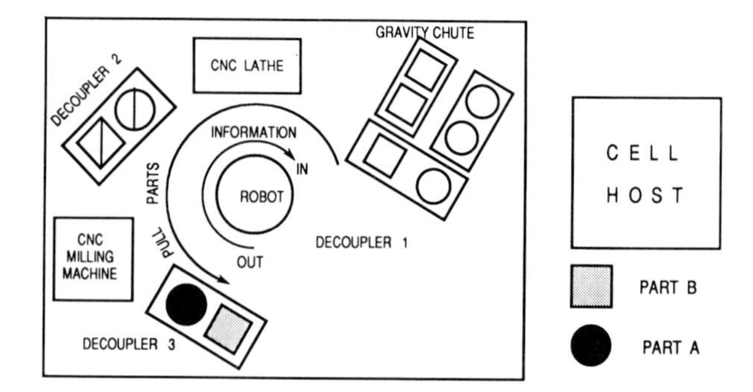


*Figure 8.2 An unmanned manufacturing cell.*

## Normal operation under 'pull' system

The operations and communications of the cell members are controlled by an IBM PC-compatible microcomputer. With no intrusions, the normal operation is that this cell-operation system first checks the status of the output port (decoupler 3) to see if one of the family parts is removed. If all family parts are in decoupler 3 (i.e. there have been no requests from the preceding cell), no cell production will occur. If one of the family parts is removed, this cell-operation system will retrieve the system program that will download the appropriate robot-moving program to move the robot arm. According to the downloaded program, the robot moves to pick up the corresponding part (which is absent from decoupler 3) from decoupler 2 and moves it to the milling machine, releases it, and goes back to the original position (called 'home'). After this part transportation, the system will download Numerical Control (NC) programs to the milling machine. The milling machine will mill the part into a final product. During the milling machine working time the host system downloads the NC part program to the controller. Then it keeps the main program running. In the next step, it moves a part from decoupler 1 to the lathe machine, because one of the part locations in decoupler 2 is empty. The same operation is running again, and the robot moves to grasp the part, then moves it to the lathe. The cell host downloads the NC program to the lathe machine, and the lathe machine follows the program to cut the part. At the same time, the robot moves back to the home position. During the lathe or mill machine working time, the robot stays in the home position waiting for a signal to move. No matter which machine finishes first, it gives a signal to the cell host. Then the cell host commands the robot to move a part from the machine to the decouplers until all three decouplers are filled with two parts, after which the whole system will stop to wait for another request.

## Implementation of SSLD

The proposed SSLD has been implemented in the unmanned manufacturing cell, which includes one robot, one CNC lathe, one CNC mill and three decouplers.

The safety devices include sensors attached to grippers, a light curtain, pressure mats, warning signals, buzzers and a barrier with an interlock. A plane view drawing is shown in Figure 8.3. The whole cell is controlled by a IBM PC-compatible computer through a PCI 2000 Input/Output (I/O) board, an 8255A PPI, 8250 UART boards and a specially designed circuit board.

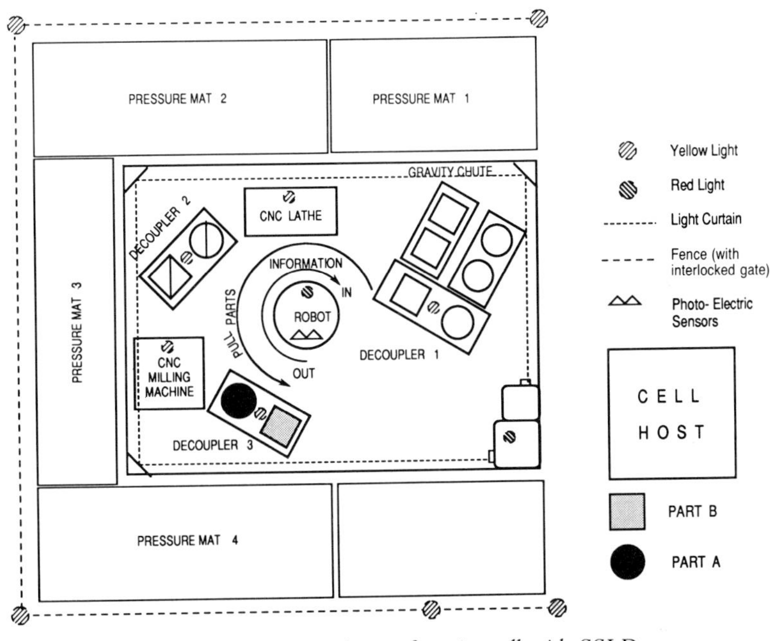


*Figure 8.3 An unmanned manufacturing cell with SSLD.*

*Hardware design*

The communication structure of the robot safety system is shown in Figure 8.4. Because this is an unmanned cell, all the processing information is controlled by a cell-information system (Lin, 1988), as shown on the lefthand side of Figure 8.4. On the other side, the cell operation is performed through a BASIC program. The operation system communicates with safety devices and warning devices by means of the PCI 2000 I/O board and a designed circuit board.

The PCI 2000 I/O board is a plug-in board purchased from the Burr-Brown Company. In this sytem, channels 8 to 15 are assigned as the input channels, and channels 16 to 23 are the output channels (Figure 8.5). The input channels only receive the Transistor–Transistor Logic (TTL) signals (i.e. 0 or 1) indirectly from sensing devices (i.e. sensors, light curtain and pressure mats), but the output channels are needed to switch the 12 V DC warning signals on and off. Therefore, a PCI Digital Opt-Isolator Module is used to convert TTL outputs from the PCI system to switch the 12 V warning signals.

The designed circuit board (Figure 8.6) provides the connections for nine relays, three flashers and a buzzer used in the safety system. If the computer

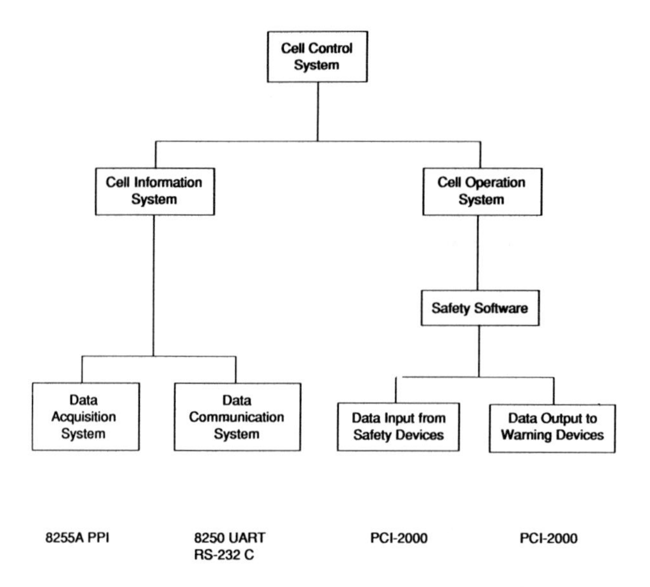


*Figure 8.4 System communication.*

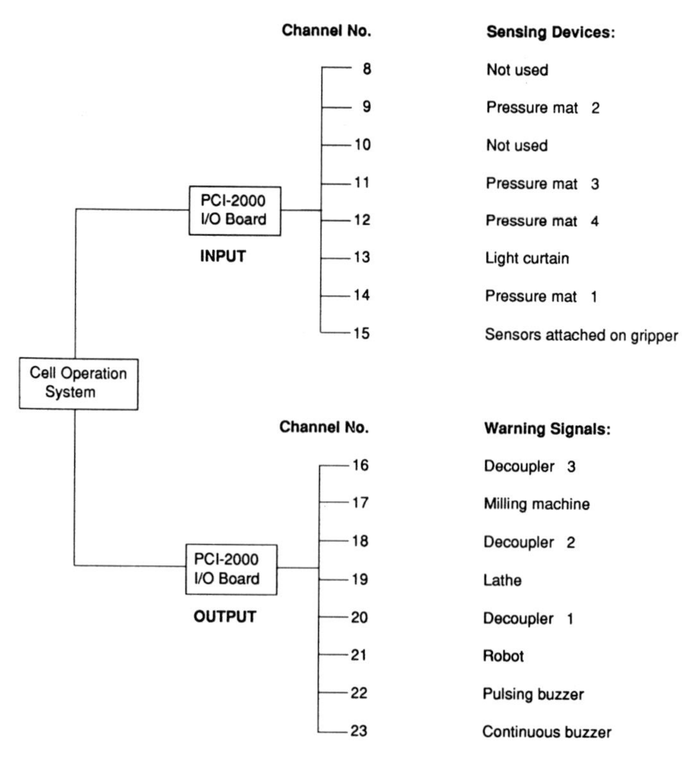


*Figure 8.5 The PCI 2000 I/O.*

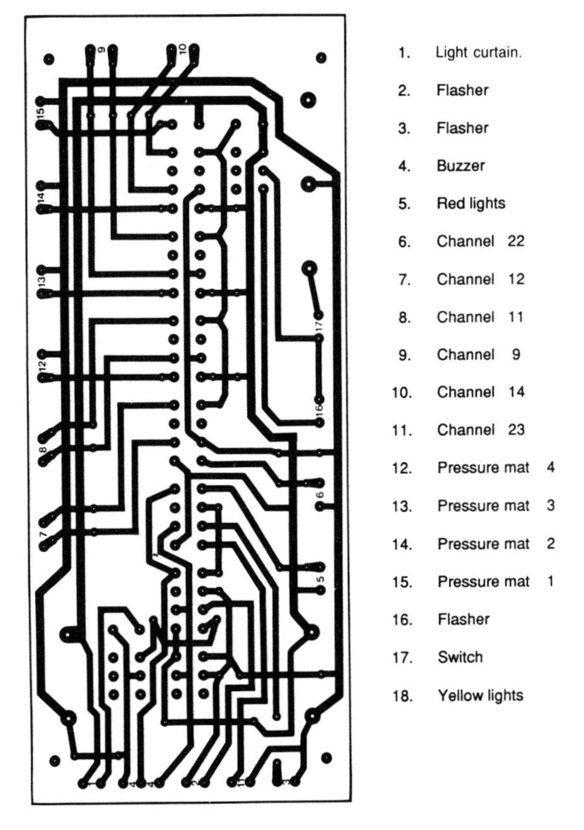


*Figure 8.6 The designed circuit board.*

is running, these devices can be triggered by sensing devices without computer control. As a matter of fact, all the sensing devices must trigger the coordinate relays first, and at that time the relays input specific signals to the PCI 2000 I/O input board. Then the computer can take further action. The information-flow process is illustrated as follows:

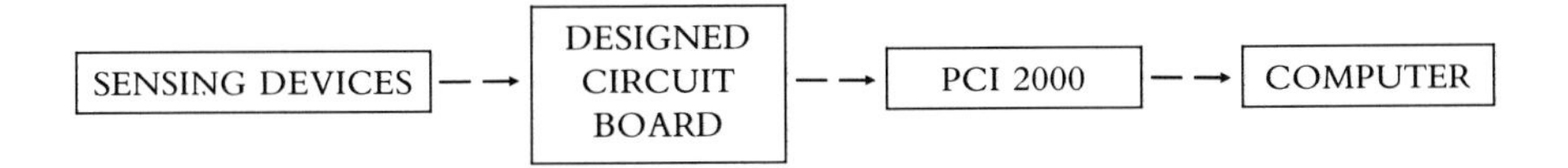


*Software design*

The software is designed to monitor the robotic system, either in robot-operational mode or robot-nonoperational mode. In nonoperational mode, the designed circuit board connected with those sensing devices can monitor the severity levels continuously to ensure that the system is in a safe condition;

otherwise, a warning signal is triggered. In the operational mode, the software and the sensing devices keep checking the robot's present condition. The status of the robot working area is displayed by three warning signals to ensure that it is not entered inadvertently. The detective function is activated for each specific zone where the robot will work once it is ready to move. Sensors attached to a gripper continue to monitor the robot skin area to prevent unexpected contacts. Once the monitoring indicates an abnormal condition, the software generates an emergency-handling procedure to ensure that no accident happens.

The whole software flow chart is shown in Figure 8.7. First, the cell host

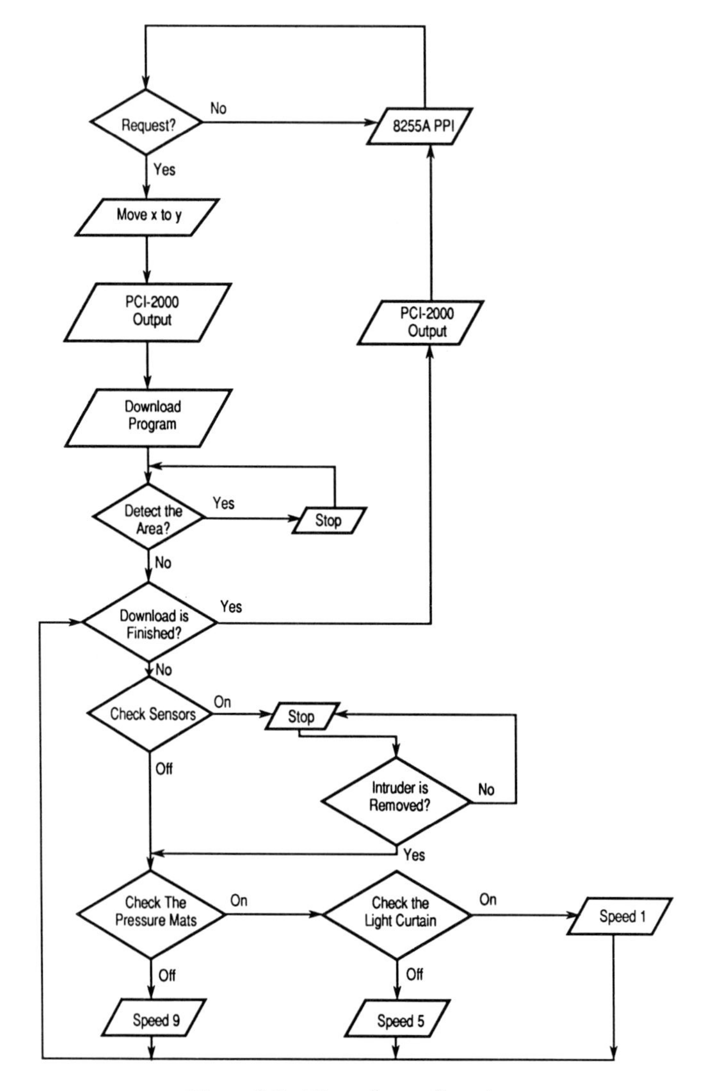


*Figure 8.7 The software flow chart.*

displays 'Request?' on the screen. Each decoupler sends a signal back to the data-acquisition system (8255A PPI). If the answer is 'Yes' (i.e. a part is requested), the main program starts sending visual signals through the PCI-2000 system to display the working area. Then the main program downloads the part-moving program to the robot controller. Before the robot moves, the detective function checks to determine whether the working area is safe or not. If the answer is 'Yes', the robot starts moving; otherwise, the robot stops and the screen presents a warning message. After the robot moves, the main program continues to check the sensors attached to the gripper. If an object approaches it, the robot stops and the screen presents a warning message. The robot will not move until the computer is reset. Before the robot moves, the system checks the light curtain and pressure mat, then assigns an appropriate speed for robot movement. When the movement is finished, the main system sends out the message to turn off all the warning signals in the working area through the PCI-2000 system. Following the six severity levels (see Table 8.2), the software is designed to accomplish the functions of monitoring safe operations with robot, machines, sensors and other warning devices, and of supervising robot motion from a safety standpoint.

## *A proposed quantitative method for evaluating safety of manufacturing systems*

In order to evaluate the safety function in a manufacturing system, a quantitative evaluation method is proposed based on the following safety theory (Cheng 1990): for machines, the sequence for accident prevention is to use hardware safety devices first and then to use software safety techniques. For human beings, applying software safety techniques is the first priority, and then the hardware safety devices should be applied next.

The quantitative method defined here is a method using numerical data to interpret the safety of the system. Three major advantages of such a quantitative evaluation method are: the magnitude of safety can be displayed clearly; the subjective factors within the evaluation procedure can be minimized; and without a great knowledge base, a person can evaluate the safety system well if the definitions in the evaluation procedure are described clearly.

The following is a safety objective function, including two variables—hardware safety devices and software safety strategies—that are used to evaluate the system safety. Each variable is expressed by multiplying a weighted value by the type of hardware or software. The formula and the detailed procedure are as follows:

$$\text{Safety Objective Function} = \text{F (Hardware, Software)}$$
$$= \sum_{1}^{n} W_i H_i + \sum_{1}^{m} W_j S_j,$$

where

$W_i$ = weighted value for hardware,
$W_j$ = weighted value for software,
$H_i$ = the type of hardware device,
$S_j$ = the type of software program,
$n$ = no. of hardware safety devices applied in the system and
$m$ = no. of software safety strategies applied in the system.

## The hardware weighing system

There are seven types of hardware safety classification categorized in three ways: improve safety, leave system unaffected and danger (Barnett *et al.*, 1989). The safety device that can improve safety all the time is the ideal, but not every safety device can really provide the safety factor exactly as designed. Besides, the effectiveness of some safety devices depends on the state of the system. For example, when the system stops, the safety device function can sometimes be stopped, too. Or some safety devices can produce dangerous conditions during specific states. The examples and the types are described in Table 8.4.

*Table 8.4 The hardware weighting system.*

| | Weight ($W_i$) | | | | |
|---|---|---|---|---|---|
| Type ($H_i$) | (2) Improve safety | (1) Leave system unaffected | (−2) Danger | Examples | $H_i$*$W_i$ |
| Type I | (1) (always) | (0) | (0) | safeguarding window on lathe machine | 2 |
| Type II | (1/2) (sometimes) | (1/2) (sometimes) | (0) | barrier, light curtain | 3/2 |
| Type III | (0) | (1) (always) | (0) | redundancy fail-safe | 1 |
| Type IV | (1/2) (sometimes) | (0) | (1/2) | interlock | 0 |
| Type V | (1/3) (sometimes) | (1/3) | (1/3) (sometimes) | seat belt | 1/3 |
| Type VI | (0) | (1/2) | (1/2) | | − 1/2 |
| Type VII | (0) | (0) | (1) | | − 2 |

$\Sigma H_i * W_i = 7/3$
*Source:* Barnett *et al.* (1989).

*Type I.* The type I device can 'always' improve the safety of the system; therefore, it has the weight '1' under the 'improve safety' item.
*Type II.* The type II device can 'sometimes' improve the safety of the system, but 'sometimes' leaves the system unaffected; thus, it has the weight '1/2' under the 'improve safety' and the 'leave system unaffected' items, respectively.
*Type III.* The type III device 'always' leaves the system unaffected; therefore,

it has the weight '1' under the 'leave system unaffected' item.
*Type IV.* The type IV device 'sometimes' can improve safety of the system, but 'sometimes' produces danger; thus, it has the weight '1/2' under the 'improve safety' and the 'danger' items, respectively.
*Type V.* The type V device has three possible conditions: 'sometimes' it improves safety of the system, but 'sometimes' it leaves the system unaffected, and 'sometimes' it produces danger. It therefore has the weight '1/3' under each item.
*Type VI.* The type VI device 'sometimes' leaves the system unaffected and 'sometimes' produces danger; thus, it has the weight '1/2' under each item, respectively.
*Type VII.* The type VII device 'always' produces danger, having the weight '1' under the 'danger' item.

The total value is the weighted scale for each specific type of safety device. A positive number means that some safety improvement can definitely be expected. A negative number should be avoided or investigated.

### The software weighing system

The software safety program used in the system can be classified into six levels. Software safety reliability is evaluated first as it is the most important criterion. The second level is to evaluate whether software safety warning signals are compatible with the hardware safety devices or not. Poor coordination of hardware and software is like a brain that cannot command muscle well. The safety-fault tolerance of software is the third level. At this level, human error tolerance in software is evaluated. High-tolerance software can direct the operator to a correct operation, also avoiding man-made tragedies. The fourth level is related to evaluation of the Zero Mechanical State (ZMS) function. Pseudo-stop always leaves energy in the system, which causes accidents, and system safety cannot be maintained. The fifth level is the evaluation of the software detection ability. The more flexible the software, the more responding time is required. The sixth level evaluates whether the system contains software predicting function. Avoiding accidents is an intrinsic ability of humans. If he or she is equipped with the knowledge to foresee problems, the operator will avoid moving into the danger zone.

The weighted value decreases following the level priority. The lower levels are weighted less. Using 1 (present) and 0 (absent) to evaluate the software, we have summarized the scale in Table 8.5. The value of the scale divided by 21 is the index of the software safety. The 100% indicates the best, and 0 indicates no software safety achievement evident.

## Discussion

A safe maintenance procedure is important to protect workers. A worker must often enter a robotic cell to perform maintenance, tool changing and simple repair

*Table 8.5 The software weighting system.*

| *Level* | *Weight* ($W_i$) | Present/absent ($S_j$) | $W_j * S_j$ |
|---|---|---|---|
| I | 6 | 1 | 6 |
| II | 5 | 0 | 0 |
| III | 4 | 1 | 4 |
| IV | 3 | 0 | 0 |
| V | 2 | 1 | 2 |
| VI | 1 | 1 | 1 |
| Total | = 21 | | = 13 |

I Software safety reliability (parallel circuits).
II Software safety warning signals (compatibility with hardware).
III Software safety fault tolerance (human error tolerance).
IV Software safety of Zero Mechanical State function (ZMS consideration).
V Software safety of detection ability.
VI Software safety of predicting function (ability to foresee).

tasks. Some of the guarding devices and warning signs or signals can be deactivated for these purposes. For example, to change a tool for a machine in the potential danger zone, the buzzer can be turned off temporarily while the robot stops moving until the tool change is completed. It is preferable to turn off all the power before performing a maintenance job. If the robot must be moved, however, it should be controlled by a remote teach pendant at a slow speed. A set of guidelines for robot maintenance has been published by the National Institute for Occupational Safety and Health (NIOSH) (Etherton, 1988).

It has been recommended in the available guidelines and standards that emergency stop buttons or deadman switches should be installed at various locations around the robotic workplace where an operator can use them as needed. The use of an emergency device usually causes the system to shut down, thus causing maximum interruption in production since system operation must be resumed through a careful restart procedure. The implementation of the SSLD will minimize the use of such emergency devices in control system malfunctions. The safety measures have been designed to function primarily in the cell control system in normal operating cycles.

Sensor technology is a rapidly growing field. Robot users need to be aware of the currently available sensors and their characteristics in implementing a robotic safety system. There are still several problems in applying sensors in a robotic cell. Most of the current sensors are still sensitive to both an object (e.g. raw material or machine) and a human. If a new sensor is developed that can detect a person while not detecting other objects, then personnel can be better protected in a robotic cell. Another problem is in the area of sensor reliability, fusion and coordination. Sensor reliability has been a persistent problem. Proper functioning of sensors is critical for monitoring the safety of a robotic cell. When redundant or supplementary sensors are used in the system, however, their information merging (fusion) and coordination become a problem. Another critical issue is how this information is used in interfacing with the manufacturing process.

Training and education supplement the technical solutions described above by making employees more aware of robotic safety problems. For example, a person will not move close to a robot if he or she knows the robot has a dwell time during operation, even if the robot is stopped momentarily (e.g. Rahimi and Karwowski, 1990). Traditional training on factory safety discipline (e.g. a lockout procedure) is also important for enforcing safety guidelines.

## Conclusion

Safety should be the first priority when designing a manufacturing cell. A bandage approach (i.e. fixing the problem after it occurs) may result in tragedy. This paper has presented and demonstrated the usefulness of SSLD at the design stage of a manufacturing cell. This design concept not only provides a system with maximum protection for the operators but also provides one that results in minimum interruption of production.

## Acknowledgment

This study was supported by Grant 1 RO1 OHO2230–O1A1 from the National Institute for Occupational Safety and Health of the Center for Disease Control and the Advanced Manufacturing Technology Center (AMTC) at Auburn University. Special thanks to Mary Jo Mykytka for the preparation of the manuscript.

## References

Bailey, R. W., 1982, *Human Performance Engineering: A Guide for Systems Designers* (Englewood Cliffs, NJ: Prentice-Hall).

Barnett, R. L. and Switalski, W. G., 1988, Principles of human safety. *Safety Brief,* **5,** 1, 1–15.

Barnett, R. L., Barroso, P. Jr., Hamilton, B. A. and Litwin, G. D., 1989, Selected principles of human safety in the workplace. *International Journal of Materials and Product Technology,* **4,** 2, 125–44.

Black, J. T., 1983, Cellular manufacturing systems—an overview. *Industrial Engineering,* **15,** 11, 36–48.

Black, J. T., 1988, The design of manufacturing cells (step one to integrated manufacturing systems). *Proceedings of Manufacturing International '88,* Atlanta, GA, April, Vol. III, (New York: American Society of Mechanical Engineers), pp. 143–57.

Black, J. T., 1988, Decouplers in integrated cellular manufacturing systems. *Journal of Engineering for Industry,* **110,** 77–85.

Bonney, M. C. and Yong, Y. F. (Eds), 1985, *Robot Safety* (Bedford: IFS Publications Ltd and New York: Springer).

Cheng, O. S. H., 1990, A Proposed Integrated Safety and Manufacturing System. Technical Report, Auburn University, AL.
Collins, J. W., 1986, Hazard prevention in automated factories. *Robotics Engineering,* **8,** 8–12.
DeGarmo, E. P., Black, J. T. and Kohser, R. A., 1988, *Materials and Processes in Manufacturing,* 7th Edn (New York: Macmillan Publishing Company).
Eastman Kodak Company, 1983, *Ergonomic Design for People at Work,* Vol. 1 (Belmont, CA: Lifetime Learning Publications).
Etherton, R., 1988, *Safe Maintenance Guidelines for Robotic Workstations.* NIOSH Technical Report, DHHS (NIOSH) Publication No. 88–108 (Morgantown, WV: NIOSH).
JISHA (Japanese Industrial Safety and Health Association), 1983, *Prevention of Industrial Accidents Due to Industrial Robots* (Tokyo: JISHA).
Jiang, B. C., Liou, Y. H., Suresh, N. and Cheng, S. H., 1990, An Evaluation of Machine Guarding Techniques for Robot Guarding (in press).
Kilmer, R. D., 1985, Safety sensor systems. In Bonney, M. C. and Yong, Y. F. (Eds) *Robot Safety* (Berlin: Springer), pp. 223–35.
Lin, J. C., 1988, 'CAD/CAM integrated modeling manufacturing cell', unpublished Master's thesis, Auburn University, Alabama.
Linger, M., 1984, 'How to Design Safety Systems for Human Protection in Robot Applications', Paper presented at the 14th International Symposium on Industrial Robots, 7th International Conference on Industrial Robot Technology, Gothenburg, Sweden.
Liou, A. Y. H., Jiang, B. C. and Suresh, N., 1989, From machine guarding to robot guarding. In Mital, A. (Ed.), *Advances in Industrial Ergonomics and Safety I* (London: Taylor & Francis), pp. 623–30.
MTTA (The Machine Tool Trades Association), 1982, *Safeguarding Industrial Robots: Part 1—Basic Principles* (London: MTTA).
Monden, R., 1983, *Toyota Production System* (Atlanta, GA: Institute of Industrial Engineers), pp. 23–7.
NIOSH (National Institute for Occupational Safety and Health), 1975, *Machine Guarding—Assessment of Need,* HEW Publication No. 75–173. (Cincinnati, OH: US Department of Health, Education, and Welfare).
National Safety Council, 1985, Robot, Data Sheet 1–717–85. *National Safety and Health News,* October, pp. 93–5.
OSHA (Occupational Safety and Health Administration), 1982, *Concepts and Techniques of Machine Safeguarding,* Bulletin 3067. (Washington, DC: US Department of Labor).
Rahimi, M. and Karwowski, W., 1990, The effect of simulated accident on worker safety behavior around industrial robots (in press).
Robinson, O.F., 1985, Robot guarding—the neglected zones. In Bonney, M. C. and Yong, Y. F. (Eds) *Robot Safety* (Bedford: IFS Publications Ltd and New York: Springer), pp. 181–8.
RIA (Robotic Industries Association), 1987, *American National Standard for Industrial Robots Systems.* (Dearborn, MI: Robotic Industries Association).
Verein Deutscher Ingenieure, 1987, *Safety Requirements for Construction, Equipment, and Operation of Industrial Robots,* VDI 2853 (Berlin: Beuth Verlag GmbH).
Salvendy, G. (Ed.), 1987, *Handbook of Human Factors,* (New York: Wiley).
Schonberger, R. J., 1982, *Japanese Manufacturing Techniques* (New York: Collier Macmillan Publishers), p. 16.
Strubhar, P. M., 1984, Robot Safeguarding. *Proceedings of the RIA Robot Safety Seminar* (Dearborn, MI: Robotic Industries Association).

# *Chapter 9*
# *Human intrusion into production danger zones: a simulation approach*

**K. K. Häkkinen[1], R. Booth[2], S. Väyrynen[3], J. Haijanen[4] and M. Hirvonen[4]**

*[1]Industrial Mutual, Risk Management Division, PO Box 12, SF-00211 Helsinki, Finland; [2]Aston University, Mechanical and Production Engineering, Health and Safety Unit, UK; [3]Oulu University, Institute of Work Science, Oulu, Finland and [4]Institute of Occupational Health, Vantaa, Finland*

**Abstract.** New technology and improved safety standards have brought about a decrease in serious machinery accidents. However, serious accidents still occur particularly during disturbances of automatic production systems. The increasing complexity of new production systems together with the great number and variety of hazard sources make safety assessment difficult. Simulation methods which today are much used in the optimum design of production systems may also be applied to a detailed a priori safety analysis. The objective of the project was to develop a simulation method by which knowledge of human actions in disturbance situations can be acquired and thereby to develop design criteria for the safety design of man–machine interfaces in computer-controlled production lines. The first experiments were conducted in order to analyse human actions and hand motion patterns when removing a faulty product from the line.

## *Introduction*

New technology combined with improved safety standards and better knowledge of the hazards of machinery has resulted in a marked decrease in the number of serious machinery accidents in many areas of industry. Fatal accidents occasionally occur, however, with automatic systems such as industrial robots. An example of a disturbance-induced fatality involving human intervention at the danger zone of a palletizing machine is given in Figure 9.1. The pallet fork struck a repairman on the head while he waited nearby for restart to check if the system would run correctly. These accidents represent a new generation of machinery accidents—the first generation appeared at the very beginning of mechanization when machines were running in routine production without guarding and protection. Now, when the machines are guarded, or they function automatically during the course of production, the accidents take place during disturbances and troubleshooting.

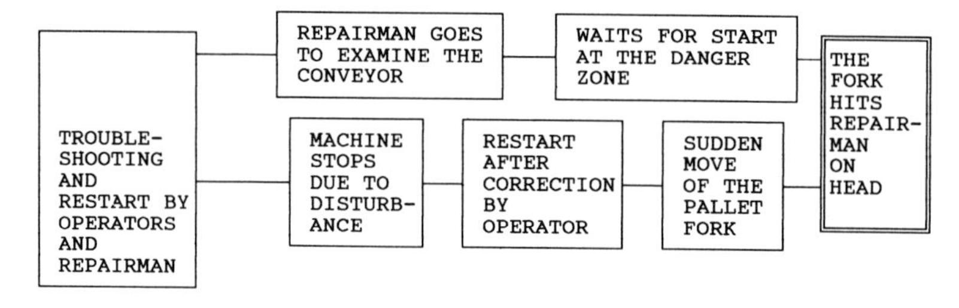


*Figure 9.1 A fatal accident involving a disturbance in an automatic palletizing line (Abbreviated from The Finnish Federation of Accident Insurance Companies, 1987).*

It is widely known that the potential for high severity injuries exists in many automatic production systems, but there are only a few accidents simply because the populations of those systems are still small and their operation times have been short. It has been estimated on the basis of the incident records available that the risk of industrial robots is at the same level as that of power presses (Helander and Karwan, 1988). In Japan, at least ten robot-related fatalities have been reported (Nagamachi, 1988). The increasing complexity of new production systems together with the great number and variety of hazard sources make safety management difficult. More information must be collected and analysed during the design and assessment of these systems and their safety requirements.

## *Disturbances as a major source of production accidents*

It has been noted, particularly in connection with accidents occurring with automatic machines, that in the design process insufficient attention has been paid to disturbance situations. These are the situations in which the operator must enter the machine's hazard zone to carry out manual operations (Harms-Ringdahl, 1985). The machines are designed only for normal and automatic running; human intervention in disturbance situations has not been taken into account.

Accordingly, many of the solutions used to prevent accidents in automatic production lines are incompatible with respect to, for example, the demands of disturbance elimination and cleaning or maintenance operations. Therefore guards are often omitted, defeated or misused (Booth, 1979). Furthermore, there is a lack of specific procedures to maintain safe systems of work in disturbance situations. An analysis of the robot safety strategies in six British companies covering 84 robots showed that in 50% of the cases the hazard control devices were systematically circumvented by workers (Jones and Dawson, 1986).

Poor management of machine disturbances is not a problem of the new-technology machines. For example, the so-called after-reach hazard in press operations has been much studied during the last few decades. After-reach is the incident which happens when the worker reaches spontaneously into the point of operation after the work cycle has been initiated in order to correct the misaligned workpiece. In spite of the several simulation studies, there is no firm consensus as to the hand speed value of human beings to be used as a design criterion. There is, however, substantial evidence that in situations where the reflex action of humans is foreseeable in the correction of disturbances, the common standard value of 1.6 m/s is usually far too low (Winsemius, 1965; Van Ballegooijen, 1979; Pizatella and Moll, 1987).

The case of the after-reach hazard demonstrates that the human performance criteria for disturbances may not be equal to those in normal and smooth production. The variations seem to be greater and the extreme values may fall far beyond any expected distribution. This is especially important when human actions are considered in connection with automated systems and robots, where the configurations of motions for both humans and machines are far more complex than those with presses and other traditional machines.

Human actions and behaviour in accidents are to a large extent governed by various situational factors (Winsemius, 1965, 1969; Corlett and Gilbank, 1978). 'A plywood sheet rattling between conveyor belts can annoy a worker enough to make him grasp it spontaneously', reported the safety manager of a Finnish plywood factory. In our field study of the safety of packaging machines, some workers occasionally 'caressed' the moving packages with their hands and thereby reached near to danger points. A closer examination showed that they tried to avoid disturbances in the stamping process by pressing down the slack covers of the boxes, or by reducing the pressure in the product queue.

More than 50% of the accidents in connection with packaging machinery occur when people are trying to correct or to cope with disturbance situations (Rieppo and Häkkinen, 1985; Defren, 1988). Missing and deficient machine guards are typical contributing factors. The packaging machines involve several types of danger. The safety devices and guards must also meet many production requirements, e.g. they must allow easy cleaning at the points of operation and they must provide visibility. Thus the guarding problems are often difficult to resolve completely.

## *The role of simulation in production safety*

The traditional method of collecting data for industrial safety studies has been to analyse actual accidents and to try to find typical accident scenarios on which to base the prevention measures. Apart from accident analysis, hazard analysis methods are also being used more commonly to ferret out the risky properties of workplaces and tasks before accidents have actually occurred. Systematic hazard analysis is usually the only method for novel automatic installations, because no accident experience is available and because there is a demand for safety from the very beginning of the new production technology. Moreover, the price of learning through accidents is now considered too high from both the human and the economic viewpoints.

According to Shannon (1975), simulation is 'the process of designing a model of a real system and conducting experiments with this model for the purpose of either understanding the behavior of the system or of evaluating various strategies for the operation of the system'. A simulation model can be an exact physical model of the system, or a scaled one presented on a computer screen or on a piece of paper. In simulation, the disturbance events can be repeated several times to provide enough material for a statistical analysis, and accurate measurements can be conducted during the simulated events. When the simulation results are combined and validated with the results of the accident and workplace studies, the production and safety experts can devise a profound safety picture of the system under development.

The situational effects in disturbances are very difficult to simulate and in many studies they are considered exceptional and totally ignored. Moreover, most simulation models have failed to demonstrate the trade-off situation between safety and productivity corresponding to that in real production environments. Our view is that incorporating the situational variables into the simulation models yields a better understanding for the safety management of disturbances.

## *Simulation method and the study design*

The objective of this study was to develop a simulation method by which knowledge of human actions in disturbance situations can be acquired and, on the basis of this, to develop criteria for the safety design of computer-controlled man–machine systems. The method is essentially based on a physical simulation model. Along with the physical model, however, a computer simulation model for the same process is developed simultaneously. These two approaches in simulation can be used to support each other.

The simulation model is developed so as to meet the following requirements:

- the test persons are not subjected to any danger or harm by the method;
- the method is realistic, reflecting a packaging process or a part of it as used in production;
- the procedure is equally applicable to study disturbances induced by machine, product or human action;
- several alternative guarding solutions are available for studying the ranking order of preference of various safety and efficiency criteria;
- the group of test persons are real operators of packaging machines;
- the method produces disturbance situations randomly: how the operator copes with them depends on his/her own 'cost–benefit' risk evaluation in that situation, which determines whether a safe or a hazardous practice is chosen; and
- human operating modes and man–machine interactions in disturbance situations are measured quantitatively using motion analysis techniques.

The study is based on the assumption that the probability of the man–machine system's safe operation depends on (a) the necessity of entering the machine's hazardous zone (frequency, urgency, distance covered) and (b) the ease of hazardous access and the degree of difficulty of safe operation.

Consequently, the simulation system is designed to manipulate the following variables quantitatively:

- frequency and urgency of the need to enter the hazardous zone;
- distance of the hazardous zone from the worker;
- degree of difficulty of entering the hazardous point, accomplished with machine guards and safety devices; and
- degree of difficulty in defeating the machine's safety devices.

## *Preliminary experiments and results*

For the development of the simulation model, the man–machine system of packaging work was selected. In earlier safety simulation studies, production-line work has seldom been considered.

The experimental production system is constructed by using four belt conveyors (Figure 9.2). The conveyors are installed in the laboratory so that they constitute a rectangle and are connected with each other by slides and a carousel. During the experiments, empty boxes are continuously moving around. The conveyor system is controlled by a programmable logic unit. Various control devices and interlocking guards are coupled with the machine operation by the computer. Moreover, the logic system counts the number of packages 'produced'. Some disturbances can be programmed such as sudden stoppages of conveyors.

The first experiments were conducted in order to analyse human actions and hand motion patterns when removing a faulty package from a line. At one conveyor there are two tunnel guards, separated by about one metre. The package entering from the first tunnel may be faulty. In that case the operator should take the product from the line before the next production step in order to avoid major damage and production stop. When the action is delayed the operator may reach into the danger zone of the second tunnel in order to save production. As the speed of the conveyor is 0.35 m/s, the operator has nearly three seconds to remove the faulty box.

The hand motions were analysed by the Selspot computerized measurement system with two cameras and the infrared LEDs fastened to the hand. The measurement was triggered automatically when the photo-cell indicator at the end of the first tunnel received a special reflection from the faulty product. Three test persons participated. Their hand motion patterns were recorded, calculated and reported. A hand motion analysis programme was developed for this experiment using the MultiLab command language provided by the manufacturer of the Selspot system.

The preliminary motion analysis results demonstrate the influence of the position and the posture of the operator during the initiation of the disturbance. Figure 9.3 illustrates typical hand speed and motion path curves when the distance between the operator and the faulty product is only 30 cm. The

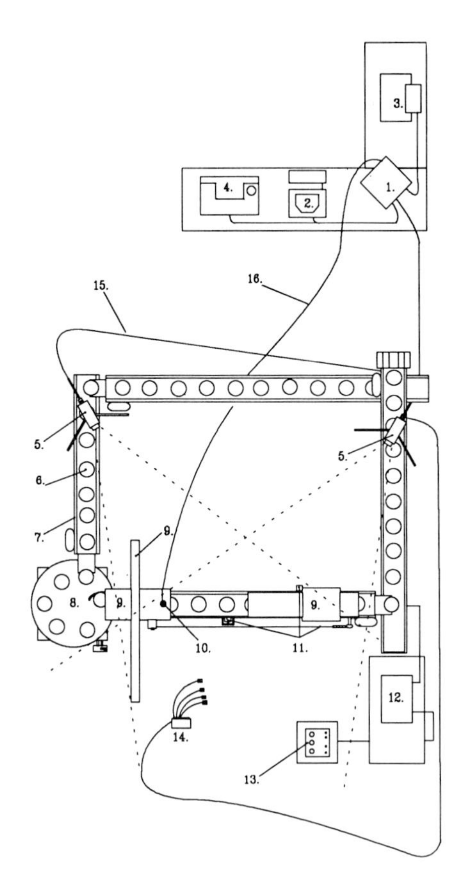

1. Computer
2. Terminal
3. Printer
4. Plotter
5. Infra-red camera
6. Package
7. Belt-conveyor
8. Conveyor table
9. Barrier
10. Light beam
11. Safety device
12. Programmable logic
13. Control panel
14. LEDs
15. Camera cable
16. Trigger cable

*Figure 9.2 The laboratory installation for physical simulation experiments.*

maximum hand speeds with varying distances from 30 to 100 cm are presented in Figure 9.4.

The hand motion usually started within 0.5 s of the faulty product emerging from the tunnel. The rapid reaction in the experiments is natural because the test persons were actually waiting for the broken package to come, unlike in real life, where people have simultaneously many goals and tasks to coordinate. In all motion paths there was some irregularity and fuzziness, which tended to

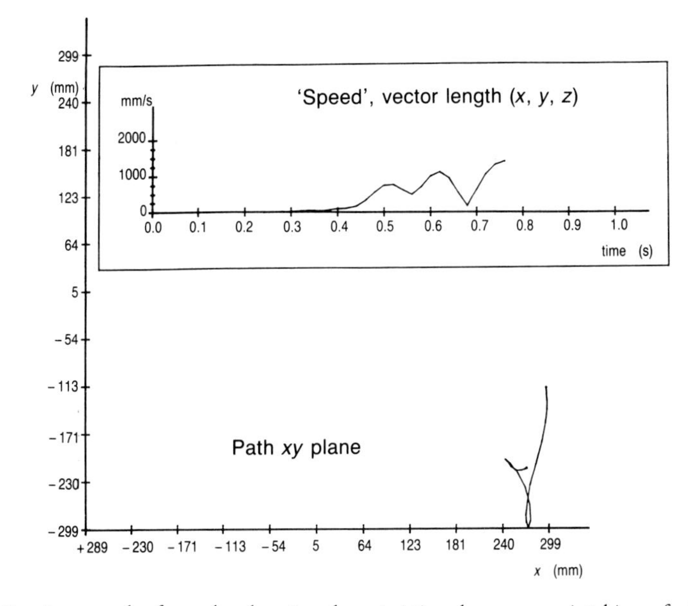


*Figure 9.3 An example of some hand-motion characteristics when a person is taking a faulty product from the experimental conveyor line starting at a distance of 30 cm from the line; speed at* xyz *level and motion path at* xy *level.*

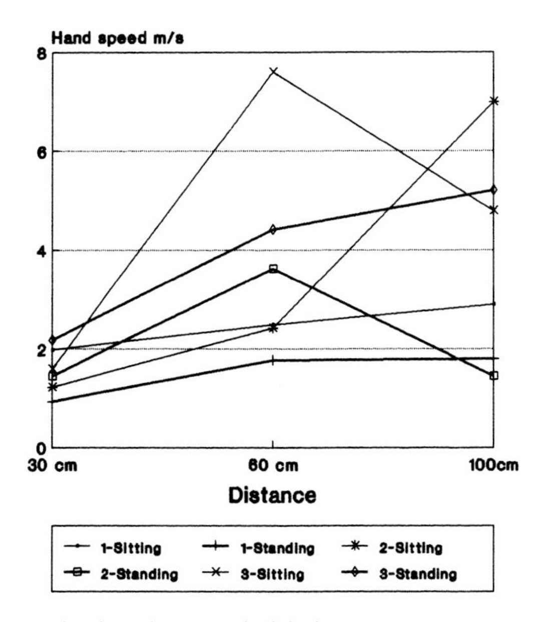


*Figure 9.4 Maximum hand speeds measured while three persons were grasping the faulty product in sitting and standing positions; their distances from the line varied from 30 to 100 cm. Average values of three measurements are presented.*

increase when the distance was lengthened. Also the speed values were widely scattered, even when the same person repeated the test at the same distance. One of the test persons maintained regularity in the motion patterns while the others showed considerable irregularity with vague motion paths and high variations of speed. The maximum speeds at short distances (30 cm) were smaller than those for longer distances (60 and 100 cm). The greatest hand speeds were over 8 m/s.

## *Discussion and directions for future work*

Realistic simulation of production disturbances in a laboratory is a controversial task, and therefore difficult. In a laboratory setting the variables should be strictly controlled and all the 'noise' factors eliminated. At workplaces, where disturbance-induced accidents happen, a complex and dynamic interaction of events and man–machine–environment elements takes place. Even in a research laboratory there are disturbing factors like telephones ringing and people walking through, looking at experiments or speaking. All those seem to cause changes in behaviour and actions, at least for some people. On the other hand, a highly controlled laboratory experiment may be unrealistic. Hence physical simulation in the laboratory is only one instrument in the safety study and incomplete by itself.

Simulation conducted in parallel with appropriate field studies may often be necessary to produce relevant new knowledge. The combination of computer and physical simulations is a promising new approach. For example the extremes of human actions in disturbances can first be demonstrated by laboratory measurements, and then these extreme values can be transferred to a computer which shows by dynamic man–machine models if there are dangerous interactions. The extreme values and vague patterns of human motion in disturbances are only one side of the coin. Actions are based on decisions which people make based on the information they receive from production and on the knowledge and experience they already have. Human intervention at danger zones is often neglected by designers, because they do not see any logical reason for workers to take such actions. There is a need for engineering models which accommodate typical—and especially exceptional and irrational—patterns of human decision-making and risk behaviour for the design of production facilities and machine guards.

The study will be continued by constructing the event trees of the possible occurrences in the following typical disturbances of modern production systems: (1) faulty or loose product on line; (2) product becomes jammed and causes line blockage; and (3) sudden unexpected motion of machine parts. Based on the event trees and the field experience collected, appropriate critical 'slices' are cut from the event trees for detailed studies by physical simulations and by a dynamic Computer-Aided Design (CAD) simulation model.

# *References*

Booth, R., 1979, The design of effective machinery guards. *Design Studies,* **1,** 1, 12–4.

Corlett, E. and Gilbank, G., 1978, A systemic technique for accident analysis. *Journal of Occupational Accidents,* **2,** 25–8.

Defren, W., 1988, Methoden zur Ermittlung der Ursachen von Störungen dargestellt am Beispiel 'Verpackungsmaschinen'. *Die Berufsgenossenschaft,* **11,** 722–7.

Finnish Federation of Accident Insurance Companies, 1987, A Fatality of a Maintenance Man at an Automatic Sack Palletizer. An investigation report (In Finnish).

Harms-Ringdahl, L., 1985, Production disturbances—a safety problem in automatic machines. In Mancini, G. (Ed.) *Analysis, Design and Evaluation of Man–Machine Systems* (Oxford: Pergamon), 231–41.

Helander, M. and Karwan, H., 1988, Methods for field evaluation of safety in a robotics workplace. In Karwowski, W. (Ed.) *Ergonomics of Hybrid Automated Systems I* (Amsterdam: Elsevier), 403–10.

Jones, R. and Dawson, S., 1986, Strategies for ensuring safety with industrial robot systems. *Omega International Journal of Management Science,* **14,** 4, 287–97.

Nagamachi, M., 1988, Ten fatal accidents due to robots in Japan. In Karwowski, W. (Ed.) *Ergonomics of Hybrid Automated Systems I* (Amsterdam: Elsevier), 391–6.

Pizatella, T. and Moll, M., 1987, Simulation of the after-reach hazard on power presses using dual palm button actuation. *Human Factors,* **29,** 1, 9–18.

Rieppo, K. and Häkkinen, K., 1985, Pakkauskoneiden työturvallisuus. (Safety of packaging machines, in Finnish). *Pakkaus,* **7–8,** 49–52.

Shannon, R., 1975, *Systems Simulation—the Art and Science.* (Englewood Cliffs, NJ: Prentice-Hall).

Van Ballegooijen, A., 1979, Messungen an Mensch–Maschine-Systemen unter Berücksichtigung der Nachgreifgefahr. (Mannheim: Berufsgenossenschaft für Nahrungsmittel und Gaststätten), *Symposium,* **6,** 79–89.

Winsemius, W., 1965, Some ergonomic aspects of safety. *Ergonomics,* **8,** 2, 151–62.

Winsemius, W., 1969, *On Single and Multipersonal Accidents.* (Leyden: Netherlands Institute for Preventive Medicine).

# *Chapter 10*
# *Hazard control of robot work area*

**R. Kuivanen**

*Technical Research Centre of Finland, Safety Engineering Laboratory, PO Box 656, 33101 Tampere, Finland*

**Abstract.** Automatic production is the main target for intelligent safety sensors. In many cases, fences and other mechanical obstacles can be too inflexible. The safety of the workers should be guaranteed with technical methods. The automation level of an intelligent safety system has to be planned carefully, because complicated systems can cause unnecessary interruptions in the process, and simple systems are of no advantage to safety and production. Intelligent safety sensors offer new ways of avoiding unnecessary defects which occur if, for instance, a stop signal is sent to a welding robot during the welding process. The sensors can also ensure the safety of persons by distinguishing them from workpieces in situations where traditional methods have turned out to be inadequate. This chapter shows that intelligent safety sensors can be adapted to production so that they become an integral part of the production system. Given that these safety systems are more non-intrusive, the workers' safety level is higher than when traditional safety devices are used.

## *Introduction*

Measures which govern robotics safety have been implemented to varying degrees in different cultures and countries. Different approaches to robotics safety reflect different procedures and philosophies regarding work. At one extreme, robot users may accept a high level of risk by operating robots without safety equipment. To avoid this situation, every attempt should be made to heighten worker awareness of safety rules and regulations. Whatever the philosophy of the workplace, the safe operation of robots should be a top priority.

Another way of ensuring the workers' safety is to make the robot system itself as safe as possible. A safe robot system limits entrance to hazardous areas during its working cycles and stops the dangerous functions of the system when a human's presence is detected in the hazard area. This not only reduces the level of risk but also reduces the level of production adaption. In this case, safety is a matter of technology. Rules and regulations are used only when it is not possible for the system itself to ensure safety.

The adaptation of safety systems to production has been studied in various research institutes. Germany, Great Britain, France, the USA and the Nordic countries have been especially active in this field. The purpose of these studies has been to increase workers' and operators' safety without making the work process more difficult. Some of the systems have also been developed as commercial products (Derby *et al.*, 1985; Kilmer, 1985; Malm, 1986; Kuivanen, 1988).

Before the adaptation to production can be attained, the safety system must use different kinds of sensors to monitor the hazard zone. The safety system's interpretation of the information from its sensors determines whether or not the situation presents a danger to the worker. Thus the hazard zone is divided into smaller subzones inside the robot's working area. The robot and the whole system stop only when necessary. These systems regulate the individual functions of the robot such as speed, movement and access to certain stations. In principle, this ability combines both the open hazard zone having a high risk level and a high level of production adaptation with the guarded hazard zone having a reduced risk level and a reduced level of production adaptation. The best result of this combination is an open hazard zone with reduced risk level and high production adaptation.

## *Basic principles*

### Control of the automation system border

The most common way to control movements of persons into or out of a system area is to control the border between the system and the environment. Encircling most systems are mechanical fences or obstacles having one or two openings which are used for entering the restricted area. In some cases either the layout of the system or the room where the system is located forms natural borders around the system. Openings can be controlled by mechanical gates or doors which are connected to the control system by limit switches or interlocking devices. Openings can also be controlled by sensors such as contact mats, light beams, capacitive sensors or passive infra-red sensors. The system detects persons either opening the gate and/or passing through the opening. It does not know, however, the direction of the movement or the number of people crossing the border.

When the borders of the system are free from mechanical obstacles, they can be controlled by sensors such as light beams or contact mats. In principle this situation creates a wide opening to the system causing the problem mentioned above. When the border is invisible, it is easy to get into the system inadvertently. This causes unnecessary breakdowns of the system operation.

### Control of worker's presence in the system area

The next step from controlling the borders of the system is to control the whole area. This can be done at three levels: area control, zone control and place control.

When using *area control*, we know whether there is one person or several persons in the controlled area but we do not know the exact location of these persons (Figure 10.1). In certain situations we do not even know if the person or persons are still in this area. We cannot always know if the signal was caused by a human being entering a controlled area or if it was dispatched, for example, in order to fool the system.

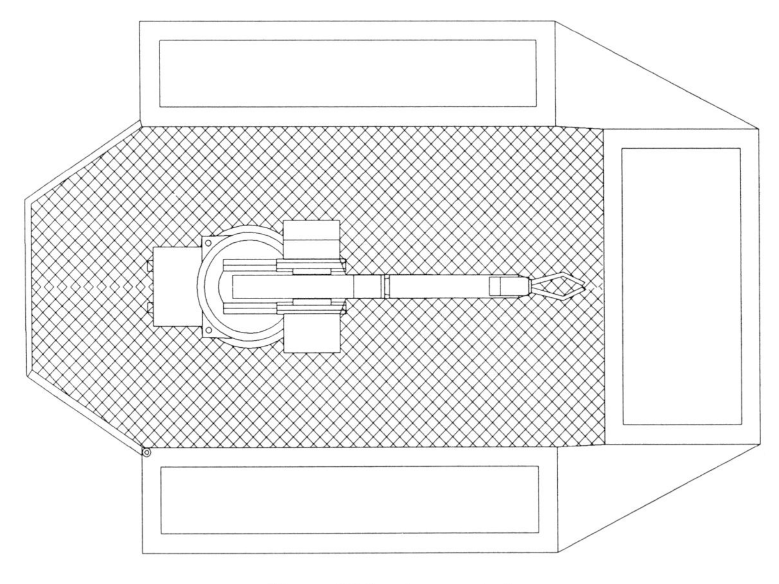

*Figure 10.1 Area control.*

*Zone control* indicates the approximate location of people in a controlled area (Figure 10.2). It also indicates the number of people in different parts of the controlled area. The easiest way to divide an area into zones is to use contact mats. Each mat monitors a specific area which constantly transmits information.

*Place control* gives exact information about the location and number of people in the guarded area (Figure 10.3). This is achieved by using grid zones or sensors in two directions. Due to the great amount of demands made on the system the control unit often needs to be computer based.

## Logical conclusions

Information about crossings of the border or presence inside the guarded area is not always enough to ensure the safety of the workers and operators. For example, in some situations, the size of the workpiece may be so large that one could not know whether it is a person or a large workpiece entering a hazardous area. Generally, these systems cover an area which is difficult to isolate completely. While using dangerous machines and processes we may need a more integrated safety system in order to control hazardous areas and their openings.

It is often possible to detect the difference between persons and workpieces. We may need 5 to 10 detections to distinguish between a workpiece and a person

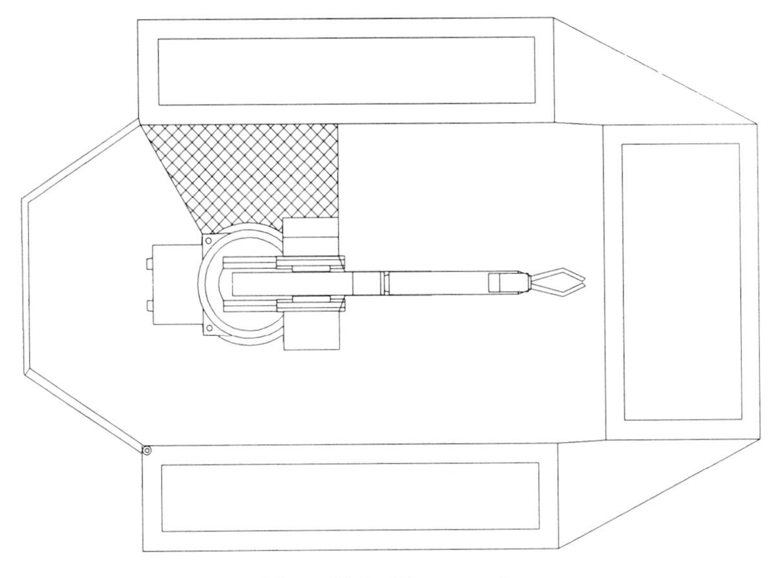

*Figure 10.2 Zone control.*

entering the area. However, when making deductions one always has to presuppose that the object crossing the border is a human being. Only after we have accurately interpreted and identified specific information from the sensors can we be assured that the object is in fact a workpiece. This type of safety

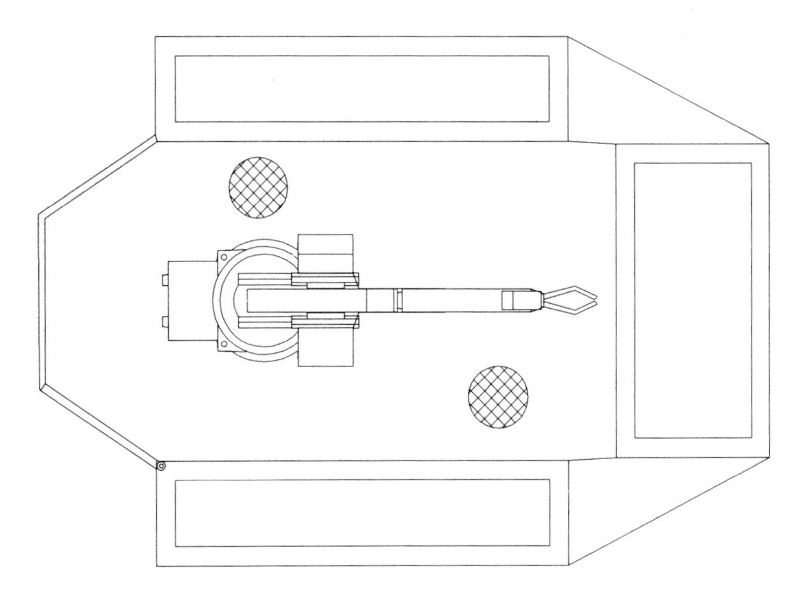

*Figure 10.3 Place control.*

equipment is based on identification of hazardous situations and collision avoidance (Figure 10.4).

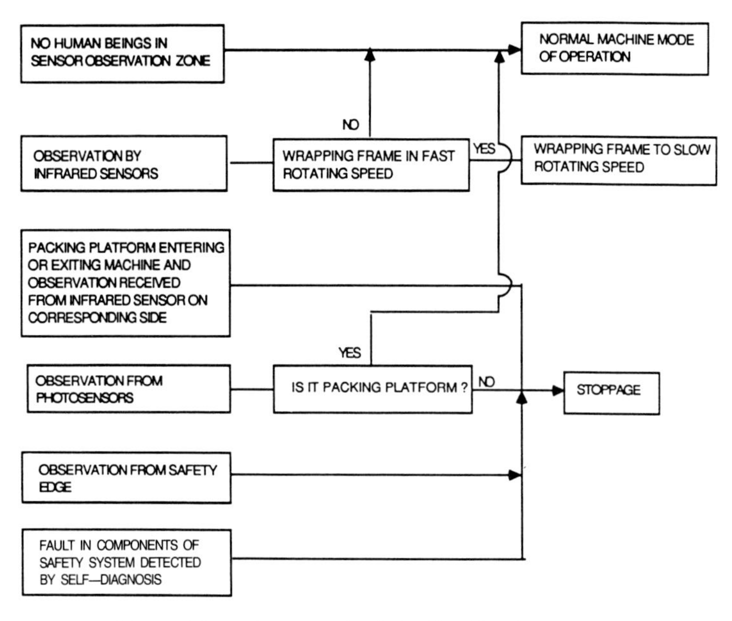


*Figure 10.4 An example of the identification process.*

## Sensors

Sensors which detect persons operate according to basic physical principles. They can, for example, detect changes in pressure, temperature, movement or electrical properties. These sensors react either to direct touch on the element or to presence in the detection area.

When these sensors are employed as safety equipment, they must fulfil certain requirements concerning their failures and reliability. For example, when the sensors are used for opening a door, even in case of malfunction there is generally no danger involved. In other words, as safety equipment these sensors must fail safely, and they must not lose their protective influence under any circumstances. A great number of presence-sensing detectors do not, however, function unfailingly. They can be easily fooled by changing properties of the objects or the area under control (Kuivanen *et al.*, 1988; Rahimi and Hancock, 1988).

### Touch detectors

Touch detectors have several methods of converting a touch into a signal. A touch either causes a direct electrical contact between two metal sheets, thus changing the electrical properties of the element's material, or becomes transmitted by a mechanism or material into a switch. On some occasions a touch can be detected by combining mechanical sensor elements with presence-sensing principles, such as light sensors in the ends of a soft reflective pipe.

In mechanical touch detectors the sensing element can be a bar, a wire, a plate or a mechanism (Kuivanen and Tiusanen, 1986). A touch on the sensing element triggers a switch or another type of sensor. Mechanical touch detectors are often designed for case-specific situations.

### Capacitive sensors

Capacitive sensors measure small changes in the capacitance between the sensing unit and the ground or between two wires or two plates of a capacitor. A capacitive sensor consists of an oscillator, adapting circuits and an amplifier. The oscillator includes the actual sensing unit which functions as a plate of the capacitor. The other plate is usually connected to the ground. When a human being or any other object that has a high dielectric constant comes near the sensor, the properties of the oscillator change. The adapting circuits then transform the changes into a form suitable for the amplifier. The amplifier can give either a signal that is proportional to the distance between the object and the sensor, or a simple on/off signal suitable for the relays (Malm, 1986). The sensor may also include an impedance bridge. All changes in the capacitance will increase the current that flows through the galvanometer (Graham and Meagher, 1985). The amplifier can receive the signal straight from the galvanometer.

The maximum detection range of a capacitive sensor is usually below 800 mm (Vartiala and Järvinen, 1982), normally about 400 mm. It is possible to lengthen the detection range, but then disturbances will become more probable (Figure 10.5). The maximum detection range for capacitive proximity switches is usually below 120 mm.

### Ultrasonic sensors

Ultrasound sensing is based on the production of a high-frequency (above 20 kHz) sound wave, transmission of this sound wave and the measurement of the time interval needed until a reflection is detected at the source. Thus it is necessary to both produce and detect ultrasound signals. The distance to the reflecting object is linearly related to the observed time delay by the speed of sound (Irwin and Caughman, 1985).

There are two types of ultrasound sensor: electrostatic and piezoelectric. In an electrostatic sensor there is a plate which has a vibration frequency that can affect the circuit. The frequency of 50 kHz is mainly used in an electrostatic ultrasound sensor. In a piezoelectric sensor there is an oscillator vibrating at a frequency caused by a voltage change in an electric circuit.

Ultrasound sensors have been used in intrusion detectors, for controlling focus in instant cameras, and for industrial gauging and ranging. In connection with automatically guided vehicles, ultrasound sensors have been applied to detect objects in front of the vehicles.

*Figure 10.5 Usage of a capacitive sensor.*

## Optical sensors

The functioning of an optical sensor is based on one or several light beams. Cutting the light beam sends the detection signal. Optical sensors with only one beam are usually called optical switches and sensors with several parallel light beams are called light curtains. The light beam is produced in a transmitter unit, and the intensity of the beam is measured in the receiver unit. The main structures of these units are shown in Figure 10.6.

Transmitter and receiver units can be in different ends of the beam, or they can be integrated into one enclosure, in which case the beam is reflected to the receiver with a special reflector. There are also sensors which detect the light

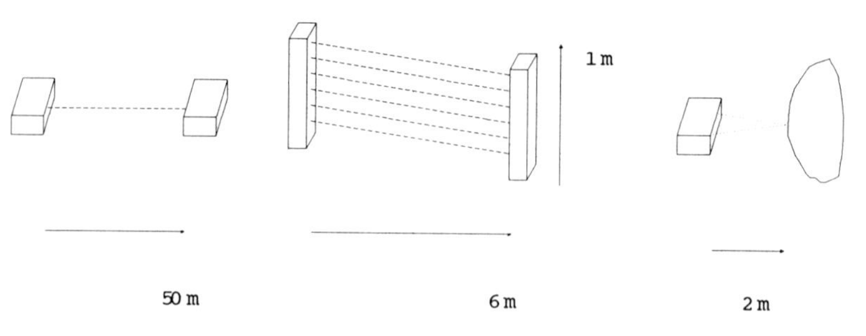


*Figure 10.6 Structure of a light sensor.*

reflected from the object, but they are not reliable enough to be used as safety sensors (Flueckiger, 1980; Health and Safety Executive, 1984; Liljeroos *et al.*, 1986).

### Vision sensors

Vision sensors are camera systems which can recognize and measure figures from pictures. When a vision sensor is used for safety purposes, it can detect a human being or certain changes in the picture. If the vision sensor perceives changes in the picture, the computer decides without recognizing figures whether it is a person that has come into the picture or not. If there are enough changes in the picture, the computer assumes that it is a person. Such a system can be realized with the help of the latest technology. The system is, however, not valid if there are mobile machines in the picture. As it often happens, a vision sensor has to distinguish people from moving machines, and that is a difficult and time consuming task for a computer (Morris, 1985; Holland *et al.*, 1986; Pugh, 1986).

### Microwave sensors

Microwave sensors are based on producing and receiving electromagnetic radiation in GHz range. There are two types of microwave sensors available: presence-sensing and movement-sensing (Doppler) sensors.

An amplitude- or pulse-modulated signal is used in the presence-sensing microwave sensor. The receiver, opposite to the transmitter, observes the energy of the received signal. When an obstacle (e.g. a human) comes between the transmitter and receiver, the received energy decreases. When the energy reaches a predetermined level, an alarm message is generated,

The movement-sensing microwave sensor is based on the principle of the Doppler effect. In this sensor type the transmitter and receiver are in the same unit. The transmitter sends fixed-frequency microwave radiation to the area to be controlled. The power of the reflected signal is directly proportional to the object's surface area. If the object's size and velocity exceed the predetermined levels, an alarm message is generated. Presence-sensing microwave sensors can be applied in the same way as optical sensors, especially when the environment is too hostile to permit reliable operation of photo-electric controls.

The use of Doppler sensors as machine safety equipment devices is not very extensive, instead they are often used as burglar alarms. The Doppler sensor detects nothing but a moving object. One has to be sure that the person who caused the alarm has left the danger area before the system is started again, because the sensor cannot detect a person standing still in the detection zone. According to the Doppler principle, the sensor is more sensitive to movement towards or away from the sensor than to lateral or tangential movement. This problem can be solved by mounting two microwave sensors perpendicular to each other on the robot.

Doppler sensors are practical only for the guarding of quiet places. If one wants to use a Doppler sensor, for example, to guard a robot's working area, the problem is how to separate the signals generated by the robot movement from those generated by a human being entering the detection zone (Kuivanen *et al.*, 1988).

## Passive infra-red sensors

Passive infra-red sensors can be used to detect changes in temperature inside a detection zone that is determined by optics. The use of a passive infra-red sensor is based on the idea that when a human being enters the detection zone, the sensor will detect the infra-red radiation emitted by the human being.

The sensor consists of optics, which collect the infra-red radiation from the detection zone, and electronics, which convert the received radiation into an electrical signal (Figure 10.7) (Juergens, 1986; Malm, 1986; Kuivanen *et al.*, 1988).

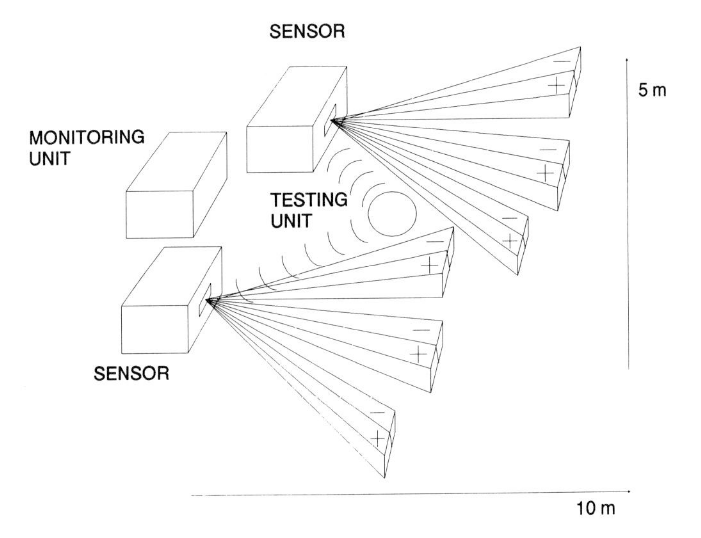


*Figure 10.7 A passive infra-red sensor.*

A pyroelectric crystal is normally used as a detector. The infra-red radiation is absorbed into the crystal which thus changes its temperature. Due to the change in temperature, the polarity of the crystal changes and an electrical charge is formed over the crystal plate. This causes a current which can be measured.

To prevent slow changes in background temperature or static sources of heat (e.g. light bulbs, electric motors) from causing false alarms, two crystals of opposite polarity are often used in the detector. If the background temperature changes, the detector senses heat on both elements. One element generates a positive signal, while the other generates a negative one. These signals cancel each other, and thus no alarm signal is generated. The situation changes when a person enters the protected area. The optical system is designed so that the person is sensed by one single element at a time. This results in a signal imbalance which generates an alarm output.

**Summary**

Touch detectors can be and are used together with safety equipment. The principle of changing a touch into a signal is easy to implement, and the functioning of the system can be tested. Carefully planned and constructed systems increase the safety level in places where simple methods can be used for the detection of human beings.

There are numerous methods and physical principles that can be used to detect human beings that do not come into actual physical contact with the sensor elements. A person can disturb or change properties in many ways, but it is difficult to generate a reliable signal on all occasions. The use of presence-sensing detectors with safety equipment is therefore not possible without careful study. All the presence-sensing principles, except some optical sensors presented in this text, have certain limitations when used as safety equipment. It is often necessary to use more than just one principle at a time in order to get sufficient reliability for safety use.

In addition to the fact that it cannot detect human beings at all times, the sensor can also be oversensitive to disturbances. Vibrations, temperature, sunlight, electromagnetic fields and other factors can cause false alarms which damage safety equipment. Safety equipment which hinders production will be turned off, and the situation will become worse than it would be without any planned safety equipment at all. Moreover, the price of the systems can be too high for general use.

## *Control of hazard zones as part of an automation system*

In workshop industries, the assumption has been made that safety and productivity are in conflict. This point of view evolved over the years because safety equipment design was often inadequate for specific situations. The top priority in production has always been a machine or equipment that functions efficiently, and poorly conceived safety systems often interfered with efficiency. When these systems made work difficult, they were often removed or bypassed. As a result, negative attitudes toward safety measures were reinforced. Today, this situation is changing. Automation systems are more reliable, and they are no longer in the prototype stage. Interest in safety has increased due to demand for better working conditions, a growing labour shortage, and higher standards of living.

Control of the robot working area as part of an automation system means that the safety measures are connected to the robot control system. The safety equipment is integrated into the system, and the robot cannot function unless all safety measures allow it to do so. In order to prevent unnecessary breakdowns of the system one must establish production adaptation. Due to some cultural influences, the idea of controlling workers by technical or other means elicits negative connotations. People regard presence-detecting sensors as being unpleasant and talk about 'Big Brother who is watching'. Nevertheless, safety measures are accepted up to a certain point. As long as safety equipment forms an integral part of a machine or a system, it will be accepted. This creates new demands on modern safety systems which control automation systems (Table 10.1).

*Table 10.1 Analysis of the tasks of automatic safety systems.*

| Task | Object or definition | Solution or sensor prinicple |
|---|---|---|
| Identity harmless situations | | Computer-based logical conclusions |
| Prevent collisions between | Man and robot<br>Robot and structures of the station<br>Robot and other machines<br><br>Two robots | Ultrasonic<br>Capacitive<br>Mechanical limits<br>Software limits<br>Software checking |
| Prevent material damage | | Touch detectors<br>Ultrasonic |
| Control the robot's ability to | Hazard stop<br>Stop<br>Slow down the speed<br>Evade obstacles | Self control<br>Combined sensor information<br>Logical conclusions |
| Monitor the movements of persons | Inside the system area<br>Entering the area<br>Leaving the area<br>Outside the area near hazardous points | Infra red<br>Ultrasonic<br>Touch detectors<br>Capacitive<br>Light<br>Microwave<br>Mechanical gate |
| Warn people about | Wrong direction of movement<br>Failures in the robot<br>Failures in the safety system<br>Wrong person too near the system | Warning light or sound<br>Signal light<br>Active obstacles<br>Enabling switch |

Controlling the hazard zone and raising the safety level require identification of the critical components and functions of the system. In the design phase of the system, this can be achieved by using safety analysis methods, such as Failure Mode and Effect Analysis (FMEA). Risks and hazards can be identified by using certain growth level analysis methods such as the Energy Barrier Analysis method. Good results can be obtained by combining these two methods and by trying to find critical failures and disturbances which can cause hazards and damage in the automatic system.

Certain mechanical devices need to be used in critical areas in order to prevent the environmental hazards from escaping into the surroundings. Automation itself cannot obstruct the passage of ultra-violet (UV) or other radiation, noise, vibrations, chemicals and electrical chocks; instead, mechanical obstacles and devices need to be utilized.

The fail-safe principle and self control are the features which always belong to the construction of intelligent safety systems. As the systems become more complicated, careful planning and testing of the safety measures become more important.

Industrialized countries lack uniformity in their safety instructions, standards and legislation in the usage and construction of safety systems (for example BS 6491, 1985). There are still countries in which intelligent safety systems are totally unknown. However, industrialized countries are developing more sophisticated safety systems.

## *Examples*

Industrial robots are used for different purposes in different countries. Spot-welding robots are very widely used by countries with major car manufacturing industries; other countries can have many assembly applications. The structure of the Finnish industry supports robotized arc welding, which is one of the most popular applications. That is why the following examples mainly include welding solutions. Arc welding in small production, however, provides a good example for safety measures due to its many elements which influence safety: light arc, smoke, programming near the machine and fast movements.

### Production-adapted safety system for a welding robot system

One truck manufacturer in Finland has several robotized welding stations for large body pieces. The robot stations consist of welding robots which have to make one long movement of about 10 to 14 m. One robot has three welding stations on each side of the welding line. While the robot is welding in one of the welding stations, the worker loads or unloads the others (Figure 10.8). The safety system must therefore allow work in one station while the robot is working in another. Large workpieces, the light arc, smoke, long and fast movements of the robot and internal transports especially have to be taken into account when designing a safety system for this use.

The developed safety system consists of sensors for detecting persons who are loading the welding stations or moving into or out of the system. In addition, it has sensors for detecting the location of the robot. These sensors give information about the area where the robot is working and in which direction the robot arm is pointed. All the functions of the sensors are ensured. In Finland, certain mechanical limit switches are officially approved for safety purposes, which explains their preference in safety equipment in this case as well. Persons are also detected with contact mats and light beams.

The first and the simplest way of ensuring the safety of this sytem is to place

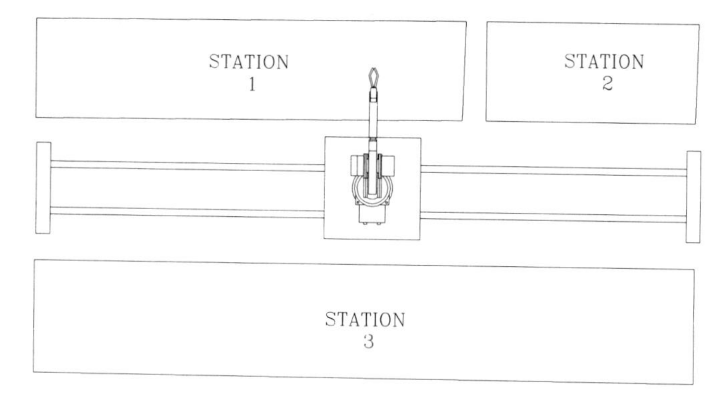


*Figure 10.8 Three welding stations in a robot system.*

fences around the system area. But due to the size of the workpieces in this situation it was too difficult to arrange a safety fence system. A second method is to use contact mats that are connected together to the safety circuit of the robot. However, this will cause the system to stop each time a person enters a welding station. Such an arrangement cannot solve the problem. The designed safety-system control unit consists of a relay logic which allows the robot to weld in a station in front of which no one is standing on the contact mat. It also prevents the robot from moving to a station where the worker is loading or unloading workpieces. Entering the vicinity of the robot stops the robot. A welding robot produces about 2.7 times the amount of smoke that a hand welder produces. The smoke is removed by a suction apparatus placed on the welding line. The walls prevent the light arc from causing an environmental hazard.

## Intelligent safety system for a welding robot system

The safety system described below has been installed at a robot welding station in a car parts factory (Malm, 1986; Kuivanen, 1988). An industrial robot with six axes welds small workpieces at three different stations. The system uses more than 180° of the robot's working area. One worker handles workpieces at all three stations. The stations form mechanical obstacles on three sides of the robot. The work is done in two shifts. The potential for danger is presented when someone goes inside the working area, near the movements of the robot. The space inside the system is very limited. Occasionally the worker must, however, go inside the system area in order to change the welding bar or rectify malfunctions. The system developed for this case is able to control problematic situations similar to those that have occurred in Japan in which the robot suddenly started to move in spite of a stop command situation.

This intelligent safety system consists of sensors for the detection of persons and the robot as well as a central unit. The system monitors the movements of the people and the robot in the work area. The main principles of the system are presented in Figure 10.9. The hazard zone of the robot welding station is divided into six subzones. The system layout is presented in Figure 10.10. The system perceives the zones occupied by the persons or robot. The information is then transmitted to the control unit which prevents collisions. The control unit can give orders to the robot at four levels. According to the situation, the robot can: operate manually; slow down its speed; stop; or hazard stop.

Thus the system controls its own function as well as that of the robot and the sensors. All types of sensor which give on/off signals can be used in the system. In this case the whole hazard zone has been covered with contact mats. Other possible types of sensor are, for example, light curtains and beams or passive infra-red sensors. However, the use of these sensors can greatly increase the price of the system and decrease its quality and reliability.

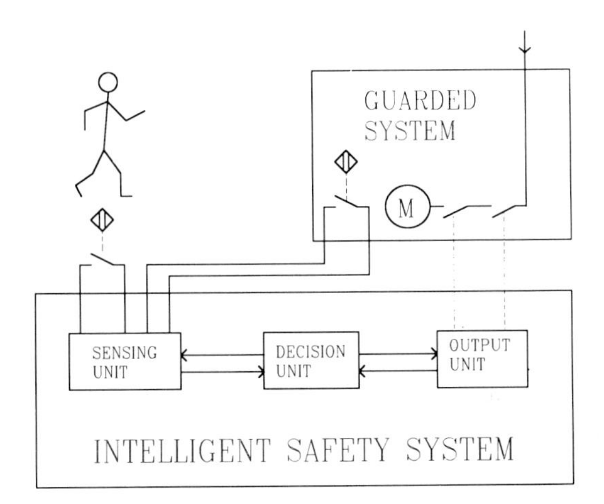


*Figure 10.9* *Basic principles of an intelligent safety system.*

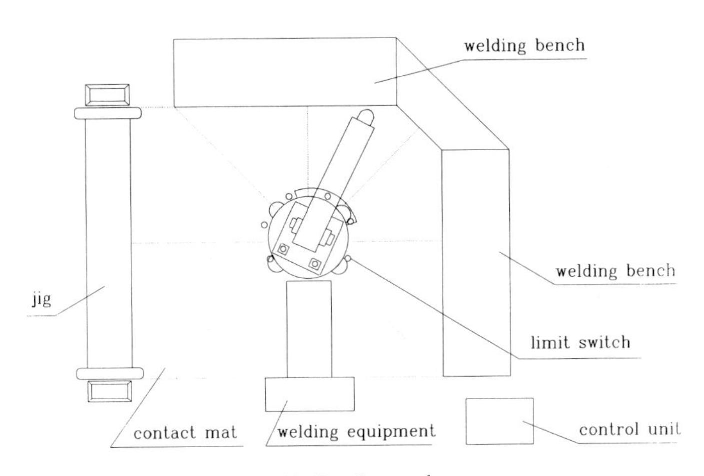


*Figure 10.10* *System layout.*

The safety system is assembled to the robot system after having been tested under laboratory conditions. Moving the system from laboratory use to production creates no difficulties because the type of robot is the same in both cases and the robot system is at the assembly phase. Most of the problems are caused by the shapes of the sensing elements. Due to the wide range of possible applications of the sensors it is very difficult to find general forms for the sensors of the various safety systems (Figure 10.11). This creates a need for the easiest possible adaptation of sensors to different kinds of conditions.

*Figure 10.11 The robot welding station.*

It was noticed at the laboratory that the reaction speed of the safety system (about 20 m/s) was too high for the robot. When the safety system gave the robot the order to stop, it took too long for the robot to give feedback after the order had been carried out, and consequently the safety system sent a new command for an emergency stop. Another type of difficulty causing a reaction in the safety system was inaccurate function of the robot control system. In this situation, the control unit had a very short functional breakdown which resulted in the safety system ordering an emergency stop due to robot failure. These reactions were corrected with time delays.

During the course of 210 working days, the safety system stopped the robot 1100 times. The order for safety speed was given 110 times. An emergency stop was ordered only three times, two of which were failures in the safety system and the other in the robot. The first failure in the safety system was caused by the adjustment of a sensor and the second was caused by the program unit which stopped due to a peak in the power-distribution network.

In the latter case it was noticed that the safety system included many unnecessary precautions and functions due to the workers' positive attitudes toward the whole system. The reactions of the operators, the foreman and the production leader towards the system during the testing period were positive. According to the operators, the safety system has helped to improve their work as it enables them to spend their time working rather than watching the robot.

When designing intelligent safety sensors one must consider all possible situations that may occur. This may cause problems because of the workers who do not even try to cooperate or who disturb the operation of the system intentionally. If the attitude to a safety system is negative, it is very difficult to prevent deliberate actions being made against the system.

The automation level of a safety system must not be higher than needed, because the possibilities of failure will increase with the automation level. If the system has a great number of functions, all of which are ensured, it can cause disturbances in the controlled system. The same phenomenon occurs, for example, in flexible manufacturing systems. The designers of the systems have reported that if all the possible precautions are taken, the system will not function at all, since some of the conditions are always in force in an extensive system. That is why the automation level of a safety system is always a compromise.

The possibility of bypassing the safety system with automatic production can be one of the key questions in production adaptation. If the safety system is designed correctly it has to stop the robot in case of failures. The bypass is acceptable when a failure in the safety system stops the whole production and the corrections cannot be made in a reasonable span of time. A reasonable time span differs from case to case. This happens, for example, if spare parts cannot be obtained without delay. If the bypass is accepted, it has to be done in a way that excludes unintentional or accidental execution and is therefore done with a key that can only be used by a foreman or other person responsible for the safety. The bypass must be performed under control, and the workers must be aware of the new situation.

Intelligent safety sensors offer new ways of avoiding unnecessary breakdowns and their consequences in a controlled system. In the case of welding, it is important that the false stop signal is not given during the welding process, otherwise the workpiece being welded can become damaged. When a safety system is adapted to a welding station, information from the arc-welding system can be used. If the light arc is on, the robot cannot move too fast, and if there are no other risks in that situation it is not necessary to stop the robot whenever someone enters the area. Another example is that the speed of the robot can be controlled with separate sensors. Too high a speed may indicate a failure in the robot and it will have to be stopped.

## Intelligent safety system for an automatic stretch film wrapping machine

In the next example the design of a safety system for an automatic wrapping machine is described. This machine is quite similar to a gantry type robot. It is totally computer controlled and has three degrees of freedom. It is possible to increase the safety level of the machine to some extent by using fences. The fences can, however, be replaced by an intelligent safety system, which maximizes the safety by protecting the input and output entrances of the machine as well (Suominen, 1988).

The aim of the sensing system is to detect the human beings entering the danger zone as reliably as possible. Since sensors are not physically able to limit the movements of human beings into the danger zone, such a zone surrounding the wrapping machine must be monitored so that the dangerous machine movement will stop as someone crosses into the zone. The sensing must also be able to distinguish between an entering or leaving platform and a human being. Passive infra-red sensors and photo-sensors are used to monitor the area surrounding the wrapping machine. The layout of the monitoring solution is presented in Figure 10.12.

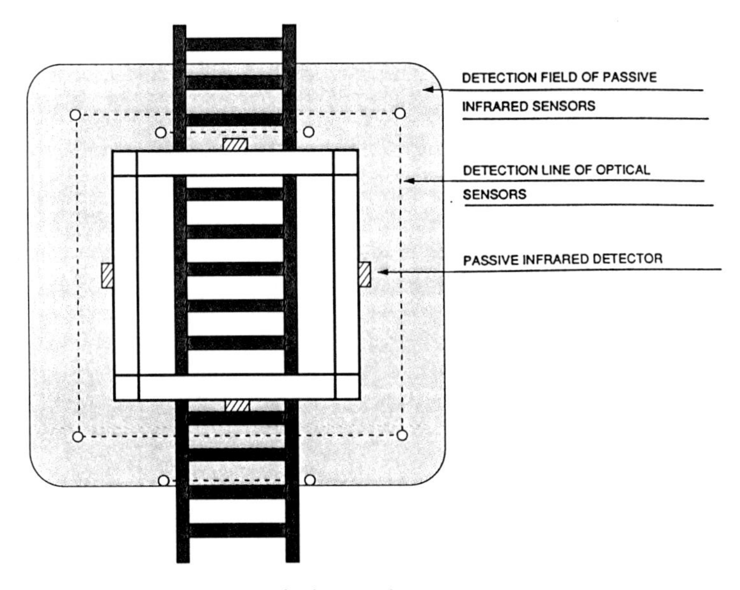


*Figure 10.12 The layout of the monitoring solution.*

Passive infra-red sensors are only sensitive to human beings; packing platforms on the conveyor do not cause false alarms. When a person enters the detection zone of the passive infra-red sensor, the speed of the wrapping frame is lowered. Only crossing the photo-sensor beam causes the machine to stop. On the conveyor sides, information from one photo-sensor only does not reveal if it is a platform or a human being that crosses the beam. Additional photo-sensors are therefore mounted onto the conveyor sides.

On the exit side of the machine, the human being and the platform go in opposite directions. The outcoming platform first crosses the inner beam and then the outer beam, whereas a human being crosses the outer beam first. In the latter case the machine stops.

On the entrance side of the machine, the situation is more difficult because both the human being and the platform move in the same direction. Now the extra photo-sensor has to be mounted outside the actual stopping line. The incoming object is interpreted as a platform if:

the conveyor is operating;
the outer photo-sensor gives the signal first;
the signal is continuous; and
the time interval between the inner and the outer photo-sensors is correct (the conveyor moves at a constant speed).

If any of these conditions is not fulfilled, the object is a human being.

The inner area of the wrapping machine is not fully monitored. The only sensor inside the machine is a mechanical safety edge in front of the stretch film rack.

The decision unit is basically very similar to the safety system of the industrial robot. A programmable logic functions as a decision unit and is duplicated with a relay logic. The programmable logic and the relay logic can independently give either a stop command or a safety speed command. In addition to the functions of the relay logic, the programmable controller takes care of certain checking and diagnostic functions and adds some features which improve the usability of the safety system.

## References

BS 6491, 1985, Electro-sensitive safety systems to machinery. Part 1. Specification for general requirements. (London: British Standards Institute).

Derby, S., Graham, J. and Meagher, J., 1985, A robotic safety and collision avoidance controller. In Bonney, M. C. and Yong, Y. F. (Eds) *Robot Safety* (Bedford: IFS Publications Ltd), pp. 237–46.

Flueckiger, N., 1980, Protecting photoelectrics from false signals. *Machine Design,* **52,** 15, 100–3.

Graham, J. and Meagher, M., 1985, A sensory-based robotic safety system. *IEEE Proceedings,* **132D,** 4, 183–9.

Health and Safety Executive, 1984, *The application of photoelectric safety systems to machinery.* Guidance Note PM 41. (Sheffield: Health and Safety Executive).

Holland, S. W., Rossol, L. and Ward, M. R., 1986, A vision-controlled robot system for transferring parts from belt. In Pugh, A. (Ed.) *Robot Sensors, Vol. 1: Vision* (Bedford: IFS Publications Ltd), pp. 213–28.

Irwin, C. T. and Caughman, D. O., 1985, *Intelligent Robotic Integrated Ultrasonic System.* SME Technical Paper MS 85–620 (Michigan: SME), pp. 19–39.

Juergens, H., 1986, *Erhöhung des Niveaus der Arbeitssicherheit in flexiblen automatisierten Fertigungsystemen (FMS),* Thesis, Technische Universität, Dresden, 150 pp.

Kilmer, R., 1985, Safety sensor systems. In Bonney, M. and Yong, Y. (Eds) *Robot Safety* (Bedford: IFS Publications Ltd and New York: Springer), pp. 223–35.

Kuivanen, R., 1988, Experiences from the use of an intelligent safety sensor with industrial robots. In Karwowski, W., Parsaei, H. R. and Wilhelm, M. R. (Eds) *Ergonomics of Hybrid Automated Systems I* (Amsterdam: Elsevier), pp. 553–9.

Kuivanen, R. and Tiusanen, R., 1986, *Kosketuksen tunnistimet turvalaitekäytössä* (Touch detectors used as safety equipment). Espoo, Valtion teknillinen tutkimuskeskus, Tiedotteita 619, 84 pp.

Kuivanen, R., Malm, T., Määttä, T., Suominen, J. and Tiusanen, R., 1988, *Safety Equipment for Human Detection*. Technical Research Centre of Finland, Research Notes 867, 35 pp.

Liljeroos, A., Silvola, M. and Sormunen, E., 1986, *Valokennojen käyttö turvalaitteena* (Fotoelectric sensors used as safety equipment). Espoo, Valtion teknillinen tutkimuskeskus, Tiedotteita 599, 30 pp.

Malm, T., 1986, *Älykäs anturi teollisuusrobotin vaara-alueen valvonnassa* (Intelligent safety system for supervising the hazard zone of the industrial robot). MSc thesis, Tampere, Diplomityö, Tampereen teknillinen korkeakoulu, sähköosasto, 64 pp.

Morris, H., 1985, Industry begins to apply vision systems widely. *Control Engineering*, January, 68–70.

Pugh, A., 1986, Robot sensors—a personal view. In Pugh, A. (Ed.) *Robot Sensors, Vol. 1: Vision* (Bedford: IFS Publications Ltd and New York: Springer), pp. 3–14.

Rahimi, M. and Hancock, P. A., 1988, Sensors, integration. In Dorf, C. D. and Nof, S. Y. (Eds) *International Encyclopedia of Robotics: Applications and Automation* (New York: Wiley), pp. 1523–32.

Suominen, J., 1988, *Intelligent Safety Sensor Systems for Controlling the Hazard Zones of Automatic Machines* (Tampere: Tampere University of Technology, Department of Electrical Engineering), 82 pp. (In Finnish).

Vartiala, H. and Järvinen, E., 1982, *Koskettamatta Toimivat Elektroniset Koneturvalaitteet* (Presence sensing electronic machine safety devices), katsauksia 53 (Helsinki: Työterveyslaitos), 51 pp.

# Chapter 11
# Robot safety research and development in Poland: an overview

**D. Podgorski[1], T. Wanski[2] and S. Kaczanowski[2]**

[1]*Central Institute for Labour Protection, ul. Tamka 1, 00–349 Warsaw, Poland and* [2]*Industrial Institute of Automation and Measurements, Al. Jerozolimskie 202, 02–222 Warsaw, Poland*

**Abstract.** This chapter presents an overview of research and development issues related to safety of industrial robotics in Poland. The first part of the chapter contains data concerning the number and types of robot manufactured and installed in Polish industry and reviews the range of scientific studies conducted on the development of robotics in Poland. In the second part the characteristics of a new Polish robot control system are presented and research and design studies on improvement of intrinsic robot safety are described. The next part outlines the results of a study directed at the design of robotic safety systems based on ultrasonic sensors, and presents the results of examination of ultrasonic sensor detection capability under static and dynamic conditions. Finally, prospects for further research in the area of robotic safety are considered.

## Introduction

The field of robotics has been developing in Poland for about 15 years. This development is characterized by an increase in the number of industrial robots manufactured and installed in Polish industry, and by qualitative progress in research and development efforts in advanced manufacturing technology. An overview of some more significant research and design studies on robotics in Poland is presented in the first part of this chapter.

Taking into consideration the importance of safety issues when introducing robots into industry, some of the relevant research programmes on robotic safety were funded through the State Office for the Promotion and Progress in Science and Technology Development. The Industrial Institute of Automation and Measurements (PIAP), Warsaw, and the Central Institute for Labour Protection (CIOP), Warsaw are the most active research institutes in this domain.

Safety problems related to industrial robot applications are divided into two main groups:

1. consideration of safety issues in the design of industrial robots—mechanical design of the robot arm, design of the robot control system hardware and software; and
2. installation of safety systems in robot workplaces for detecting human presence inside or near the robot dangerous movement zone.

Since both types of problem are equally important to achieve the appropriate safety level in robot workplaces, research on robotic safety in Poland is conducted in these two directions almost at the same time. Results of the studies on design of safe industrial robots are described in the second part of this chapter, while the results of studies devoted to safety system development are presented in the third part.

## *Development of robotics in Poland*

Robotics, as a branch of science and technology, has been developing in Poland since the mid 1970s. In 1977 licence for manufacturing IRb robots was purchased from the Swedish company ASEA. These robots were manufactured up to 1988 under the same trademark, and from 1989 under the name of IRp. The type designation has been altered due to intrinsic changes in the robot manipulator design and development of the new robot control system.

Besides the IRb robots, several other types of robot of original Polish design have been manufactured. These pneumatically-, hydraulically- and electrically-driven robots are used mostly for workpiece manipulation (i.e. pick-and-place tasks), spray painting and spot welding. At present about 100 robots a year are manufactured in Poland. The growth in number of industrial robots installed in Poland in the period 1986–1989 is shown in Table 11.1 (Morecki, 1990).

*Table 11.1 Number of industrial robots in Poland 1986–1989.*

| Year | 1986 | 1987 | 1988 | 1989 |
|---|---|---|---|---|
| Number of robots | 380 | 410 | 470 | 520 |

Most of the robots used in Polish industry are manufactured in Poland. The most popular foreign robot manufacturers are UNIMATION (USA), CLOOS (Germany) and COMAU (Italy). The main application areas of industrial robots are arc and spot welding, and machine-tool loading and unloading operations. Other more sophisticated robot applications include: 1) plasma spraying, 2) polishing, 3) machine-part regeneration with aluminum padding, 4) sintering details from iron powder and 5) assembly of small electrical equipment. Several specific robot parts and certain equipment used in robot applications are also designed and manufactured in Poland. These include electric drives with DC and AC motors, harmonic drives, automatic gripper change equipment, various types of sensors, industrial vision systems, welding equipment especially designed for robot technology and rotary positioners for welded workpieces.

# *Safety aspects of industrial robot design*

## Intrinsic robot safety

It is generally accepted that the introduction of robots into industry can improve working conditions by elimination of workers from dangerous and monotonous jobs (Wilson, 1982; Parsons, 1987). But robots create new hazards and many accidents due to industrial robots have been reported (Carlsson, 1985; Sugimoto, 1985; Nagamachi, 1988). Since malfunctions of robot system components can result in potential hazards for personnel interacting with robots, (Khodabandehloo *et al.*, 1985; Sugimoto and Kawaguchi, 1985), it is particularly important to build intrinsic safety into the robots. The intrinsic safety features include robot reliability, are inherent in the design of the robot and its control system and do not concern the use of external safety equipment (Akeel, 1983).

The basic approach to achieving an acceptable level of intrinsic robot safety is the fail-safe design of the robot system. The fail-safe system can be explained as a system that will fail to a safe state in the event of a failure of any component of the system.

The typical method used for the robot fail-safe design is the application of safety monitoring measures. These measures should be implemented as additional subsystems, devices, circuits or sensors incorporated in the robot control unit and the manipulator in order to monitor operation of the robot system independently. If any robot failure or improper working conditions are detected by any of these measures, the robot movements should be stopped.

## The control system for IRp-6/60 industrial robots

Recently, a modification of the control system for IRb robots manufactured under the ASEA licence was made by the Industrial Institute of Automation and Measurements. The construction of this system, now called IRp, allows for two types of robot arm trajectory control: point-to-point (PTP) and continuous path control (CP) (Jablonski *et al.*, 1986).

The new control unit enables simultaneous control of up to six robot axes and up to three external drives with DC motors for external equipment (including robot track motion). It also enables the collection of analogue and digital external signals, and the control of a number of bistable-driven external devices. The new system makes it possible to integrate the robot control unit in a flexible manufacturing system by means of a set of additional built-in modules.

The hardware of the control unit is based on a digital automation system with a 16-bit central processing unit. This hardware is modular and consists of various electronic modules. The software of the IRp controller has a modular structure and makes the development of application programs relatively easy.

Each drive system for the robot axes (as well as for external motors) consists of an axis position controller and power controller panels. The axis position controller is based on an 8-bit processor. This enables a qualitative development of a control unit and the realization of a number of new robot functional features, such as:

1. PTP and CP trajectory control with direct velocity declaration (mm/s);
2. tool centre point (TCP) trajectory programming in various coordinates: joint oriented, cartesian, or cylindrical;
3. TCP trajectory programming with linear or circular interpolation;
4. adaptive control with the use of various external sensors (proximity, tactile, etc.);
5. preparation of an application programs library with the use of magnetic tapes and disks; and
6. possibility of off-line programming.

The application program is prepared by means of a hand-held teach pendant. This microprocessor-based module is equipped with an alphanumeric display (2 rows with 40 characters each), a set of fixed-denotation keys and a set of soft keys (the current function of these keys is shown on the display just above them). The pendant communicates with the control unit by means of a standard RS232-C interface. The operator is informed about program instruction parameters and robot status through the information provided by the alphanumeric display.

The teach pendant enables editing of application programs: insertion and deletion of instructions, change of numerical parameters, etc. The multi-step robot velocity correction during automatic execution of the application program is also possible by means of a teach pendant. The pendant is equipped with a joystick for manual control of robot movements during programming. The joystick has three degrees of freedom: left–right inclination, forward–backward inclination and knob rotation. Movements of external axes can also be controlled by this joystick.

## Safety of IRp-6/60 robots

The following circuits and technical solutions have been incorporated in the IRp control unit in order to increase the level of the robot intrinsic safety:

1. hold-to-run control on the teach pendant;
2. watchdog within the teach pendant;
3. watchdogs in axis position controllers;
4. circuits for on-line signal bus checking; and
5. brakes on manipulator axes.

A hold-to-run switch prevents unintentional robot movement which can be caused by incidental pushing of the joystick by the robot programmer or by dropping the pendant to the floor. Electrical signals from the joystick initiate robot movements only when this switch is pressed.

The aim of incorporating watchdog circuits in the design of the teach pendant and the axis position controllers is to neutralize the majority of possible faults of electronic system components. The selected output signal of the monitored system triggers the watchdog periodically when the control program of the system is executed properly. If the program goes out of its regular routine for any reason, the watchdog will produce an alarm signal. In the case of the teach pendant, the watchdog alarm signal resets the internal microprocessor system and forces the pendant control program to jump to the beginning of its regular routine. The output signals of watchdogs in axis position controllers cause the robot to stop immediately due to a significant possibility of unpredictable axis movements.

The robot control unit is equipped with a control panel to monitor several internal parameters. Part of its hardware is used by the central processor unit to check signal buses by means of control data transmission in order to stop the robot when any failures or disturbances are detected.

Electrically-driven brakes on several robot manipulator axes prevent undesirable robot movements in case of emergency stop or switching off the power supply.

The solutions described above have a direct influence on robotic safety. But there are also other technical means which have an indirect influence on safety of robotic workstations. These means include: dialogue programming with the help of the teach pendant; multi-action start system for the application program; possibility of connecting the external stop devices; and reduction of the robot speed when the pendant is removed from its normal place in the control cabinet.

The multi-action start system requires the consecutive pressing of two distinct pushbuttons to prevent starting the application program by mistake. The above described solutions reduce the possibility of hazards and accidents on robot workstations. These are the basic means of safety for operators and programmers who are those most exposed to hazards in the robotic workplace.

## Further improvement of IRp robot safety

At present, research on increasing the safety level of the IRp robots is being carried on at the Industrial Institute of Automation and Measurements. The purpose of this work is to establish a set of technical and organizational means for increasing robot safety, i.e. design rules (regarding hardware and software), methods of supervising various subsystems within the robot control system, new proposed robot functional features (e.g. safety stop function and emergency withdrawal function), recommendations regarding robot workstation organization and operator training programmes.

Within the scope of this project the following main research issues have been identified: 1) identification of hazards at robotic workplaces (including analysis of workplaces in Polish industry); 2) method of estimating the safety level of a robot or a robot workplace; and 3) development of robot structures taking into consideration the highest technically and economically achievable safety level.

Several technical solutions have been proposed for these safety problems. The most important ones deal with the self-test system incorporated in the control unit, and safety performance of the drive and measurement system. The idea of the self-test system structure is shown in Figure 11.1.

Reliability of the drive system is essential for robot construction and

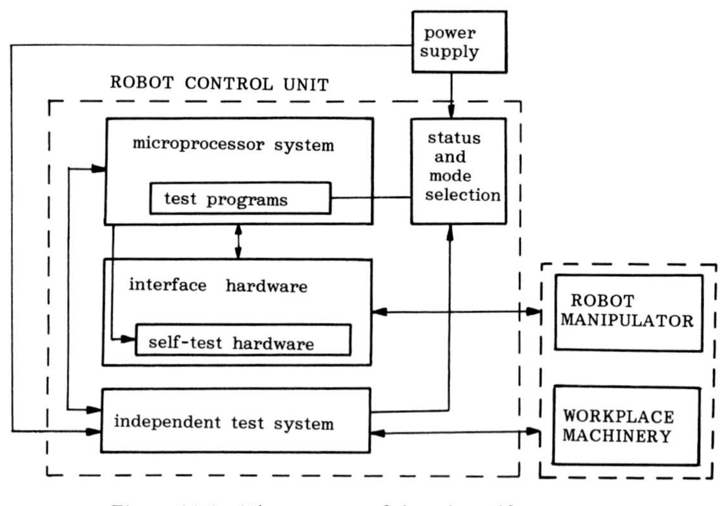


*Figure 11.1 The structure of the robot self-test system.*

application. Almost every failure in this system may cause a hazardous situation—a rapid, unpredictable robot arm movement. The most effective prevention method is the use of a redundancy technique for monitoring the drive system operation. Implementing redundancy in the design of the robot drive system is illustrated in Figure 11.2.

Several other solutions will be introduced in the IRp-6 robot design. These safety solutions include:

1. development of the start–stop system with external commands (e.g. external generation of 'safety stop' signal) in automatic and manual operation modes;
2. a robot environment monitoring system (power supplies, peripheral machinery, gate interlocks, etc.);

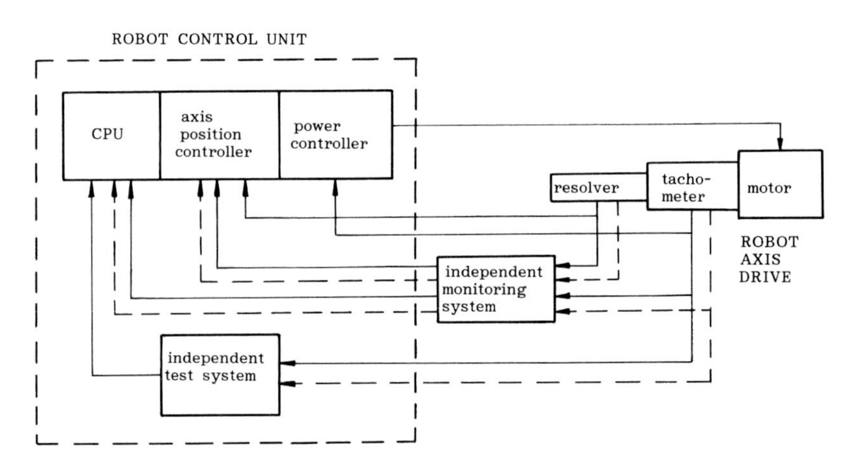


*Figure 11.2 The idea of redundancy technique application in the design of the robot drive system.*

3. self-test procedure (performed in each robot start-up routine and during normal operation);
4. several velocity ranges for manual robot movement control with the joystick; and
5. emergency withdrawal function.

The first two solutions can be easily implemented by expanding the bistable input/output system together with introducing some changes into the robot control software.

The self-test techniques should be used as widely as possible to check the proper operation of modules and circuits of the robot control system. Tests started after each power-up can check the operation of all system blocks because there are no practical time limits for the test execution. Since tests executed periodically during normal operation of the robot system should not decrease the efficiency of the system, they should be relatively short and check only the most important subsystems.

The possibility of changing the movement velocity range controlled with the joystick makes the programming easier, especially in situations where high positioning accuracy is necessary. It can also decrease the hazard level caused by inadvertent programmer errors.

An emergency robot withdrawal function has been proposed by Wanski (1983). The purpose of this function is to take the robot arm back automatically from a dangerous position to a place defined as 'safe position'. The withdrawal can be activated by the operator, external safety sensor, or by a special sensor (e.g. robot-gripper temperature sensor or a robot-arm load sensor) for collision avoidance. The emergency withdrawal function should have higher priority compared with any other safety stop command with the exception of the hardware-based emergency stop. Examples of robot end-effector trajectories during emergency withdrawal are shown in Figure 11.3.

Additional intermediate or end-safe positions can be introduced to the robot control unit memory during preparation of the application program by the robot operator. These positions are ignored during normal execution of the program. After activation of the withdrawal signal, the application program starts to be executed backwards. From this time on, the normal trajectory points are omitted and only instructions with emergency positions are executed, until the end-safe position is reached.

The emergency withdrawal function depends upon a new organization of hardware and software resources in the control system for prevention of hazards and accidents. The solutions described above constitute a complex set of technical measures that is being introduced to increase the level of intrinsic safety of the IRp robots.

## *Development of safety systems for robotic workplaces*

### Safety sensors

One of the ways to increase the safety of personnel interacting with industrial robots is to equip robot workstations with human presence detection systems.

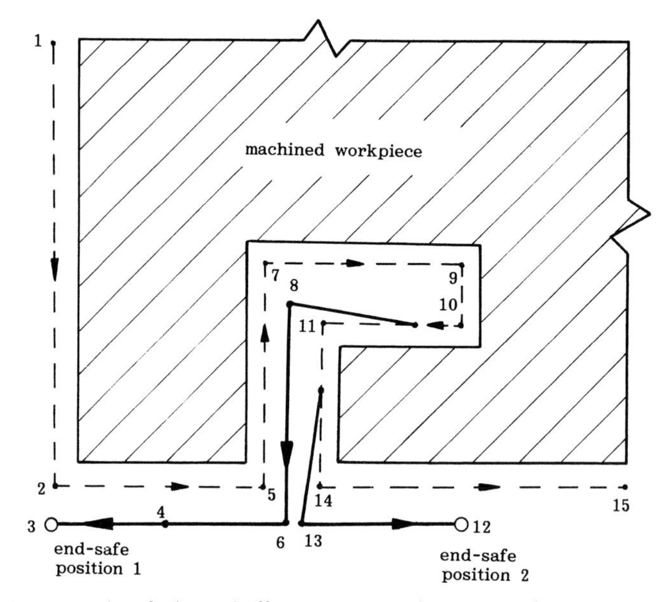


*Figure 11.3 Examples of robot end-effector trajectories during normal program execution (dashed line), and during emergency withdrawal (continuous line).*

These safety systems can be divided into three main levels according to coverage of the workspace regions around the robot (Kilmer, 1985).

First-level safety systems have the most widespread application in robot workplaces. These systems provide perimeter detection of persons crossing the workstation boundary by means of interlocked physical barriers, light curtains or floor safety mats. This type of safety system does not protect the personnel during programming the robots or during robot maintenance when close interaction with robots is required. Therefore, it is important to study the development and application of second-level safety systems which would be able to detect human presence within and around the robot movement zone. Moreover, third-level systems are needed to detect human presence near the contact surface of the robot arm.

Examination and assessment of sensor technology from the point of view of sensor operational reliability at the robotic workstation environment should be the first phase of the safety system development process.

There are various sensors which can be applied to the design of safety systems at the second and third levels. The presently available sensor technologies include infra-red, microwave, capacitance, ultrasonic and machine vision. The possibilities of their application for robotic safety have been reviewed and discussed by Derby *et al.* (1985), Rahimi and Hancock (1988) and Kuivanen *et al.* (1988). Each of the technologies mentioned above has some advantages and disadvantages, and proper selection of the sensor type is an important issue to assure an acceptable level of operator safety. The necessity of providing the most reliable detection of human presence, and the sensor immunity to false triggers caused by industrial environment factors, should in particular be taken into consideration during safety system development. Costs and technical limitations of the safety system design and manufacture are also important criteria in the selection of the most suitable sensor type.

The first research project on the safety system at the second and the third levels was carried out in Poland in the early 1980s. The system developed consisted of active infra-red sensors mounted on the robot arm. Laboratory tests on the active infra-red safety system operation used various kinds of obstacles (including a mannequin of adult human size). The results proved that the system effectiveness was sufficient to detect these objects at safe distances from the robot arm (Wydzga, 1981).

## A study of the application of ultrasonic safety sensors

Further research in the area of robotic safety systems development focused on the possibility of ultrasonic-sensor application. In order to evaluate the usefulness of ultrasonic sensors in robotic safety systems design, a laboratory model of a safety system was developed during the first phase of the study (Podgorski, 1989). A block diagram of this model is shown in Figure 11.4.

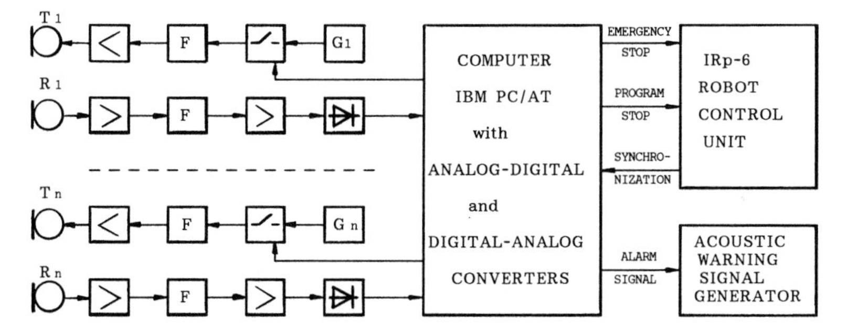


*Figure 11.4 Block diagram of the model of the safety system (T, transmitter; R, receiver; G, generator; F, band-pass filter).*

The safety-system model incorporates ultrasonic sensors which can be attached in various places to the arm of the IRp-6 robot. Each sensor consists of two piezoelectric transducers: a transmitter and a receiver. The transmitter emits short ultrasonic pulses at a rate adjusted from 10–50/s, while the receiver detects echo pulses reflected from objects placed in the detection zone.

The operation of this safety system is controlled by the computer software which is divided into three main procedures. The first one, which is run during teaching the robot, allows the presence of any objects in the sensor detection zone when the system performance is independent of the robot task (generic mode). When the robot operator is outside the robot workstation, and the robot is in the automatic mode, the second procedure is run. In this procedure, the currently measured distances between the sensor and reflecting objects are stored in the computer memory during the first automatic cycle of the robot program. During execution of the next cycles of the robot program, the third procedure is run, and the data currently obtained from the ultrasonic sensors are compared with previously determined values stored in a computer memory (mapping mode).

The detection capability of ultrasonic sensors was investigated in the generic mode by means of experiments with the use of obstacles of various shapes and dimensions and made of different materials. Two types of piezoelectric transducer were employed for ultrasonic sensor design: MPU-1 working at 40 kHz and RU-200 working at 200 kHz. The former are manufactured in Poland by UNITRA-CERAD and the latter are manufactured by Siemens AG in Germany.

The comparison of shapes and dimensions of detection zones for both types of ultrasonic sensors is shown in Figure 11.5. Points of detection zone boundaries were determined by the position of a sphere of 15 cm diameter covered with a wig of natural slightly curly hair. This obstacle simulating a human head was chosen because the head is the part of human body most sensitive to injuries, and human hair has a relatively low ultrasonic radiation reflectivity coefficient.

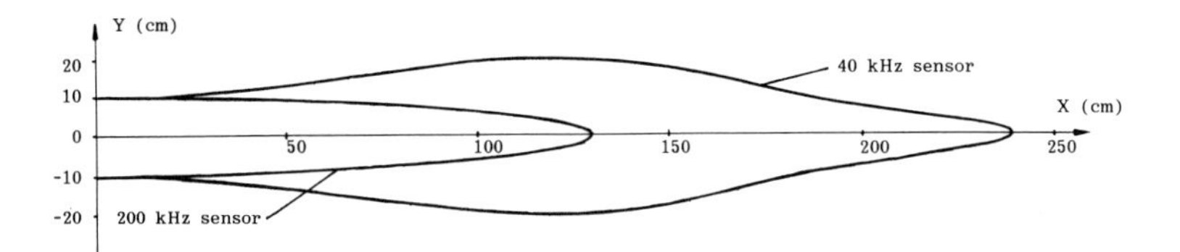


*Figure 11.5 Comparison of detection zones of 40 kHz and 200 kHz ultrasonic sensors.*

The detection zone of the 40 kHz sensor presented in Figure 11.5 is about twice as long as the detection zone of the 200 kHz sensor. Similar proportions were found by comparing detection ranges for these sensors with obstacles of other shapes and dimensions. This is caused by greater attenuation of acoustic energy in air for higher frequencies (since this attenuation increases approximately in proportion to frequency squared).

Though detection ranges for the 200 kHz ultrasonic sensors are smaller, their application to safety system design can be motivated by better immunity to false triggers caused by external noise sources of the industrial environment. Measurements of ultrasonic background noise associated with various manufacturing operations reported by Bass and Bolen (1985) show that ultrasonic sensors for robotic applications have an adequate signal to noise ratio above 100 kHz. Operations with laser etching and high-velocity fluid or air spraying are not recommended where ultrasonic noise can even disturb sensors with a working frequency of up to several hundred kHz.

Since the attenuation of ultrasonic radiation in air depends on temperature and humidity, the influence of these climatic factors on detection ranges of ultrasonic sensors was measured in a microclimatic chamber. The results of these measurements showed that detection ranges decrease when temperature or humidity increases.

Under normal climatic conditions occurring at most robot workstations in industry (temperature: 15–25°C, humidity: 40–60%), changes in detection ranges of ultrasonic sensors do not exceed a few percent; but when the safety system is used in more fluctuating climatic conditions, the changes in detection ranges can reach even 30%. In such cases, the influence of temperature and humidity on system safety performance should be predicted and eliminated by means of compensation of echo signal amplification in the receiving channel of each sensor.

In order to determine the appropriate arrangement of ultrasonic sensors on the robot arm, a series of measurements of distances between a mannequin (of adult size) and the robot arm stopped by the safety system were performed. Combinations of the following factors were applied during these measurements:

1. robot arm movement speed (variable from 0.3–1.5 m/s measured for the end-effector in maximum arm extension);
2. repetition frequency of pulses emitted by the ultrasonic sensors (variable from 10–50 Hz);
3. distance between the mannequin and the robot base main axis (variable from 40–160 cm);
4. distances between places of attaching the sensors and the robot base main axis (variable from 0–90 cm); and
5. angles between the sensor detection zones and the surface of the robot arm (variable from 15–60°).

The optimum arrangement of four ultrasonic sensors on the robot arm determined as the result of these measurements is shown in Figure 11.6. The distances between the front surface of the mannequin trunk and relevant points on the surface of the robot arm stopped by the safety system are presented as functions of the distance between the mannequin and the robot base main axis.

The performed measurements also revealed that the adequate ultrasonic pulse repetition rate should be about 20–30 times/s. When the repetition rate was lower, in some cases the robot would be stopped too late and consequently struck the mannequin.

The examination of ultrasonic sensors conducted in the first phase of the CIOP study (1989), showed that these sensors have suitable detection capabilities under static and dynamic conditions, so they can be applied in the generic safety systems. But further research is still required to obtain evaluation of the sensor feasibility for mapping robotic safety system design.

## *Prospects for future research*

Safety consideration is still an important trend of industrial robot design in Poland. It can be expected that intrinsic safety will have to be built into various robot hardware subsystems which will be designed in future. Further evolution of robot intrinsic safety systems will include application of artificial intelligence

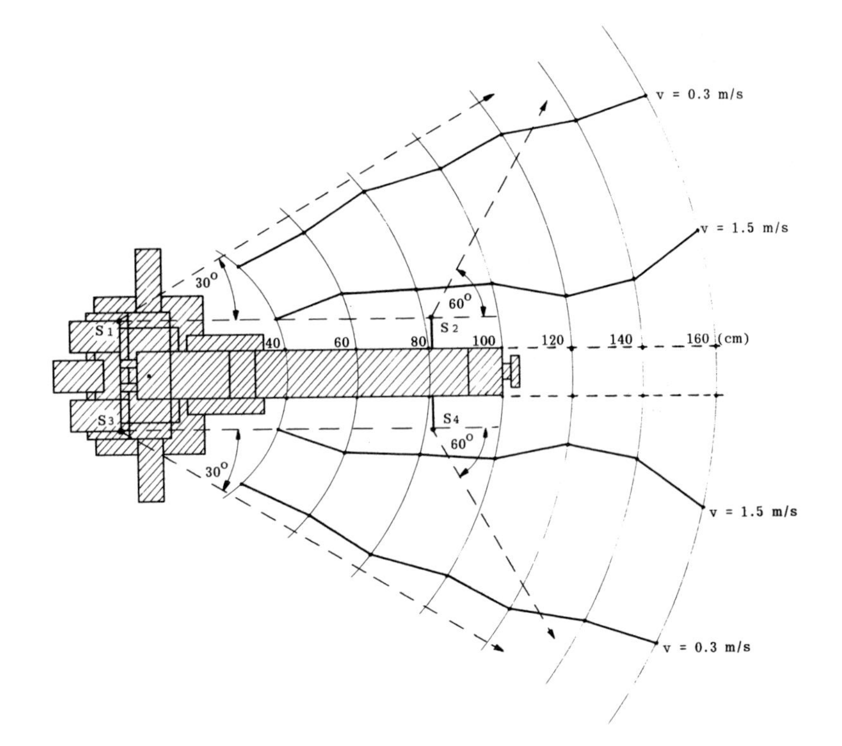


*Figure 11.6 Distances between a mannequin and the robot arm stopped by the safety system measured for optimum arrangement of four ultrasonic sensors (S1—S4) and for minimum and maximum robot movement speed.*

methods for robot software design. The redundancy and self-diagnostic techniques are going to be more widely applied in future robot constructions. This will be particularly important for mobile robots and for robots with vision-sensing systems.

With regard to the development of safety systems for detecting humans in robot workplaces, further evaluation of ultrasonic sensor feasibility will be performed. Hardware and software of the developed safety system will be modified in order to secure the compensation of temperature and humidity, and better synchronization with the robot control unit. Additional tests are needed on the safety-system operation when the mapping mode is applied. The reliability of sensors and safety system controller will be investigated in the framework of fail-safe design methods.

## Acknowledgments

The research presented in this chapter was sponsored by grants from the State Office for the Promotion and Progress in Science and Technology Development,

Central Programs of Research and Development: 'Industrial Robots' (project no. 100), and 'Protection of Man in the Working Environment' (project no. 121/IV).

## *References*


Akeel, H. A., 1983, Intrinsic robot safety. *Professional Safety,* 12, 27–31.

Bass, H. E. and Bolen, L. N., 1985, Ultrasonic background noise in industrial environments. *Journal of the Acoustical Society of America,* **78,** 6, 2013–6.

Carlsson, J., 1985, Robot accidents in Sweden. In Bonney, M. C. and Yong, Y. F. (Eds) *Robot Safety* (Bedford: IFS Publications Ltd and New York: Springer), pp. 49–64.

Derby, S., Graham, J. and Meagher, J., 1985, A robot safety and collision avoidance controller. In Bonney, M. C. and Yong, Y. F. (Eds) *Robot Safety* (Bedford: IFS Publications Ltd and New York: Springer), pp. 237–46.

Jablonski, P., Pachuta, M. and Wanski, T., 1986, The new control system for IRb-6 and IRb-60 robots. In *Proceedings of the United Nations Economic Commission for Europe Conference* 'Industrial Robotics', Brno, Czechoslovakia, February.

Khodabandehloo, K., Sayles, R. S. and Husband, T. M., 1985, Safety integrity assessment of robot systems. In Quirk, W. J. (Ed.) *Safety of Computer Control Systems* (Oxford: Pergamon Press), pp. 13–20.

Kilmer, R. D., 1985, Safety sensor systems. In Bonney, M. C. and Yong, Y. F. (Eds) *Robot Safety* (Bedford: IFS Publications Ltd and New York: Springer), pp. 223–35.

Kuivanen, R., Malm, T., Määttä, T., Suominen, J. and Tiusanen, R., 1988, Safety Equipment for Human Detection. Research Notes No. 867. (Espoo: Technical Research Centre of Finland).

Morecki, A., 1990, Present state and future possibilities of industrial robotics, *Scientific Papers of the Institute of Technical Cybernetics of the Technical University of Wroclaw,* 82/37, 5–15.

Nagamachi, M., 1988, Ten fatal accidents due to robots in Japan. In Karwowski, W., Parsaei, H. R. and Wilhelm, M. R. (Eds) *Ergonomics of Hybrid Automated Systems I* (Amsterdam: Elsevier), pp. 391–6.

Parsons, H. M., 1987, Robotics and the health of workers. In Noro, K. (Ed.) *Occupational Health and Safety in Automation and Robotics* (London: Taylor & Francis), pp. 196–211.

Podgorski, D., 1989, Electronic Safety System for Protecting Workers on Robot Workstations (Central Institute for Labour Protection, Warsaw)—unpublished report in Polish.

Rahimi, M. and Hancock, P. A., 1988, Sensors, integration. In Dorf, C. D. and Nof, S. Y. (Eds) *International Encyclopedia of Robotics: Applications and Automation* (New York: Wiley), pp. 1523–31.

Sugimoto, N., 1985, Systematic robot-related accidents and standardization of safety measures. In Bonney, M. C. and Yong, Y. F. (Eds) *Robot Safety* (Bedford: IFS Publications Ltd and New York: Springer), pp. 23–9.

Sugimoto, N. and Kawaguchi, K., 1985, Fault-tree analysis of hazards created by robots. In Bonney, M. C. and Yong, Y. F. (Eds) *Robot Safety* (Bedford: IFS Publications Ltd and New York: Springer), pp. 83–98.

Wanski, T., 1983, Emergency withdrawal—a new robot function. *Pomiary, Automatyka, Kontrola,* 7, 243–5.

Wilson, R. D., 1982, How robots save lives. In *Proceedings of Robots VI Conference,* Detroit, MI, USA, March.

Wydzga, S., 1981, The Device Protecting the Industrial Robot from Injuring the Human (Industrial Institute of Automation and Measurements, Warsaw)—unpublished report in Polish.

# *PART III*
# *DESIGN AND IMPLEMENTATION*

# *Introduction*

The section on robot design and implementation reflects a set of conceptual stages in planning design and implementation of current and future robotic systems. In the opening chapter of this section, the author considers the application of socio-technical systems to generate feasible designs of human-oriented robotic systems. The concept of human-centred technology is fully explained. Methodological issues are outlined in the light of prospective designs aimed to optimize the human–robot performance, using operator skills and knowledge.

In Chapter 13, the authors discuss strategies for human error reduction in advanced manufacturing systems that incorporate robots as subsystems. A generic list of incidents and accidents and the ways in which these reports can be analysed are discussed. In this review, the reliability of robotic systems is linked to the technical and human errors as the root causes of system stoppages (i.e. lack of system productivity). In terms of system safety, the authors recommend work design, work layout and operator training as critical components of a successful robotic system.

An important issue in human–robot interface is the design of robot teach pendants. Chapter 14 contains a comprehensive account of design issues related to remote-control units for industrial robots. In particular, the need for standardization of robot pendants has been emphasized. In a related area, the authors in Chapter 15 address the human factors issues in the design of a force-reflecting hand controller for a six-degree-of-freedom cartesian coordinate telerobotic system.

The final two chapters of this book reflect the growing interest in implementing robots in task domains other than industrial environments. The issues of human–robot integration in service robotics are discussed in Chapter 16. The field of service robotics challenges the designers to address the issues of human compatibility, safety and utility. The primary thrust of this chapter is to show how robotic technology can augment human capabilities. Several research programmes in the area of service robotics are reviewed and the relevant human–robot integration problems are discussed. Finally, Chapter 17 contains an application of robotics in the construction industry. Robot economics and management issues are discussed in view of current developments in optimum robot assignment to various construction tasks. A computer model for such considerations is elaborated.

# Chapter 12
# Design of human-centred robotic systems

**P. T. Kidd**

*Cheshire Henbury Research and Consultancy, Tamworth House, PO Box 103, Macclesfield, SK11 8UW, UK*

**Abstract.** The design of a human-centred robotic system is considered. The main design issues addressed are the choice of robot and the layout of the robotic cell. A number of methodological problems were encountered during the project. These problems hindered consideration of the people issues during the initial phase of the design work. The relevant design criteria that should have been considered are identified and discussed. The application of social science knowledge in a prospective manner to generate feasible design options is demonstrated. This way of using social science criteria is compared with the more normal retrospective evaluation method. The difficulties of considering human aspects during the design of a human-centred robotic system are discussed and the lessons learnt from the project are enumerated.

## Introduction

In recent years there has been a notable increase in the application of the label human-centred to various types of system. Unfortunately this has resulted in some confusion about what constitutes a human-centred system. The label now seems to be applied to systems designed using the more conventional approach of the social science community and also to any system which has some characteristic that makes it more acceptable for people.

It is clear from original proposals (Rosenbrock, 1989), however, what constitutes a human-centred system. One important concept underlying human-centred systems is the development and use of skill supporting and skill enhancing technology rather than skill substituting technology. Also, the human-centred approach places emphasis on the prospective application of social science criteria in the design process, rather than on retrospective evaluation or on a parallel design approach. Human-centred design methods are also interdisciplinary because they are concerned with the issues that lie between disciplines. More conventional retrospective evaluation and parallel design methods are more multidisciplinary, because they tend to ignore these interdisciplinary issues.

A human-centred robotic system can therefore be characterized by four key features: 1) the technology is designed to be skill supporting and skill enhancing; 2) social science knowledge is applied in the design in a prospective manner to generate initial design proposals; 3) social science knowledge is used to shape the technology; and 4) the design process focuses attention on issues which lie between disciplines (i.e. between technology and the social sciences).

The design of human-centred robotic systems requires that human, financial and technical issues be considered simultaneously when shaping the technology. However, whilst financial and technical issues pose few problems for systems designers, the relevant human considerations can be difficult to deal with. The reason for this is that social science insights should be used in a prospective manner to shape feasible design options. This requires technologists to have a broad range of skills and knowledge in both the technical and the social sciences. It is also necessary for technologists to internalize the judgement criteria used by social scientists. It is not very common to find technologists who have these skills, or who have internalized the judgement criteria used by the social science community.

As a consequence, the design of robotic systems tends to focus on those issues which are governed by legislative requirements. Primarily this means safety, which can have wide ranging implications for the design of the robot and the layout of the cell, as well as for the design of associated control and programming systems (Swedish Work Environment Fund, 1987). It is often the case, therefore, that the relevant social science issues are not identified and do not get considered until after initial design proposals have been formulated by technologists. These initial design proposals are shaped using technical and financial criteria, and no account is taken of the relevant social science issues. In the case of robotic systems emphasis is then placed almost entirely on the important issue of safety. Effort is then focused on making the technical system safe.

This technology-led approach, followed by retrospective evaluation and social science input is not very satisfactory if the evaluation shows that major design changes are needed. Technologists are often reluctant to make major design changes. This is especially true if a substantial amount of design effort has already been expended, or if further design changes would introduce delays in the project work programme. It is likely in these circumstances that the social science issues, other than safety, would be ignored.

The work described in this paper concerns the application of social science to the design of a human-centred robotic system. The system, a manufacturing cell, included a number of major elements: robots, various machine tools, computers and other forms of workhandling devices. The design and implementation of this cell was undertaken as part of the European Strategic Programme for Research and Development in Information Technology (ESPRIT) project 1199 (Human-centred CIM).

A number of technologists undertook to design a human-centred robotic system, but the way in which the majority of these technologists went about this task was unsatisfactory. Their main concern was with the rather obvious, but nevertheless important, question of safety. They also conceded that another social science issue was the need for social interaction between people working in the cell. Task allocation was also identified as one further important design issue.

This chapter describes how social science considerations can be used to influence the initial selection of robots and the layout of robotic cells. The conclusions address some of the important lessons that can be learnt from the design project. This focuses on the method that should be used to design human-centred robotic systems and the problems of interdisciplinary design. The implications for human–robot interaction are also addressed.

## *Human-centred technology*

The possibility of designing human-centred technology was proposed by Rosenbrock (1977), although at the time this terminology was not used. The human-centred label was added later, probably by Brödner (1982) who used the phrase anthropocentric (which means human-centred).

Rosenbrock (1977) observed that in designing new technology, designers are faced with a choice. A machine such as a computer can collaborate with the skill of the user, making it more effective and more productive, and allowing it to evolve into new skills in relation to new facilities and new theoretical insights. Alternatively, it is possible to use the computer in a way that rejects the user's skill and ability, and attempts to reduce to a minimum his/her contribution to the work process. The same comment can be made about robots, numerically-controlled machine tools, office automation, and other computer-based machines.

To demonstrate how computers could be used to support human skills, Rosenbrock initiated an interdisciplinary research project to develop a Flexible Manufacturing System in which people would not be subordinate to the machines. He established a research team consisting of one engineer, one computer scientist and one social scientist. The objective was to use social and psychological insights and knowledge, in a prospective rather than in a retrospective way, to design new technology.

The implication of this approach is that technological developments should not just be evaluated for their impact on people, but rather that technological ideas should be shaped by social science insights in order to produce a better outcome. This interdisciplinary project is fully described in Corbett (1985), Kidd (1988a; 1990a) and Rosenbrock (1989). One of the important insights that emerged from this work was an understanding of the nature of human-centred technology and the basis on which human-centred technologies and systems should be designed.

In recent years it has been argued that technical and economic benefits can accrue from making better use of people (see e.g. Craven, 1986). This argument is often based upon the idea that human skill is required as a means of compensating for the inadequacies of a system, i.e. the human is required to override the system when something goes wrong, or to correct computed data, or to compensate for poor quality information, or to do those tasks which for technical and economic reasons cannot be automated. This has led to a call for joint optimization of the human and technical aspects of systems to achieve systems that integrate human capabilities with those of machines (e.g. Kamall *et al.*, 1982). Such systems have been called hybrid systems (Karwowski *et al.*, 1988a; Rahimi *et al.*, 1988), but some people have mistakenly referred to them as human-centred systems (e.g. Craven, 1986).

Hybrid systems have been conceived as being a necessary step along the road to fully automated systems. Human-centred systems were conceived on a different basis. One of the primary aims was to change the path of technological development, such that fully automated systems would no longer be seen as the ultimate goal. Hybrid systems are designed on the basis that human skill is still required. A human-centred system should be designed on the principle that human skill is a desirable attribute, which should not be designed out of systems. Design of human-centred systems does not, therefore, proceed on the basis that human skill is still required, but rather on the basis that human skill should be required (Kidd, 1988a).

There is a clear divergence of philosophy relating to the issue of whether systems should be designed on the basis that human skill is a necessity or on the basis that human skill is desirable. The second of these encompasses the first, but not vice versa. Which option is pursued has fundamental implications for the technology that is developed. It is also relevant to the problematic situation that arose during the course of ESPRIT project 1199 among those people addressing the design of the human-centred robotic system.

## *ESPRIT project 1199*

ESPRIT (European Strategic Programme for Research and Development in Information Technology) is a ten-year research programme funded by the European Community on a shared cost basis with industry (Kidd, 1990b). ESPRIT project 1199, entitled 'Human-Centred CIM' was a research project which developed out of the work undertaken by Rosenbrock and his co-workers (Rosenbrock, 1989). Part of the ESPRIT project was concerned with developing working prototypes of the various software tools resulting from Rosenbrock's work.

The main purpose of the ESPRIT project was to build a number of demonstration systems. The objective was to prove that Computer-Integrated Manufacturing (CIM) systems which are based on the use and development of human skills and abilities can be economically more effective than CIM systems which are designed to minimize or to eliminate the role of people. Several implementations at user sites were undertaken as part of the project. One of these demonstrations involved testing prototype systems developed from the work of Rosenbrock and his co-workers. Further details about this ESPRIT project can be found in Ainger (1988, 1990), Kidd and Corbett (1988), Hamlin (1989), Havn (1989, 1990), Rosenbrock (1989), Kidd (1990b).

One of the planned ESPRIT demonstrations was a robotic cell for a factory engaged in the manufacture of small gas turbines. The factory in question is operated along conventional Tayloristic lines (see Rose, 1978). It was planned that the cell would operate on a very different basis (see below) within this traditional environment. This in itself was not a very wise move. Similar situations have resulted in the complete isolation of the alien manufacturing system from the rest of the factory (see Schott, 1990).

### Composition of the manufacturing cell

The robotic cell proposed for this site consisted of two computer-numerically controlled turning centres, a cell control computer, a work bench for inspection of machined parts, and other more minor items of equipment. The use of robots and other types of workhandling systems was proposed as a means of automating certain tasks.

It was proposed that the cell should be manned by skilled machinists rather than semi-skilled machine minders. These machinists were to be given responsibility for programming the turning centres and other automated equipment such as the robots. The machinists were also to be given responsibility for meeting manufacturing targets set by the factory planner and to be made responsible and accountable for the quality of their own work. This meant that both during the programming and operation of the turning centres, the machinists would need to be able to gain access to control panels mounted on the machine tools.

## *Development of initial design proposals*

The first step in the design process should be to define the major design decisions. This is not difficult in this case and in this respect the project did start off on the right track. The major design decisions are choice of robot type and the layout of the cell.

The next step should be to identify the relevant social science considerations before any design work starts. This did not happen on the ESPRIT project. A number of social science issues were eventually identified (i.e. safety, need for socialization, and allocation of functions), but not before the design options had started to be considered and opinions formulated.

The final step should be to generate feasible design options using the social science criteria in a prospective way. More detailed design work is then required to evaluate the feasible options. In this later phase of the design process further decisions can be made. More of the design details will also gradually begin to emerge at this stage.

In technical design, many technical and financial criteria are utilized by technologists. However, these criteria are rarely stated explicitly, but tend to be absorbed through training and experience. They form part of the tacit knowledge base of conventional design practice. Hence decision making during design involves a substantial amount of subjectivity and intuition (Kidd and Corbett, 1988).

When designing human-centred robotic systems, social science criteria should be used in the same way as these subjective technical and financial criteria. It is unfair and unreasonable for technologists to apply subjective criteria, while expecting social science experts to justify every proposal. This is one of the important messages of the human-centred approach to the design of technology: equality of social criteria with technical and financial criteria.

## Identification of relevant social science criteria

The first and most obvious consideration is that of safety (see Parsons, 1986; Rahimi, 1986; Rahimi and Hancock, 1986; Swedish Work Environment Fund, 1987; Karwowski *et al.*, 1988a, 1988b). In the UK there are fairly severe legal requirements placed upon employers to provide a safe working environment. These legal requirements are enforced by a Health and Safety Executive which has the power to order immediate shutdown of unsafe equipment.

The safety requirements for industrial robots are given in a set of guidelines developed by the UK Health and Safety Executive. At the time the project was being undertaken these guidelines had been published in draft form for industry comment (UK Health and Safety Executive, 1986). These guidelines are very thorough and detailed. As a result, robots in UK factories operate within fenced off areas to prevent unauthorized access. Numerous safety devices, interlocks, etc., have to be installed to ensure that the robots are deactivated should anybody enter the fenced off area. Only specially trained and authorized personnel are allowed within these areas.

The allocation of tasks is the next area that needs to be considered. The social science literature is full of discussions concerning allocation of tasks. A typical approach to this problem might involve an assessment of the attributes of both human and robot in order to achieve optimal completion of the tasks (Kamall *et al.*, 1982; Parsons and Kearsley, 1982). Another approach is a systems-based one which takes into account anticipated tasks, product design, allocation between people and robots and iterative improvement in product design (Ghosh and Helander, 1986).

Since the cell is a manned system and much work will have to be done in the vicinity of the cell, the noise emanating from the workhandling system is an important factor. Robot noise level must therefore be taken into account. The techniques used for human–robot interaction must also be considered. These include robot teaching via force torque sensors (Hirzinger, 1982; 1983), computer-based programming methods and teach pendants (Rahimi and Karwowski, 1990).

All of the above considerations are in fact fairly standard design issues. Industrial psychology, however, has identified a number of other important issues which are relevant to the design of this system. These can be established quite easily by examining any standard textbook on industrial psychology or the social psychology of work (e.g. Brown, 1954; Argyle, 1974).

Socialization has been established as one of the most obvious things that employees seek from their working environment. As one of the factors affecting quality of working life, provision and opportunities for social interaction should be taken into account when designing the system. Barriers to social interaction should also be considered. There is an obvious link here with robot noise levels.

Stress control is also another important issue relevant to the design of the system. This is a wide ranging factor which covers such things as the physical environment, job content, control over the work process, etc. There is an obvious link here with robot safety. Stress control is also linked to socialization and hence to the opportunities that exist for colleague support. Opportunity for colleague support is therefore a further issue that should be addressed.

Colleague support is linked to opportunities for socialization, but it is also related to the structure of the work group. Group structure is primarily concerned with size of work groups and communication and information structures. This, therefore, is another issue that should be considered.

In systems where robots are used training is clearly an important issue. Training is therefore linked with safety and stress control. However, training is also related to the need for self development which is another issue that industrial psychology has identified as being important to the quality of working life (Martin, 1983). Training of personnel is an issue that should be addressed, but in relation to the actual design of the system, the opportunity for on-the-job training is more important.

## Prospective social science design to generate feasible options

The role that the robot takes in the robotic cell was derived from the nature of the human-centred approach. Technology should be developed and used in a way that allows human skill and ability to be more productive and more effective. In this case the reason for using a robot was to free the machinists from the more routine tasks that need to be undertaken. This leaves the machinists with more time to undertake skilled work and cell management tasks. This has major implications for the overall layout of the cell and for the choice of robot.

Although one of the objectives of using robots was to free the machinists from more routine tasks such as loading and unloading the turning centres, etc., there were also other reasons for automating the workhandling tasks. Automated workhandling, for example, would also contribute towards reducing manufacturing throughput times.

A number of workhandling systems could be used for the tasks that need to be undertaken. In the human-centred robotic cell considered, human and robot have to work in close collaboration. This poses more severe safety problems than is normally the case with systems that are used to replace or to minimize the human element.

The transfer of parts in and around the cell can be achieved fairly simply by the use of a conveyor system. Such conveyor systems can vary from a simple manually controlled on/off type of conveyor, to a high precision automatic computer-controlled system. The more advanced type of conveyor systems incorporate facilities for the merging and diverging of paths, reversing, 90° take off, stop control, and precise presentation of items at desired speeds and time intervals. Such complex conveyor systems can of course be quite expensive if they are used extensively throughout the cell. From the safety perspective the large space requirements of conveyors make them a potentially significant hazard, but one that can be easily identified and dealt with.

Loading and unloading of the turning centres can be undertaken by dedicated workhandlers attached to the machines. These workhandling systems provide a limited number of movements. Thus any components must be placed within precise target areas at the right time if they are to function automatically and correctly. As the movements of such a device are restricted, cyclic and precisely timed, the hazards associated with their actions are easily observed and restricted to a limited area of operation. The operational restrictions of dedicated workhandlers make them less versatile than robots, but they are also less expensive.

The other option considered was various types of robot. These, being more versatile than conveyors and dedicated workhandlers, could be used both for loading and unloading the turning centres, and transferring parts around the cell. There are of course different categories of robot. Gantry and fixed-base robots were considered as options for this particular problem. These two types of robot, however, have different safety implications.

A workhandling system employing a gantry robot can be designed such that the robot can only enter the working environment at specified locations. Physical barriers can be fitted at these points to prevent access of the loading arm to the working area when there would be danger to personnel. A fixed-base robot on the other hand carries out its operations within an envelope, which must (at least in the UK) be fenced off to prevent unauthorized access. This type of robot, therefore, potentially places severe restrictions on the movement of personnel within and around the cell.

### *Allocation of tasks*

Task allocation in this design example is derived from the role of the robot, which is to free the machinist from routine work. This means loading and unloading the turning centres, and transferring materials around the cell. No other role was perceived for the robot.

This role, however, has implications for the design of the robot's workhandling tools. Some of the parts to be handled by the robot were rather flimsy metal rings of varying diameters. The thickness of these rings could be as small as 2 mm. Batch sizes varied from tens to hundreds, and the more flimsy rings had larger batch sizes. It was vitally important, therefore, if the robot was to fulfil its role of relieving the machinists from repetitive tasks, that the robot be supplied with grippers that could accommodate the complete part range. It was especially important that the robot be capable of handling the more flimsy rings since these were manufactured in relatively large numbers.

The requirement that robot noise level be considered eliminates pneumatically activated and driven robotic systems. This is especially relevant to the gantry system, because the cheaper type of gantry robot is pneumatic. The noise level resulting from venting air valves would not be acceptable in this manned system, where the machinists are working in close proximity to the robot. The noise would also be a barrier to effective socialization and communication. If a gantry robot is used, it should therefore be driven and activated by an alternative type of drive, for example, by electric servo-motors.

### *Socialization and associated issues*

The next requirement to consider is that of socialization and the associated issues of stress control, colleague support, group structure and opportunities for on-the-job training. All of these suggest a minimum manning level of two people.

It is known that group size can have important effects. Large groups tend to become fragmented into smaller, less effective and cohesive subgroups. Absenteeism is also more significant in larger working groups and satisfaction also seems to be lower. Smaller groups are therefore to be preferred. The recommended group size is between five and ten people but there is no way in which this recommended range can be achieved in the example considered. This suggests that the idea of creating a robotic cell of the form being considered was ill-founded and that a larger cell would have provided a better situation.

Group structure is another issue which should be taken into account. It is known that in simple problem-solving situations a highly centralized 'wheel' structure is better than 'circle' or 'all channel' structures. In more complex situations, however, circle or all channel structures are better (Argyle, 1974). In a system that is only manned by two people with relatively equal capabilities, there is no need for a supervisory figure. So-called boundary management tasks, i.e. dealing with the rest of the factory, can be shared. When considering the major design issues in this particular case, group structure appeared to have little or no importance.

### *Human–robot interaction*

The device used for human–robot interaction is an important issue. Some systems such as the teach pendant require close approach to the robot. The design of human-centred systems attempts to provide users with more than one way of undertaking tasks (Kidd, 1988a; Kidd and Corbett, 1988). For safety purposes it is desirable to keep human and robot apart as much as possible. For this reason it was proposed that robot programming facilities would be provided on the cell computer. A secondary programming tool such as a teach pendant was also proposed for the situations where programming needs to be undertaken in close proximity to the robot.

Given that the cell would be manned by two machinists and possibly one other person such as an apprentice, the relevant requirements, that is, opportunities for socialization, stress control, colleague support and on-the-job training, can, in principle, be met. However, the choice of robot and the layout of the system can hinder or hamper the achievement of these objectives.

### Cell layout

Socialization, stress control, colleague support and on-the-job training can be hindered, for example, if the machines are positioned the way shown in Figure 12.1, where the turning centres are positioned back to back. This layout has cost advantages because only one robot is required to serve both machines. The robot can also be used to effect parts transfer between machines. The two machinists are, however, separated from each other, and are also out of sight of each other. The only point in the system where they are in close proximity

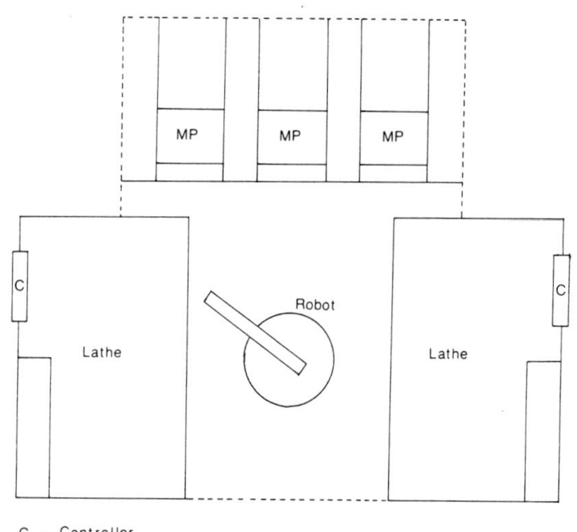


*Figure 12.1 Infeasible design option.*

to each other is in the area where the inspection bench and cell computer are located. If they wish to communicate with each other while working at the turning centres, one or both of them need to leave their workstations. This may not always be possible, depending on the work. This layout also suggests that it will not be easy to keep an overall eye on what is happening in the cell. The arrangement is in addition not very convenient if one of the machinists is providing training for the other (for example a new recruit). This arrangement is a good example of an infeasible design option which need not be considered further.

An improved situation can be brought about by the arrangement shown in Figure 12.2. A more intimate working environment can be created, however, by using the layout shown in Figure 12.3. This arrangement of the machines provides two points where the machinists are working in reasonably close proximity to each other. Communication is much easier, and the machinists are in close proximity to all the machines. It is thus easier to keep an eye on what is happening overall, and much simpler to satisfy the requirements for socialization, stress control, colleague support and on-the-job training. The disadvantage of this layout is that two fixed-base robots and a conveyor are required to serve the turning centres, as illustrated in Figure 12.3. This is clearly more expensive than the layout shown in Figure 12.1. Costs may be reduced, depending upon the price of the fixed-base robots used, by using a gantry robot as shown in Figure 12.4.

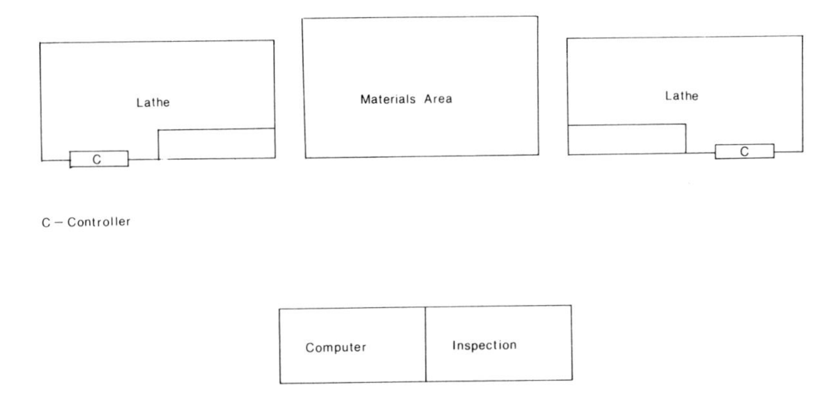


*Figure 12.2 Improved design option.*

Initial prospective application of social science considerations therefore suggests two feasible design options, those shown in Figures 12.3 and 12.4. Both provide safe working environments without the need to restrict movement of the machinists in and around the cell. The machinists are in close proximity to each other for most of the time and can easily support each other. It is also possible to keep an overall eye on what is going on in the cell.

The option shown in Figure 12.2 is much less satisfactory. This particular

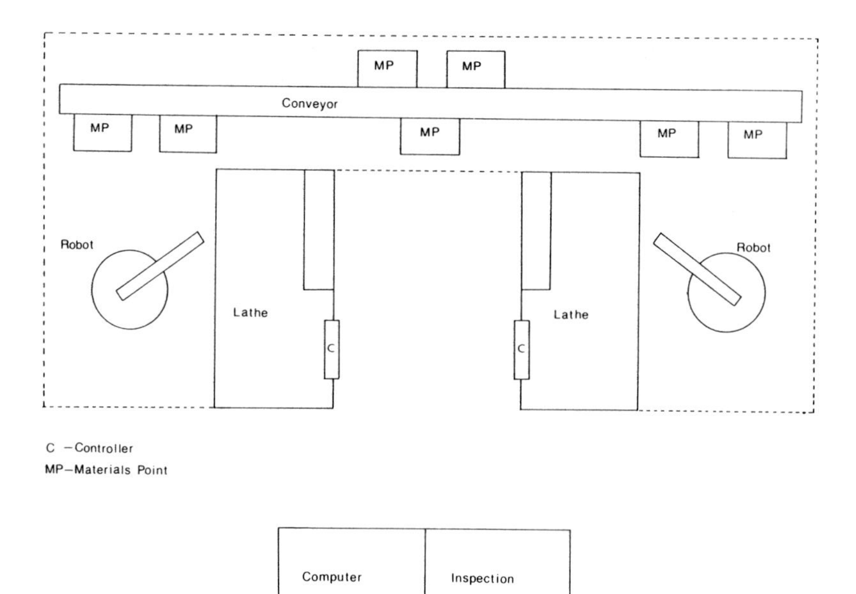


*Figure 12.3 Feasible design option.*

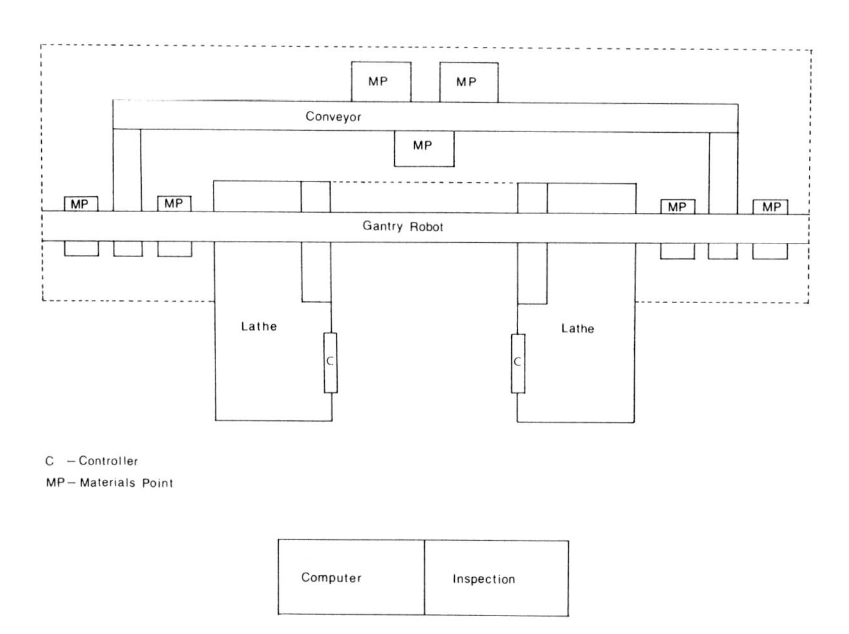


*Figure 12.4 Feasible design option using a gantry robot.*

option could employ two fixed-base robots, one serving each machine tool, or one gantry robot to serve both machine tools. Initial considerations suggest that the layout in Figure 12.1 is not satisfactory and should not be investigated further.

### Detailed design

Having established two feasible design options more detailed design work can then be undertaken. This aspect of the design process is less problematic. If one can generate feasible design solutions taking social science issues into account in a prospective way, then more detailed social science analysis can be done without much problem. The primary reason for this is that using social science in a prospective way requires technologists to internalize the judgement criteria that social science experts use. This is the essence of successful interdisciplinary design (Jones, 1981), but it is more difficult than asking other experts to make retrospective evaluations.

What follows is more routine and less creative. It is the convergent part of the design process in which the details of the design are built up. It is at this point in the design process that detailed checklists and more conventional social science methods of analysis can be used.

Thus a wide range of tools and techniques can be employed at this point. Computer simulations can be used to check that the manning levels are adequate and that the machinists are not overloaded. Material and part movements around the cell can be predicted and mapped to establish that the layout is satisfactory and that the robots can fulfil the desired function. Movements of personnel can also be predicted and mapped to establish if the machinists are not tied to their machines. Hazard analysis can be undertaken to identify safety risks so that appropriate safeguards can be devised.

## *Conclusions and discussion*

Quality of working life and humanization of work are becoming important issues in Europe. There is, therefore, no logic in developing and using technologies such as robotic systems in a way that attempts to replicate the skills of the people who will have to use the system, if this leads to unsatisfactory work. Moreover, the human-centred philosophy offers the potential of a better way in which to introduce new technologies. It can also avoid creating the need for skills that do not exist. It can result in systems which allow older skills to evolve into newer skills, while still leaving scope for the exercise of these older skills.

Although the future is unknown and largely impossible to predict, the philosophy of human-centred systems recognizes that the vision of the unmanned factory is nothing more than a dream. The unmanned factory is also a concept that is based upon Tayloristic thinking about the role of people in manufacturing (see e.g. Bolwijn *et al.*, 1986; Kidd, 1988b). It is also based on an outdated belief that competitive advantage can be derived primarily from investments in technology and automation (Kidd, 1989; 1990a; 1990c).

In Europe there are indications that the 1990s will see an expansion of interest in human-centred systems. Some of the ideas have already been applied with success in industry (Ainger, 1990), and efforts are being made to change research programmes such as ESPRIT to include the human-centred approach (Kidd, 1990d). There is nothing to be gained, therefore, from pretending that projects involving engineers and social scientists do not encounter problems and conflicts. There is, on the other hand, much to be gained by industry and researchers, by learning from the mistakes of those who have participated in such projects.

The work attempted by the group designing the human-centred robotic system was a response to the changing nature of manufacturing, but the work that was actually undertaken was not too successful from a human-centred perspective. While no design method could have resolved all the difficulties experienced on ESPRIT project 1199, it is clear in retrospect what should be done when designing a human-centred robotic system.

The first step in the design of a human-centred robotic system should be to define the major design decisions. In the case considered, the major design decisions are choice of robot type and the layout of the cell. The next step should be to identify the relevant social science considerations before any design work starts. This was not attempted until later in the project, after design work had already started.

It turns out that it is not difficult to identify the relevant social science issues. The main factors that should be considered include: (1) the underlying reasons for using robots and the role of the robots; (2) task allocation between people and robots; (3) safety; (4) stress control; (5) opportunities for colleague support; (6) opportunities for socialization; (7) group structure; (8) robot noise levels; (9) training; and (10) the nature of the devices used to program and control the robots.

The final step should be to generate feasible design options using the identified criteria in a prospective way. More detailed design work should then be undertaken to evaluate the feasible options. In this later phase of the design process further decisions can be made and more design detail will gradually begin to emerge.

What happened on ESPRIT project 1199 was that some technologists working on the project decided what the major design issues were. They came up with initial design proposals that were shaped by technical and financial judgement criteria. They then proceeded to develop the detailed design. At this stage they presented the design proposals to the social science specialists for retrospective evaluation. They did not, however, even stop to allow this evaluation to take place, and ploughed on with their work.

This paper has shown that it is not difficult to find a wide range of social science issues that should be considered in the design of a human-centred robotic system. It is not claimed that the list is comprehensive, just that it is possible to use the knowledge in a prospective way if given the chance.

One of the major problems with an interdisciplinary approach to the design of robotic systems is the way in which social science criteria and design methods are used. Social science considerations will not have a major impact if they are not used in a prospective way to guide the subconscious mental processes from which design solutions emerge.

It has rightly been said that the application of social science in a sequential way is unsatisfactory. When design options are shaped by technical and financial considerations and then retrospectively evaluated for the social science implications, the end result is invariably unsatisfactory. Social science issues tend to become secondary issues and major design changes are not made. This is because it may be too difficult to do so, or because the social science critique cannot be operationalized into an alternative design. These criticisms of the conventional social science design approach are, however, only secondary ones. The main point is this. If social science criteria are not used in a prospective manner, in the same way that technical and financial judgement criteria are used, then new ideas, unforeseen possibilities, and new and original ways forward will not be forthcoming.

This last point is an extremely important one with implications for human–robot interaction. If robots are just perceived as reliable and faster replacements for people, then the vision of human–robot interaction is that of people serving the needs of robots and compensating for the inadequacies of these devices. If robots are seen as tools to support skilled workers, as a kind of assistant, then human–robot interaction becomes potentially much richer and deeper. In the former case, emphasis is placed on producing a user-friendly robot (e.g. Schulman and Olex, 1985), while in the latter case the emphasis should be on developing a cooperative robot (e.g. Scibor-Rylski, 1986).

Designing a cooperative robot is fundamentally different from designing a user-friendly robot. There is a need to think about how robotic systems are designed, what they do, and how they function. It is necessary to consider the deep system characteristics of the robot technology, in other words, the design should consider interfacing in depth.

This paper has primarily addressed the question of how to design human-centred robotic systems. The actual technology of the robotic system (i.e. how it functions) also needs to be addressed in more detail. This is an issue for future research.

## *References*

Ainger, A. W. S., 1988, Computer integrated manufacturing: the human centred approach. *Technology in Action, January, 29*–31. (Available from BICC Plc, Devonshire House, Mayfair Place, London W1X 5FH, UK).

Ainger, A. W. S., 1990, Aspects of an experimental industrial application of a human-centred CIM system. In *Proceedings IEE Colloquium on The Human Factor in CIM* (London: IEE), pp. 3/1–3/5.

Argyle, M., 1974, *The Social Psychology of Work* (Harmondsworth: Penguin).

Bolwijn, P. T., Boorsma, J., van Breukelen, Q. H., Brinkman, S. and Kumpe, T., 1986, *Flexible Manufacturing. Integrating Technological and Social Innovation* (Amsterdam: Elsevier).

Brödner, P., 1982, Humane work design for man–machine systems: a challenge to engineers and labour scientists. In *Proceedings of IFAC Conference on Analysis, Design and Evaluation of Man–Machine Systems* (Oxford: Pergamon), pp. 179–85.

Brown, J. A. C., 1954, *The Social Psychology of Industry* (Harmondsworth: Penguin).

Corbett, J. M., 1985, Prospective work design of a human-centred CNC lathe. *Behaviour and Information Technology,* **4,** 3, 201–14.

Craven, F. W., 1986, A human-centred turning cell. In Lupton, T. (Ed.) *Proceedings 3rd International Conference on Human Factors in Manufacturing* (Kempston: IFS Publications Ltd).

Ghosh, B. M. and Helander, M. G., 1986, A systems approach to task allocation of human–robot interaction in manufacturing. *Journal of Manufacturing Systems,* **5,** 1, 41–9.

Hamlin, M., 1989, Human-centred CIM. *Professional Engineer,* **2,** 4, 34–6.

Havn, E., 1989, CIM and integration of work. In Halatis, C. and Torres, J. (Eds) *Computer Integrated Manufacturing, Proceedings of the 5th CIM Europe Conference* (Kempston: IFS Publications Ltd), pp. 145–53.

Havn, E., 1990, Designing for cooperative work. In Karwowski, W. and Rahimi, M. (Eds) *Ergonomics of Hybrid Automated Systems II* (Amsterdam: Elsevier), pp. 35–42.

Hirzinger, G., 1982, Robot teaching via force-torque sensors. In Trappl, R. (Ed.) *Cybernetics and Systems Research, Proceedings of the 6th European Meeting on Cybernetics and Systems Research* (Amsterdam: North-Holland).

Hirzinger, G., 1983, Direct digital robot control using a force–torque sensor, *Proceedings IFAC Symposium on Real Time Digital Control Applications,* Guadalajara, Mexico.

Jones, J. C., 1981, *Design Methods. Seeds of Human Futures* (Chichester: Wiley).

Kamall, J., Moodie, C. L. and Salvendy, G., 1982, A framework for integrated assembly systems: humans, automation and robots. *International Journal of Production Research,* **20,** 4, 431–48.

Karwowski, W., Parsaei, H. R. and Wilhelm, M. R. (Eds), 1988a, *Ergonomics of Hybrid Automated Systems I* (Amsterdam: Elsevier).

Karwowski, W., Rahimi, M., Nash, D. L. and Parsaei, H. R. (Eds), 1988b, Perception of safety zone around an industrial robot. In *Proceedings of the Human Factors Society 32nd Annual Meeting* (Santa Monica: Human Factors Society), pp. 948–52.

Kidd, P. T., 1988a, The social shaping of technology: the case of a CNC lathe. *Behaviour and Information Technology,* **7,** 2, 193–204.

Kidd, P. T., 1988b, Human and computer-aided manufacturing: the end of Taylorism? In Karwowski, W., Parsaei, H. R. and Wilhelm, M. R. (Eds) *Ergonomics of Hybrid Automated Systems I* (Amsterdam: Elsevier), pp. 145–52.

Kidd, P. T., 1989, Systems based approaches to CIM: questions of method, competitiveness and profitability. In *Proceedings IEE Specialist Seminar on The Systems Engineering Contribution to Increased Profitability* (London: IEE), pp. 7/1–7/7.

Kidd, P. T., 1990a, Human factors in CIM: a European perspective. In *Proceedings IEE Colloquium on The Human Factor in CIM* (London: IEE), pp. 2/1–2/7.

Kidd, P. T., 1990b, Human factors, CIM-Europe and the ESPRIT research programme. *International Journal of Industrial Ergonomics,* **5,** 105–12.

Kidd, P. T., 1990c, Organisation, people and technology: towards continuing improvement in manufacturing. In Faria, L. and van Puymbroeck, W. (Eds) *Proceedings 6th CIM Europe Conference* (London: Springer), pp. 387–98.

Kidd, P. T. (Ed.), 1990d, *Organisation, People and Technology in European Manufacturing. Interdisciplinary Research for the 1990s. A Report Prepared by Cheshire Henbury for the European Commission FAST Programme* (Brussels: Commission of the European Communities).

Kidd, P. T. and Corbett, J. M., 1988, Towards the joint social and technical design of advanced manufacturing systems. *International Journal of Industrial Ergonomics,* **2,** 305–13.

Martin, T., 1983, Human software requirements engineering for computer controlled manufacturing systems. *Automatica,* **19,** 755–8.

Parsons, H. M., 1986, Human factors in industrial robot safety. *Journal of Occupational Accidents,* **8,** 25–47.

Parsons, H. M. and Kearsley, G. P., 1982, Robotics and human factors: current status and future prospects. *Human Factors,* **24,** 5, 535–52.

Rahimi, M., 1986, Systems safety for robots: an energy barrier analysis. *Journal of Occupational Accidents,* **8,** 127–38.

Rahimi, M. and Hancock, P. A., 1986, Perception-decision and action processes in operator collision avoidance with robots. *Proceedings of the 19th Annual Conference of the Human Factors Association of Canada,* Richmond (Vancouver), pp. 119–22.

Rahimi, M. and Karwowski, W., 1990, A research paradigm in human–robot interaction. *International Journal of Industrial Ergonomics,* **5,** 1, 59–71.

Rahimi, M., Hancock, P. A. and Majchrzak, A., 1988, On managing the human factors engineering of hybrid production systems. *IEEE Transactions on Engineering Management,* **5,** 4, 238–49.

Rose, M., 1978, *Industrial Behaviour. Theoretical Developments Since Taylor* (Harmondsworth: Penguin).

Rosenbrock, H. H., 1977, The future of control. *Automatica,* **13,** 389–92.

Rosenbrock, H. H. (Ed.), 1989, *Designing Human-Centred Technology: a Cross-Disciplinary Project in Computer-Aided Manufacturing*. (London: Springer).

Scibor-Rylski, M., 1986, 'Yes-man'—a cooperative robot workstation. In Lupton, T. (Ed.), *Human Factors* (Kempston: IFS Publications Ltd), pp. 65–14.

Schott, E. S., 1990, Flexible work practices for flexible manufacturing systems. In *Proceedings IEE Colloquium on The Human Factor in CIM* (London: IEE), pp. 6/1–6/3.

Schulman, H. G. and Olex, M. B., 1985, Designing the user-friendly robot: a case history. *Human Factors,* **27,** 1, 91–8.

Swedish Work Environment Fund, 1987, *Robot Safety. A Manual on the Art of Creating Safe, Practical Workplaces in FMS Systems.* (Stockholm: Swedish Work Environment Fund).

UK Health and Safety Executive, 1986, Industrial Robot Safety. Draft for comment (Bootle: UK Health and Safety Executive).

# *Chapter 13*
# *Human error reduction strategies in advanced manufacturing systems*

**B. Zimolong and L. Duda**

*Department of Psychology, Ruhr University Bochum, Postfach 10 21 48, 4630 Bochum 1, Germany*

**Abstract.** Error and reliability of human operators in advanced manufacturing systems is the focus of this chapter. With a continuing increase in incorporating robots into manufacturing systems, there is a need for an accurate statistical method of incident and accident reporting. Human factors issues of robot accident causation and mitigation are discussed as important causes of robot stoppage. Other contributing factors such as invisible danger zones and unpredictable robot motions are also highlighted. Various strategies that may be used to reduce human error in complex robotic systems have been reviewed. The last part of this chapter explains how workstation design, task/job allocation and management styles may help improve robot system reliability and safety.

## *Introduction*

The Robotics Institute of America (RIA) defines a robot as a manipulator designed to move material, parts, tools, or specialized devices through variable programmed motions for the performance of a variety of tasks. The Industrial Robot (IR) is controlled by a software program that enables a robot moving at various speeds along many axes (up to eight) to reach a predetermined location.

Robots are capable of storing and executing the program data needed to perform a more or less complex sequence of motions that allows them to interface with other machines and humans. According to the complexity of the tasks undertaken, four types of robot are distinguished:

1. pick-and-place units (nonservo-robots) only allow for two positions per axis, namely starting point and end point of a motion sequence that typically remains unchanged over a longer production period;
2. a servo-controlled robot can, in addition, execute different programs that guide it to certain movements and stops anywhere within the machine's specified limit;
3. point-to-point robots are initialized with a basic program but allow for further modifications during the later course of operation; and
4. continuous robots in contrast are programmed on a time basis rather than fixed coordinates, thus allowing for a great variety of smooth continuous motions (Bullinger *et al.*, 1987).

Intelligence characteristics are fundamental to future robot developments. These are the methods of teaching and communicating with the robot, sensing and decision making abilities, and learning and adapting abilities. For instance, robots with vision capabilities are needed for inspection purposes, while a robot with no sensing ability is sufficient to perform a predetermined, repetitive task.

The rapid growth of robot distribution in the last decade is illustrated in Table 13.1. It should be noted that the Japanese figures are somewhat problematic because of the differing definition of what is counted as an IR. Thus Busch and Imken (1986) pointed out that while the British Robot Association estimated a population of 16 500 Japanese robots at the end of 1986, other sources calculated 44 000 or even 70 000 units. However, all authors agree that Japan's worldwide number one position (with regard to absolute frequency) is unquestioned.

*Table 13.1 Worldwide robot population in the 1980s.*

| Country | End of 1983[1] | End of 1984[2] | End of 1989[3] |
|---|---|---|---|
| Japan | 16 500 | 30 000 | 180 000 |
| USA | 8 000 | 15 000 | 37 000 |
| FRG | 4 800 | 6 000 | 22 400 |
| Sweden | 1 900 | 5 000 | 3 500 |
| Italy | 1 800 | 1 500 | 10 000 |
| England/UK | 1 750 | 2 000 | 5 900 |
| France | 1 500 | 2 000 | 9 000 |

*Sources:* (1) Busch and Imken (1986), (2) Wolovich (1987), (3) IPA (1990).

In the same manner Wolovich (1987) reported an estimated number of 30 000 IRs in Japan at the end of 1984 while noting that this figure excludes an almost equal number of more rudimentary, non-servo-controlled, pick-and-place manipulators. He also stressed that while Japan and the USA account for almost two-thirds of the worldwide robot population, it is Sweden that 'ranks first, by far, in the number (about 25) of robots in use for every 50 000 segment of the population, making it the most highly automated country in the world from this point of view' (Wolovich, 1987, p. 10).

The most recent statistics concerning the number of IRs in use worldwide were made available by the Fraunhofer-Institut für Produktionstechnik und Automatisierung (IPA, 1990). Figures indicating the different tasks undertaken by IRs in West Germany were also provided (Table 13.2). For a consideration of the characteristics and consequences of changes in organizational structure and management caused by advanced manufacturing technology see Tynan (1985).

The tremendous development rate of the world robot population has necessitated the application of the principles of human factors in such areas as the role of the human in robotic systems, organizational and job-design issues and ergonomic design and safety principles (Noro and Okada, 1983; Ostberg and Enqvist, 1984). While Bullinger *et al.* (1987) address robots from a general

*Table 13.2 Tasks of industrial robots in West Germany (end of 1989).*

| Tasks of IR | Number |
|---|---|
| Tool handling | |
| Assembling | 4201 |
| Spot welding | 4055 |
| Path welding | 3790 |
| Painting and coating | 1542 |
| Deburring | 115 |
| Other | 671 |
| Workpiece handling | |
| Work-tool machines | 2302 |
| Die casting | 770 |
| Forging | 557 |
| Other | 3817 |
| Research, testing and training | 575 |
| Total | 22 395 |

*Source:* IPA (1990)

human factors point of view, this article aims at safety issues of robot and flexible manufacturing systems.

A robot system consists of three elements: man, the industrial robot and a communication network. The robot can be designed and programmed to perform a variety of material handling, positioning and processing tasks that can be of use for a number of work activities such as welding, painting and coating, assembling and handling of workpieces, or loading of other machines.

Humans involved in robotic systems can be classified into four groups:

1. programmers and setters who, in order to teach the required tasks to the robot and to initialize the unit, often work within the immediate motion range of the machine;
2. robot operators and inspectors who, depending on the performed job, more or less often interact with the robot during its automated mode of production;
3. maintenance and repair personnel who have to enter the work area for regular check-ups or due to breakdowns; and
4. any other intruder such as visitors, passers-by or curious persons.

The hazardous situations all of these might encounter are collision between human and robot, trapping of a part of the human body by the robot, unintended and unexpected movements of the robot, and the robot losing grip of a tool or the workpiece (Percival 1983; 1984). On a more detailed level, the safety risks personnel are exposed to can be determined by a task analysis. Programmers, for example, must first enter the data specifying the desired robot movements, usually with the help of a remote-control unit. This can be done by pure key programming, where each key or switch corresponds to a specific function or movement, or by key programming using the menu technique, i.e. most choosable commands are additionally specified by text and possible parameters appearing on a display. If text programming is used, no prompting by the menu is provided, rather the programmer independently enters and edits the relevant coordinates which are then integrated into the executive program and displayed. Finally the robot can be taught with a play-back routine where the teacher actually guides the robot arm prototypically to perform the desired motions which are then stored in the robot's memory.

Robot teach pendants are the primary device for interactive human–robot interfacing. Cousins (1988) provided a list of ergonomic shortcomings in the design of the pendants. The result of the 'Human Engineering Design Criteria Study' cited by Rahimi and Karwowski (1990a, p. 64) provided a good 'snapshot' view of the American National Standards Institute (ANSI) design standard of pendants to come. In order to execute and check exact adjustments, the programming as well as a trial run is usually undertaken in the direct vicinity of the robot. Thus possible dangers in the course of this activity are:

the programmer might enter a wrong direction, thereby causing the robot to move towards him/her;
the programmer might enter a speed which is too fast; and
the programmer might over or underestimate the speed at which a robot is working or misjudge the direction of its motion.

An appropriate method to carry out a safety-related task analysis is the Questionnaire of Safety Diagnosis (QSD) developed by Hoyos and co-worker (see, for example, Ruppert *et al.*, 1990). Results of a QSD study carried out in a robot system are reported in Hoyos and Zimolong (1988).

## *Hazards in robot systems*

### Incident and accident reports

Common experience certainly supports the face validity of linking human error and accident probability. Yet there is little quantitative, direct evidence available showing how errors have led to accidents in robot installations. This is partly due to the fact that uncorrected errors are relatively infrequent and accidents are rarer still, while on the other hand most error analysis has been directed at product defects as consequences, rather than at accidents (Parsons, 1986). Generally speaking, the occurrence of an error can lead to one of these consequences:

the error remains unnoticed;
the error can be compensated by the system;
the error leads to a machine breakdown and/or system stoppage; or
the error leads to an accident.

But not every human error that results in a critical incident will cause an actual accident, therefore the further distinction of the following outcome categories is appropriate (Swain, 1985):

*Unsafe incidents.* Any unintentional occurrence that may or may not result in injury, damage, or loss;
*Accident.* Unsafe event resulting in injury, damage, or loss;

*Damage incident.* Unsafe event, which only resulted in some kind of material damage;
*Near accident.* Unsafe event, in which injury, damage, or loss was fortuitously avoided despite a 'close call'; and
*Accident potential.* Unsafe events, which could have resulted in injury, damage, or loss, but, owing to circumstances, not even a close call was experienced.

Thus from a sequencing point of view (Zimolong and Hale, 1989) an accident is only one of the several outcomes of a man–machine interaction under hazardous conditions; near accidents and damage incidents are much more common.

The following summary of available data regarding critical incidents with industrial robots is partly based on an analysis performed by Jiang and Gainer (1987). The authors classified accidents gathered from four different sources:

1. Carlsson (1984, 1985) provided data from the computer file Information System for Occupational Injuries (ISA) maintained by the National Board of Occupational Safety and Health in Sweden. The circumstances of 36 accidents which occurred between 1979 and 1984 were described in great detail; Carlsson pointed out, however, that some injuries were missing and in some cases the definition of IR was questionable.
2. Nicolaisen (1986, 1987) also analysed a total of 18 accident reports from Sweden (not West Germany, as Jiang and Gainer (1987) assumed). Furthermore, he calculated the degree of risk to various groups of personnel using IRs (see Table 13.3).
3. Japanese Industrial Safety and Health Association (JISHA) (1983) reported on a survey covering 190 Japanese plants with 4341 robots (with 47.8% of higher level than playback robots). For the years 1978–1982 a total of 11 accidents and 37 'unsafe acts' were indicated (Nagamachi, 1986).
4. NIOSH (1984) investigated in detail the events surrounding the death of a worker in a die cast company in the USA.

Jiang and Gainer (1987) grouped those individuals injured by robots into three categories: robot operators, maintenance workers and programmers/teachers. Nicolaisen (1987) distinguished a fourth group, staff for clearing stoppages, whereas Seeger (1985) allowed for 'unknown' and 'only material damage' (see Table 13.3).

In a questionnaire survey by Gotoh (1985) with 1027 Japanese workers who interact with or work near robots, 46% indicated that they have experienced a close encounter with robots; of these, 36% said it was during teaching and

*Table 13.3 Groups at risk.*

| | | Accident distribution by group (%) | | |
|---|---|---|---|---|
| Source | Accidents | Robot operators/ line workers | Maintenance personnel | Programmers/ teachers |
| Gotoh (1985) | 472 | 34 | 30 | 36 |
| Jiang and Gainer (1985) | 32 | 72 | 19 | 9 |
| Seeger (1985) | 47 | 45 | 17 | 17 |
| Nicolaisen (1987) | – | 13 | 30 | 57 |

30% during repair, adjustment or inspection.

At first glance the figures indicating the degree of risk to various groups differ rather surprisingly from study to study. It should be noted, however, that no detailed descriptions of the robots under question were provided, whereas it seems highly likely that for example the complexity and sophistication of the machine will heavily influence the potential amount of programming as well as maintenance and repair work, thereby enlarging the hazard exposure of the relevant personnel groups (Socher and Wolowczyk, 1982). Thus such technical design differences as type of robot and task to be accomplished, type of programming mode, maintenance intervals and availability (stoppage times) might explain the dissimilar findings. In order to give an overview of some typical accident circumstances several examples grouped by worker classification are listed in Table 13.4.

*Table 13.4 Accident examples of operators, maintenance workers and programmers.*

| Accident group | Injury | Period off work |
|---|---|---|
| Machine operators | | |
| 1. Operator was to exchange device at a plant consisting of two units while the first unit was at rest. He failed to notice that the other unit was still in operation. | Hand | >7 days |
| 2. Operator tried to remove wrongly positioned workpiece while plant was running. Resumption following removal of stoppage. | Hand | 1–7 days |
| 3. Operator wanted to remove wrongly positioned workpiece but inadvertently activated an external switch. | Finger | 2 days |
| 4. Worker wanted to remove workpiece whilst plant was running because magazine was full. | Hand | >7 days |
| Maintenance workers | | |
| 1. Maintenance worker wanted to start up robot plant with the help of a colleague following repairs. Because of a misunderstanding the robot started up and squeezed hand. | Finger | 12 days |
| 2. Maintenance worker made a trial run at the robot following change of valves. Unexpected movement of one shaft since air was present in the system. | Hip, leg | >7 days |
| 3. Maintenance worker was in movement area of a robot with installation running (workpiece handling: insertion into a degreasing plant). | Bones | 12 days |
| Programmers | | |
| 1. During trial run of an IR the inspector was to hold still/counterstop the equipment during movement. | Finger | >7 days |
| 2. Whilst a machine in an IR installation was started up, the IR (programmed) also started up. | Arm | 10 days |

*Source:* Nicolaisen (1987)

## Available statistics

Available statistics of reported incidents in IR systems are summarized in Table 13.5. In a study by Jiang and Gainer (1987) the reported incidents were grouped according to effect (pinch and impact accidents) and degree of injury, with the results indicating that pinch-point accidents (trapping a part of the human body by the robot) seem to be of a more serious nature than impact accidents. In addition Nicolaisen (1985) listed the results of a 14-day survey of eight IR workplaces in Sweden dating from 1981. All in all 24 critical situations were recorded, with four leading to injuries (one of them resulting in a seven-day absence from work whilst the seriousness of the other ones can only be vaguely estimated due to lack of detailed information), and one causing material damage.

*Table 13.5 Incident types and accidents caused by IRs.*

| Source | Number and type of reported incidents | Injuries Total | Fatal No. (%) | NFLW No. (%) | NFNLW No. (%) |
|---|---|---|---|---|---|
| Jiang and Gainer (1987) | 32 accidents | 32 | 3(9) | 24(75) | 5(16) |
| Nicolaisen (1985) | 24 critical situations | 4 | – | 1(25) | 3(75) |
| Nagamachi and Anayama (1983) | >500 unsafe acts | 142 | – | 51(36) | 91(64) |
| Nagamachi *et al.* (1984) | 179 abnormal stoppage cases | – | – | – | – |

NFLW, Non Fatal Lost Work; NFNLW, Non Fatal No Lost Work.

Nagamachi and Anayama (1983) surveyed 1200 workers in six Japanese companies using robots and other advanced production systems. 41.5% of the employees had experienced near accidents, and 11.8% reported injury accidents with 4.2% resulting in loss of work time by causing a shutdown. In a follow-up study Nagamachi *et al.* (1984) analysed abnormal stoppage cases by studying maintenance reports and interviewing maintenance personnel over a period of one year at a company using 11 IRs. He reported 179 cases leading to a total stoppage time of 301 h. No figures regarding injuries were given.

Obviously the figures given in Table 13.5 are of only very limited use for a comparison due to lack of detailed information that would be needed to judge the reported frequencies in relation to common hazard criteria. The number of accidents that occurred cannot be interpreted without information on hazard exposure, i.e. the number of persons who worked in the hazardous area and the length of time they worked.

Ideally speaking one should in fact account for every chance of an accident occurring so the mean number of robot accidents/incidents should be calculated per possible encounter with a human. Of course this ought to be done separately for the different groups concerned (programmers, operators, maintenance personnel), whose specific risk potential—as noted before—supposedly varies depending on the kind of robot installed.

Because of extraordinary methodological difficulties involved in determining such accident potentials (Hoyos and Zimolong, 1988), the number of employees or work hours is often applied as a rough substitute for measures of hazard exposures, e.g. the 1000-man-rate. Yet many of the available studies do not report the information necessary to calculate such a ratio. Moreover, in some cases, neither the number of robots nor the time period covered is mentioned.

Some of the results reported by Parsons (1986) are similarly rudimentary. Sugimoto and Kawaguchi (1983) summarized a survey by the Japanese Industrial Robot Association. Over an unspecified time period (presumably sometime in the early to mid-1970s) three workers were killed, four or more injured, and there were 18 near accidents. An earlier report by Carlsson *et al.* (1979) covers a 30-month survey of 270 workers in the Swedish steel industry between 1976 and 1978. There were seven to eight accidents each year requiring sick-listing, a rate of about one accident per 45 robots per year.

The conclusion to be drawn from an extensive literature survey is that available data of reported incidents in IR installations are poor to meagre, i.e. only a few original reports on accidents, critical incidents and abnormal stoppage cases have been reported up to now. Authors often rely on second-hand information and on the reinterpretation of available statistics. There is also a lack of information on supplementary hazard data such as number of employees exposed, work hours of personnel in robot systems, or even number of robots in a work system. As a consequence, it is unclear how dangerous work with robots really is with respect to such basic concepts as number of accidents/injuries per 1000 employees, per one million work hours, or per number of robots.

## Accident causes

The 32 accidents scrutinized by Jiang and Gainer (1987) were grouped into four categories representing accident causes (human error, workplace design, robot design and other) with the possibility of an accident having more than one cause, thus summing up to a total of 40 (see Table 13.6).

The authors gave no explicit definitions of their causal categories, but apparently human error was referred to if a worker was not simply engaged in the sequence of events surrounding the accident, but made some sort of mistake which in fact can be identified as at least partly responsible for causing the injuring contact between robot and human. Examples are a 'robot started up by colleague while maintenance worker in the work area' or 'robot started up as a result of operator mistakenly hitting switch while in the robot area'. Both of these were also classified under 'workplace design', indicating that this category was used if the accident happened due to poor layout of the work area, e.g. insufficient guarding/fencing of the robot envelope and unnecessary interfacing of programmers and operators with the machine. 'Robot design', on the other hand, was applied as a causal group if the technical layout of the machine itself led to an accident. For example, if due to a lack of an appropriate valve, compressed air put an already stopped robot into motion again.

In 24 of the 32 cases the cause-and-effect analysis led to a successful determination of a definite accident cause. Although human error is the primary cause in 13 of these 24 incidents it is emphasized that in all of these cases adequate safeguarding would have prevented the presence of the injured person. Similarly 'strict adherence to suggested standards or guidelines could have prevented, in nearly all cases, the accidents presented in this study . . . Clearly the largest cause of accidents is inadequate, poor or non-existent safeguarding methods' (Jiang and Gainer, 1987, pp. 40–1).

In order to accomplish a comparative overview at least to some extent, we tried to group the data provided by Nicolaisen (1985), Nagamachi *et al.* (1984) and Seeger (1985; cited in Derichs, 1986) into the causal categories 'human error', 'technical design' (which includes both workplace and robot design), and 'work organization'. The last of these indicated that an accident happened because insufficient planning of the running work process resulted in workers having to interfere while the installation was in operation, e.g. in order to readjust some robot component, program the machine or remove remains of cast metal from the workpiece. It should be noted, however, that Seeger supplied no information about human-error related causes, whereas Jiang and Gainer as well as Nagamachi *et al.*, did not refer to failures within the area of work organization.

Contrary to the common notion that human errors lead to 80–90% of all accidents, and technical and organizational factors account for the rest, the reverse is true for accident causes of IR. If Table 13.6 is examined closely, one can see that the share of human-error causes ranges from 12 to 40%, while the technical design factors play a part in at least 66% and up to more than 90% of the reported incidents.

*Table 13.6 Accident causes of IR. Note: More than one cause per accident.*

| | | Accident causes (%) | | | |
|---|---|---|---|---|---|
| | | | Technical design | | |
| Source | No. of accidents | Human error | Workplace design | Robot design | Work organization |
| Nagamachi *et al.* (1984) | 179 | 12 | 88 | | – |
| Nicolaisen (1985) | 24 | 21 | 92 | | 30 |
| Seeger (1985) | 47 | – | 26 | 40 | 34 |
| Jiang and Gainer (1987) | 40 | 41 | 63 | 22 | – |

Nicolaisen (1987) listed weak points of robots in the areas of design (again including both robot and workplace design), planning (with regard to spatial layout, work organization and machine linkage as well as safety devices), and operational procedure, where the last of these summarized inadequate handling of weak points and unsafe working procedures as well as poor training of workers (see Table 13.7).

*Table 13.7 Weak points in robot safety.*

Design
1. Simple errors in control lead to dangerous system states
2. Unsafe gripper design (especially with power failure)
3. Inadequate strength of cables/hoses; poor laying
4. Component failure (mechanical) leads to dangerous system states (valves, position sensors)
5. Inadequate protection against environmental influences (weld spatter, dust, swarf, temperature, electromagnetic radiation)
6. Failure of basic safety devices (e.g. emergency shutdown switch)
7. No protection against unintentional activation of operating elements (knocking against, leaning on, dropping)
8. Poor ergonomic design (robot, hand-programming device, operating desk) increases the likelihood of incorrect operation

Layout
1. Poor spatial arrangement (confusion; possibility of collisions)
2. Poor organization of work (particularly when clearing stoppages and programming)
3. Unsafe or confused linkage (interfaces between individual machines)
4. Inadequate safety devices (faulty emergency shutdown circuit; insufficient guards: gaps, too low, too close to hazard points)

Operational procedure
1. Inadequate removal of weak points (no measures set up for removal; no feedback to design and layout)
2. Permitting working procedures which are counter to safety (particularly during stoppage clearance)
3. Inadequate training of personnel

*Source:* Nicolaisen (1987)

Additionally he stressed the need for cooperation among the concerned groups (manufacturers, first users, designers etc.) as well as the necessity of keeping in touch with future developments in robot technology in order to overcome the correlating upcoming safety problems.

## *Incidents in advanced manufacturing systems*

### Technical availability

In the metal-working industry modern Computer Numerical Control (CNC) lathes are distinguished from CNC workstations, which allows for different jobs such as drilling, cutting, grinding etc., thereby reducing expensive transportation and tool-change times. If several of these workstations are linked via a central transport system, this is regarded as a Flexible Manufacturing System (FMS). In order to specify the tasks performed at CNC workplaces, Konradt and Zimolong (1990), using a questionnaire, analysed the time proportions of operational tasks at 11 conventional CNC machines and eight CNC workstations located at four plants in Germany. The obtained percentages for different task groups are shown in Table 13.8.

At the CNC machines the largest share is held by supervisory control tasks

*Table 13.8 Task-time analysis of 11 conventional CNC machines and 8 CNC workstations. The 8-hour shift equals 100%.*

| | Percentage of task time | |
|---|---|---|
| Task of operators | CNC machines | CNC workstations |
| Supervisory control | 25 | 17 |
| Fitting | 18 | 18 |
| Inspection | 16 | 20 |
| Handling | 10 | 12 |
| Program correction | 10 | 6 |
| Programming | 8 | 19 |
| Repair | 7 | 4 |
| Maintenance | 6 | 4 |

*Source:* Konradt and Zimolong (1990)

that sum up to a quarter of total working time, whereas at the CNC workstations inspection tasks amount to the largest single group with 20%. The time spent with programming tasks is more than twice as high at the workstations (19%) compared with the CNC machines, while there is no difference with regard to fitting tasks (18% at both workplaces). Maintenance and repair work sum up to 14% at the CNC machines but only 8% at the more complex workstations. No direct computations of unsafe incidents (near accidents, accidents, damage incidents) are available with respect to FMSs. Instead an indirect estimation of risk must be undertaken. It is well known that preventive maintenance and repair tasks hold the most risks for FMS personnel. Since these tasks diminish with increasing technical reliability of the system, the computation of the system's reliability can serve as an indicator of the hazard potential still present to the workers.

According to Büdenbender and Scheller (1987) the reliability and productivity of FMSs can be characterized by two figures: technical availability and actual utilization. Technical availability is the complete production capacity (= 100%) reduced by technically- and hardware-caused stoppage time, e.g. burn-in phase, program transfer, tool change fault, gear shift fault, breakdown of workpiece supply. If one further subtracts the organizationally-caused stoppage time (e.g. missing raw material, cleaning) the final result is the amount of 'actual utilization' time. Availability data of several FMS studies are given in Table 13.9.

Büdenbender and Scheller (1987) recorded the relevant data in two plants in Germany manufacturing engine parts over a period of 338 and 360 h, respectively. In a study by Wiendahl and Springer (1986) three FMSs were observed, with corresponding figures being calculated for two of them. A survey among 17 users of FMSs in five West European countries was reported by Shah (1987), indicating a range of actual utilization between 67 and 95%. A number of availability figures ranging from 67% (for a complete workstation) up to 95% (for a single control unit) are summarized by Schneider and Diehl (1988), who themselves recorded data from 76, 77 and 79 machines, respectively, in the years 1984 to 1986 and calculated an overall average of 96% for technical availability.

*Table 13.9 Technical availability and actual utilization of Flexible Manufacturing Systems (FMSs).*

| Study | Number of FMSs observed | Total time observed (h) | Technical availability (%) | Actual utilization (%) |
|---|---|---|---|---|
| Büdenbender and | 1 | 338 | 91 | 84 |
| Scheller (1987) | 1 | 360 | 93 | 84 |
| Wiendahl and | 1 | 156 | 75–92 | 55–76 |
| Springer (1986) | 1 | 83 | 70–98 | 64–92 |
| Shah (1987) | 240 | – | – | 67–95 |
| Schneider and Diehl (1988) | 76–79 | – | 96 | – |
| Vossloh (1988) | 28 | – | 85–98 | – |
| Reithofer (1987) | 18 | 4.000 | 89–98 | 66–78 |
| | 9 | 2.454 | 98 | 88 |

A similar figure is given for 102 and 43 machines, respectively, in workstations observed in 1985 and 1986. Vossloh (1988) estimated that by then CNC lathes had reached a level of 85 to 98% technical availability, while workstations ranged from 75 to 90%. Apparently these figures were supported by his own recordings of 28 lathes over 2.5 years as well as by analyses undertaken by other authors. Based on an extensive study with data recorded in a number of companies Reithofer (1987) supplied—among other findings—a detailed account of breakdown figures for different machine components as well as whole workstations. Not surprisingly he noted that as the number of functional units increases the average technical availability decreases from up to 100% for a single component group to 80% for the complete workstation. Similar figures are calculated when FMSs with different degrees of complexity are compared: the more complex the plant, the lower its technical availability (e.g. dropping from 98 to 88%). On the other hand, the actual utilization rate (at least in those cases where the appropriate numbers are provided) can be higher for more complex units (e.g. rising from 66 to 78%) due to a decrease in organizationally-caused stoppage time from 18 to 6%.

In summary it can be stated that over the last decade the technical progress achieved in the planning, design and set-up of FMSs has led to a high level of technical availability, whereas at the same time the probability of breakdowns increases as the number of functional units increases.

## Causes of stoppage

The most frequent causes of stoppage observed by Büdenbender and Scheller (1987) are listed in Table 13.10. A further analysis of the technically-caused stoppages revealed that 48% of the technically-caused and even 62% of the organizationally-caused stoppages lasted less than 10 min, amounting to only 13 and 16%, respectively, of the total technical and organizational stoppage time. Similar figures are given for the second plant.

In their summary Büdenbender and Scheller stressed the fact that direct technical stoppages only summed up to a 2% loss of total production capacity,

*Table 13.10 Stoppage causes observed in FMS.*

| Causes of stoppage | Description |
|---|---|
| Technical | Component groups |
| | electrical |
| | pneumatic |
| | mechanical |
| | Control software |
| Systemtechnical | Warming up of the system at the beginning of the early shift |
| Organizational | Shortage of raw material |
| | Program improvement and correction |
| | Machine–workpiece coordination |

*Source:* Büdenbender and Scheller (1987).

indicating a high reliability of the observed machines. Apparently more attention should be paid to indirect technical stoppages that are associated with the linkage of several units (e.g. breakdown of workpiece supply) because such problems can easily lead to a stoppage of the complete system.

Wiendahl and Springer (1986) noted the breakdown potential of stoppages as an after-effect of linkage. Although the 40 technical stoppages of the linkage system recorded at one FMS over a period of 84.7 h lasted altogether only 2.8 h, they resulted in 6.8 h production loss. Again it was found that the large share (around 50%) of short failures (up to 10 min) summed up to only a small proportion (3–12%) of the total stoppage time. Almost all these short breakdowns were repaired by the line workers, with an overall average of one technical failure per hour. This led Wiendahl and Springer to the conclusion that technical availability and actual utilization of FMSs depend (at least in the near future) to a large extent on the qualification and engagement of the operating personnel. On the other hand a number of causes (resulting in one case in a total loss of 13.8% utilization time) could be traced back to conceptual and organizational failures, indicating the necessity for a stronger consideration of causes and consequences when planning and designing FMSs. Shah (1987) summarized his results by stating that the largest proportions of stoppage time are due to machine breakdown and organizational shortcomings (Table 13.11).

Besides the already reported results, Reithofer (1987) also recorded every machine interruption occurring at an FMS consisting of 18 units during one shift (observation time 697.5 h). Time lost due to short technical stoppages (i.e. those the machine operators could repair themselves) and organizationally-caused stoppages added up to 85% of the total stoppage time, with an average interval between interruptions of 40 min (see Table 13.11). Thus again it is obvious that automatic production without human presence and observation would not have been possible.

Furthermore, the degree of utilization is overestimated if short technical stoppages remain unrecorded. From a safety point of view it also becomes apparent that despite an automated manufacturing mode there remains a considerable portion of time during which workers interact with the hardware equipment. In the study of Reithofer (1987), for example, the direct interaction

*Table 13.11 Stoppage times of FMS: results of a survey by Shah (1987) among 17 companies using FMSs in five West European countries and by Reithofer (1987) observing 18 FMS in one West German plant.*

| | |
|---|---|
| Shah (1987) | |
| No. of units observed | 240 |
| Overall capacity | 100% |
| Actual utilization | 86% |
| Time lost (%) due to | |
| machine breakdown | 5 |
| organizational faults | 4 |
| control breakdown | 2 |
| transport system breakdown | 2 |
| handling device breakdown | 1 |
| Reithofer (1987) | |
| No. of units observed | 18 |
| Observation time 697.5 h | 100% |
| Actual utilization time | 75% |
| Time lost (%) due to | |
| organizationally-caused stoppages | 11 |
| short technical stoppages | 10 |
| technically-caused stoppages | 4 |

time of workers was 97.65 h for repair, which is 14% of the observation time of 697.5 h. Actually this interaction time can serve as a quantitative expression for the amount of work at a high-risk level due to the nature of the repair work under severe time constraints. Additional risks emerge from the complexity of the integrated manufacturing system; however, as mentioned before, no direct observations of accidents, near accidents, or critical incidents are available up to now to back up the practical evidence.

## *Human aspects of robot systems*

### Invisible danger zone and unpredictable motions

The implementation of industrial robots may reduce the potential for accidents involving workers in hazardous environments e.g., performing a welding task in an automobile production line. They may also introduce new hazards associated with the robot. The occurrence of various accidents and near misses indicates that robots in themselves, or in their operational process, are sources of potential hazards to humans and to equipment. The differences between robots and traditional automated machines are listed in Table 13.12. Dangers which originate from the interaction of human and IR are caused by the invisible danger zone, the unpredictability of motions, possible malfunctions in the physical structure or other control errors of the IR.

One of the important characteristics of robots is their flexibility in spatial movement. The difference between robots and traditional automated machinery is that the former are designed for 3-dimensional flexible moves compared with

*Table 13.12 Comparison of system characteristics of IR and conventional machines.*

| Industrial robots | Conventional Machines |
|---|---|
| Simultaneous movement in several, up to 'N' axes | Usually only simultaneous movement in few (1–2) axes |
| Free programmability of the speeds of every separate axis | Pre-set fixed speed |
| Free programmability of direction of movement of every separate axis (free spatial movement) | Fixed movement pattern (pre-set routes) |
| Very large range of movement compared with the volume of the appliances | Range of movement usually smaller than volume of machine |
| Range of movement overlaps the position of other machines, parts of buildings etc. | Scarcely any overlapping |

*Source:* Nicolaisen (1985)

the 2-dimensional fixed moves of machinery on a specific track. The robot's immediate motion sphere as well as the area out of the sphere are part of the danger zone, which workers tend to overlook due to the robot's ability to move the arm freely (see Table 13.12). While robots of all types require a wide area of movement, workers have difficulty in deciding how far the danger zone extends. In addition, the hazardous area is not limited to the far end of the motion sphere, but reaches beyond that line. Workers often receive injuries by thrown workpieces or material.

Unpredictability of movements is a consequence of different types of idle-time of the robot. Waiting can be divided into three classes: the complete stop with main power turned off; the conditional stop period, waiting for the next movement; and the deceptive stop in which a certain position is servo-controlled for a fixed period of time. Workers who confuse one type of idle-time with another are likely to step into the danger zone of the robot.

One Japanese fatal accident took place while the robot was under a conditional stop (Nagamachi, 1986). Other hazards are related to malfunctions of IRs. Malfunctions may originate from electronic, electric, hydraulic or mechanical elements of the physical structure of the robot. They may lead to unexpected movements or may pose a hazard in themselves, e.g. high voltage in the system. External noise can also cause abnormal movements of robots different from the programmed movements.

## Coping with hazards

There are few field studies of ergonomic issues in the human–robot interaction (Kemmer, 1984; Hoyos and Strobel, 1985; Karwowski *et al.*, 1988b; Rahimi and Karwowski, 1990b). One of the exceptions is the study by Nagamachi (1986). He investigated the aspects of safety distances between robots and workers as a function of motion speed and waiting times of the IR under which unsafe behaviour occurred.

One question was how workers would estimate the feasibility to pass under the mechanical arm of a commercially manufactured cartesian-coordinate robot. Subjects were required to estimate the feasibility with regard to different speeds of movement and the robot's three uniaxial motions *X, Y, Z*. As expected the results indicate that the speed of the *X*-axis (back and forth motion) tends to be underestimated and the feasibility overestimated, while the others (*Y*, up and down motion and *Z*, right and left) tend to be almost uniform in evaluation. The *X*-motion towards and away from objects, e.g. oncoming and disappearing cars in traffic, are by far the most difficult movements to perceive. Another problem was temporary stopping time of the robot and the risk of picking up a workpiece which dropped under the mechanical arm of the robot. Results are shown in Figure 13.1.

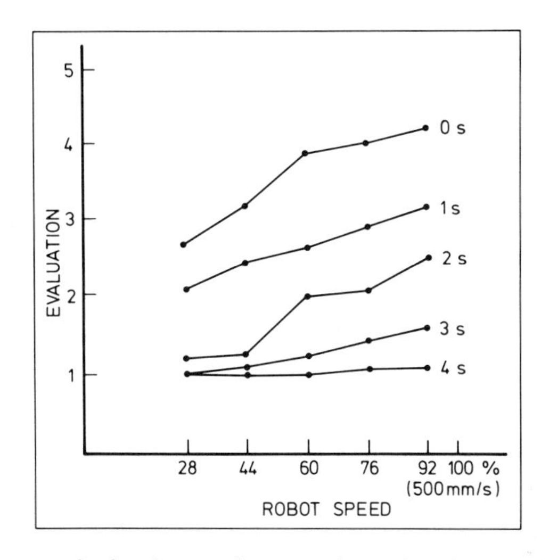


*Figure 13.1 Evaluation of safety distances between robot and worker as a function of speed and waiting times of robot (source: Nagamachi, 1986).*

Under the condition of no delay (continuous speed), for all speed conditions, subjects were aware of the danger. However, for speeds of 22 cm/s and less, subjects tended to have a more favourable attitude to picking up the workpiece. Under the condition of 3 s delay, at all speed levels, the tendency to pick up the piece was evident although some relative changes due to speed could be seen. Under the condition 4 s delay, speed hardly influenced subjects' risk assessment. Subjects' evaluation under this delay condition was '. . . able to pass under the arm of the robot quite easily' (Nagamachi, 1986, p. 13).

Karwowski *et al.* (1988b) also investigated robot idle-times during which the robot paused for 5, 10, 15 and 20 s intervals chosen in random order. The dependent variable of the experiment was the period of time the subjects waited before they decided that the robot stop was a malfunction rather than a programmed stop. Those subjects who viewed a simulated accident—the robot struck a mannequin and knocked it to the ground—did wait longer for their decision to enter the work envelope. The percentage time increase was about

33%, from a mean waiting time of 16.85 s to 25.10 s. Under both conditions the waiting time of several subjects was too short to avoid a collision when the robot would have started to move after the longest programmed pause of 20 s. From the results of both experiments, however, one cannot draw a conclusion as to what the optimum idle-time for the IR should be. Karwowski *et al.* (1988b) suggested that the programmed idle-time of robots be less than 16 s; however, as the results of Nagamachi (1986) showed, even a 4 s delay may be inappropriate. Probably human judgment of the possibility of a programmed stop or a malfunction of the IR depends on the distribution and variance of the idle-times observed and not so much on an absolute waiting time interval to be defined.

Nagamachi (1986) also investigated subjective safety distances of subjects in relation to waiting time and motion speed of the robot. The variation in waiting times of the robot (0 to 4 s) did not influence the evaluation of the safety distance, whereas a clear relationship between motion speed and safety distance could be established. Subjects approached closer to the robot as the speed was reduced and they stayed further away from the robot at higher speeds. In any case subjects approached too close, sometimes extremely close, thereby ignoring any safety standard. Subjects seemed unaware of the chance of an unexpected movement caused by a failure or breakdown of the robot, which in effect very seldom takes place.

Rahimi and Karwowski (1990b) showed that the selection of safe speed by workers for a robot depends on the size of the robot. This result is in accord with the notion that larger robots produce a hazard perception of a more dangerous motion in space. For the IRs under study, subjects preferred a maximum safe speed of 65 cm/s for the small robot (General Electric P 50) and 40 cm/s for the large robot (GE MH 33).

The findings clearly indicate human limitations while coping with hazards caused by robots. Although it is not quite clear if the results could be generalized to programmers, line workers and repair personnel due to limitations of the sample studied (subjects were undergraduate and graduate students), the findings call for technical and behavioural measures to prevent dangers, which originate from unexpected movements and underestimation of the danger zone.

## *Safety recommendations*

### Safety philosophies

Within the field of safety four safety goals are generally distinguished (Hoyos and Zimolong, 1988). The most desirable goal is naturally the elimination of the hazard itself, for example, substitution of high speed by low speed. If this is not possible, the potential danger should be kept separated from the human with the help of appropriate safety technology, e.g. fencing off the robot envelope to prevent direct contact between the operating robot and the worker. Another example is supervisory control of the robot from outside the danger zone. It is safest and most effective for the human to exercise supervisory control through computer-based networks, by designing an industrial robot system so that the operators are kept physically away from the robot. The concepts, models and applications of supervisory control are presented by Bullinger *et al.* (1987). However, potential injury to the set-up and service personnel continues to exist during maintenance and repair work.

If, however, the hazard cannot be eliminated, recourse can be taken to the direct protection of the human body. The individual worker might be forced to wear protective gear or to make use of other appropriate devices such as a safety plug which must be pulled off the door by the worker entering the danger zone so that a colleague closing the door by mistake cannot restart the robot.

Finally, safety psychology as a fourth set of accident prevention measures can be applied with the goal of influencing behaviour so that hazards will be avoided or controlled. Accordingly employees' qualification and danger awareness can be improved by information (e.g. warning signs indicating hazards) and training.

From an international perspective it becomes apparent that different strategies exist in terms of whether to rely on the individual worker's self-responsibility or rather impose adequate technical and organizational measures, thereby enforcing safe occupational behaviour. Occupational safety measures in the USA and Sweden are predominantly aimed at the individual worker. With the help of a warning sign, for example, the robot operator is reminded of his/her personal responsibility for consulting the instruction book and following its content (Derichs, 1986). In contrast most European and Japanese safety experts stress the need to establish adequate technical measures that will prevent an accident beyond the instance when a worker has made an error or willingly violated a safety regulation. 'The Japanese recognize that even with training and visual warning there will always be risk takers in any workplace, therefore, the Japanese standards call for a 'watchman' to prevent workers from entering the robot work area while the robot is in operation.' (Jiang and Gainer, 1987, p. 42).

When comparing national safety regulations Busch and Imken (1986) came to the conclusion that—despite congruencies in terms of the safety standards aimed at—a number of clear differences exist. In a publication from 1984 the RIA, for example, considered only unexpected and unintended robot movements as dangerous, while in the West German VDI regulations expected movements, too, are referred to as dangerous. Moreover the RIA stressed the user's responsibility and accordingly gave much credit to the training of workers. Correspondingly it suggested simple safety fences and warning signals that are meant to draw attention when the robot work area is entered without turning the machine off. The German regulations as well as the Japanese technical guidelines for industrial robots from 1983 on the other hand emphasize the management's responsibility. Therefore safety measures referring to the normal mode of operation concentrate on guarding access to the work area. In addition the Japanese guidelines determine content and duration of the workers' training.

Another example of such a safety strategy is the recent draft for the VDI regulations No. 2854 concerning the safety requirements for automated FMSs.

Priority in terms of space and order is given to the specification of technical measures such as fencing installations, type of control switches, or displays to support fault diagnosis in case of machine breakdown. Comparatively little attention is given to measures directly aimed at improving the users' safety behaviour. For instance, only some scarce remarks are made concerning the use and installation of warning signals and safety signs, and no details of workers' teaching and training are provided.

The VDI regulations might be valued as more detailed and precise in technical respects, whereas the RIA recommendation could be considered more profitable to the user due to its clear descriptions. However, because of the different basic safety philosophies it makes hardly any sense to evaluate any single set of instructions in an international comparison.

Generally speaking the safety behaviour of workers is affected by the following variables (Hoyos and Zimolong, 1988):

*Work design*, which includes technical and ergonomic measures applied in the area's work site, equipment and work environment. Their implementation will reduce the likelihood of unsafe action.

*Work structure*, consisting of topics such as task design, task allocation, responsibility and ergonomic layout that all affect work motivation, which in turn leads to the performance of safe (or unsafe) behaviour. Work structure encompasses organizational measures aimed at improving certain modes of operation as well as the general organizational structure of a firm. For example, informational deficits (e.g. on how to restart a system after repair) that can turn a decision process into a safety-critical situation can be counteracted by setting up information networks. Additionally, interest in questions regarding work safety can be activated by creating organizational structures that allow for participation, or by changing forms of cooperation, e.g. the establishing of safety circles (Hackman and Oldham, 1976; Hoyos, 1987).

*Instruction, information and training*, which in turn are closely connected with a number of individual psychological variables. Work safety is dependent on the extent to which employees act appropriately in safety-critical situations. The ability to choose the most appropriate way to act as well as the implementation of one's choice are important in this connection. The perception and recognition of hazards as well as the over- and underestimation of risks are not only influenced by experiences in previous work situations, but can be further shaped by training (Zimolong, 1985). Qualification which will enable the employee to avoid typical errors in critical situations can be directed at fresh and routine performances. Further means of training include the teaching of decision heuristics and the enhancement of bias awareness, which will facilitate decision-making processes (Slovic, 1987). Other key qualifications are the ability to cooperate within a firm and to coordinate work tasks successfully with the simultaneous control of the dangers resulting from the acute hazard potential.

In the following sections, recommendations adapted to the specific circumstances of human–robot interaction are grouped according to the aforementioned factors. Recommendations are adopted from Nicolaisen (1979),

Jürgens (1983), Kemmer (1986), Nagamachi (1986), Parsons (1986), Jiang and Gainer (1987) and Rahimi and Karwowski (1990a). For a discussion of education and training of engineers in automated manufacturing systems see Rathmill (1983). Safety recommendations for FMS are listed in Sheehy and Chapman (1988), Karwowski *et al.* (1988a), Karwowski and Rahimi (1989) and VDI (1989).

## Safety measures

The measures given here range from workplace design (e.g. safety fences) and robot design (e.g. gripper easily removable for repair) to suggestions concerning certain high-risk modes of robot operation such as the programming and teaching of an IR (Table 13.13).

### *Instruction, information and training*

Listed here are recommendations dealing with the installation of warning signs as well as some generally applicable training principles (Table 13.14). The more specific qualification and training requirements of personnel will of course depend on the kind of robot, the functions served by the robot, and the particular production methods employed.

### *Motivation and change of attitude*

Appropriate means of increasing the motivation to practise safe behaviour include incentives which must correspond to the employees' social and physical needs. Positive feedback to promote safe behaviour and negative feedback to reduce unsafe behaviour should help to increase the awareness and significance of safety at work. In addition, an employee will be motivated to pay attention to safety if work safety is propagated as a high-priority goal by his/her firm. Also, the management style practised by supervisors strongly influences employees' attitudes toward work safety (Kleinbeck *et al.*, 1990).

Safety recommendations applicable to industrial robots, which mainly affect motivation and positive attitude towards safety, include among others:

signs which should depict the hazardous situation and either (1) both the behaviour that would incur an accident and the accident outcome or (2) both the behaviour that would avoid an accident and the non-accident outcome;
severe penalties which can be imposed on any worker who violates robot safety procedures; and
information and training based on a regular schedule.

Motivation and positive attitudes towards safe behaviour severely depend on the efforts of the organization and management philosophy. No 'one best way' for the improvement of safety motivation can exist due to different incentive systems of the organizations. As a result, measures must be tailored according to the specific needs and philosophies of companies.

*Table 13.13 Safety recommendations applicable to industrial robots: work design and work structure.*

Safeguarding

1. The robot area should be enclosed by an iron fence to prevent intruders, with appropriate considerations regarding location and height of safeguarding
2. In the case of a robot line where it is difficult to set up a safety fence, there should be a system where the worker pushes a start button distant from the robot after setting the workpiece
3. The robot working area should be painted to distinguish it from safe areas
4. Shuttle mechanisms or feeder devices can be implemented to prevent direct contact between the operating robot and the worker

Starting and stopping

5. The door through which the robot working zone is entered should have an interlock safety plug: a worker who enters the danger zone must pull a safety plug off the door and pocket it, thereby turning off the power
6. For a machine where it is difficult to install a safety plug, a safety mat should be installed in front of it
7. Robots should be installed with a flexible mechanical stopping device
8. A signal should be installed to indicate whether the robot is waiting or not

Maintenance and repair

9. Maintenance personnel (and programmers) who must function within a robot enclosure might wear a device that would return a signal to the robot's safety sensor indicating human presence
10. The jig attached to the 'robot hand' should be able to be repaired safely and easily outside the fence
11. Each robotic work envelope into which a person might intrude should be visible from locations where others work

Programming and teaching

12. The teaching should preferably be under supervisory control from outside the robot envelope
13. If the programming has to take place in close vicinity to the robot, the worker engaged in teaching should turn a switch on the robot controller to 'teach' and hang his/her name tag on the switch
14. When the control is switched onto manual programming it has to be ensured that movements of the robot can only be initialized by the programmer with the help of the control keys. One possible way of ensuring this: the programmer should remove a safety plug at the door when going inside. The robot has such a device that inserting this safety plug into it enables the robot to be moved only in the teaching mode at teaching speed
15. Another worker should observe the first worker's teaching behaviour from outside the robot working area and assist him/her in his/her teaching task
16. The robot speed is reduced to a maximum of 16 cm/s
17. Only touch-type keys should be used on the control unit
18. The control unit must have an emergency switch-off as well as a programming button of the 'dead-man' type. The design of the teaching box should be the same for different makes

Error reduction

19. Interface design: both the software and hardware must be designed so error likelihood will be minimized
20. Test and evaluation to detect specific errors; analyses aiming at error prediction using for instance THERP or FTA
21. Safety device checking: all safety devices should be checked daily or at the start of each shift

*Table 13.14 Safety recommendations applicable to industrial robots: information and training.*

Information
1. Warning signs should depict the hazardous situation and flashing lights can be installed to indicate a robot is in the automatic mode of operation
2. Procedural information may have to be developed and tested in the first place; furthermore the information must be presented so that workers can understand it

Training
3. A person without special training in robots should not touch any robot or robot controller
4. Both initial and periodical training should include the following ideas: if the robot is not moving do not assume it is not going to move; if the robot is moving slowly, do not assume it will continue to move slowly; if the robot is repeating a pattern do not assume it will continue; and maintain a respect for what a robot is and can do
5. Simulation, e.g. a simulated maintenance man (a dummy) enters the robot's work envelope while it has paused and is struck when the robot starts up again
6. Refresher training, that could include both simulation or only information

# *References*

Büdenbender, W. and Scheller, T., 1987, Flexible Fertigungssysteme in der Praxis. *VDI-Z,* **129,** 10, 22–8.

Bullinger, H. J., Menges, R. and Warschat, J., 1987, GROSS—Graphisches Roboter-Simulationsprogramm. *Automatisierungstechnik,* **35,** 12, 476–82.

Busch, K. and Imken, B., 1986, Die Entwicklung der Robotertechnik und die Sicherheitsphilosophien in verschiedenen Industrieländern. In Rheinisch-Westfälischer TÜV (Hrsg.) *Industrieroboter und Arbeitssicherheit* (Essen: Verlag Rheinisch-Westfälischer TÜV), S.6–9.

Carlsson, J., Harms-Ringdahl, L. and Kjellen, U., 1979, *Industrial Robots and Accidents at Work* (Stockholm: Royal Institute of Technology).

Carlsson, J., 1984, *Robot Accidents in Sweden 1979–1983* (Solna, Sweden: ISA).

Carlsson, J., 1985, Robot accidents in Sweden. In Bonney, M. C. and Yong, Y. F. (Eds) *Robot Safety* (Bedford: IFS Publications Ltd and New York: Springer), pp. 49–64.

Cousins, S. A., 1988, Development of human engineering design standard for robot teach pendants. In Karwowski, W., Parsaei, H. R. and Wilhelm, M. R. (Eds) *Ergonomics of Hybrid Automated Systems I* (Amsterdam: Elsevier), pp. 429–36.

Derichs, H., 1986, Stand der Sicherheitstechnik und ihre Umsetzung in die Praxis aus der Sicht des Sicherheitsingenieurs. In Rheinisch-Westfälischer TÜV (Hrsg.) *Industrieroboter und Arbeitssicherheit* (Essen: Verlag Rheinisch-Westfälischer TÜV), S.14–20.

Gotoh, M., 1985, Occupational safety and health measures taken for introduction of robots in automobile industry. In Noro, K. (Ed.) *Occupational Health and Safety in Automation and Robotics* (London: Taylor & Francis), pp. 399–417.

Hackman, J. R. and Oldham, G. R., 1976, Motivation through the design of work: test of a theory. *Organizational Behaviour and Human Performance,* **16,** 250–79.

Hoyos, C. Graf., 1987, Einstellungen zu und Akzeptanz von unsicheren Situationen: Die Sicht der Psychologie. In Bayrische Rückversicherung (Hrsg.) *Gesellschaft und Unsicherheit* (Karlsruhe: Verlag Versicherungswirtschaft), S.50–65.

Hoyos, C. Graf. and Strobel, G., 1985, Das Gefährdungspotential des Programmierers von Industrierobotern. *Die Berufsgenossenschaft,* **4,** 194–8.

Hoyos, C. Graf. and Zimolong, B., 1988, *Occupational Safety and Accident Prevention. Behavioral Strategies and Methods* (Amsterdam: Elsevier).

IPA (Fraunhofer Institut für Produktionstechnik und Automatisierung), 1990, *Statistische Mitteilungen* (Stuttgart: IPA).

JISHA (Japanese Industrial Safety and Health Association), 1983, *Prevention of Industrial Accidents due to Industrial Robots* (Tokyo: JISHA).

Jiang, B. C. and Gainer, C. A., 1987, A cause-and-effect analysis of robot accidents. *Journal of Occupational Accidents,* **9,** 27–45.

Jürgens, G., 1983, Arbeitsschutz und Industrieroboter. *Sicher ist Sicher,* **34,** 7/8, 326–32.

Karwowski, W. and Rahimi, M., 1989, Work design and work measurement: implications for advanced production systems. *International Journal of Industrial Ergonomics,* **4,** 185–93.

Karwowski, W., Rahimi, M. and Mihaly, T., 1988a, Effects of computerized automation and robotics on safety performance of a manufacturing plant. *Journal of Occupational Accidents,* **10,** 217–33.

Karwowski, W., Rahimi, M., Nash, D. L. and Parsaei, H. R., 1988b, Perception of safety zone around an industrial robot. *Proceedings of the Human Factors Society—32nd Annual Meeting.* Anaheim, California, Vol. 2, pp. 948–52.

Kemmer, K.-H., 1984, Arbeitssicherheit beim Einsatz von Industrierobotern. *Die Berufsgenossenschaft,* **9,** 550–7.

Kemmer, K.-H., 1986, Kollege Roboter eine Gefahr? *Sicher ist Sicher,* **2,** 80–4.

Kleinbeck, U., Quast, H. H., Thierry, H. and Häcker, H., 1990, *Work Motivation* (Hillsdale: Lawrence Erlbaum).

Konradt, U. and Zimolong, B., 1990, Die Analyse von Diagnosestrategien bei Wartungs- und Instandsetzungsarbeiten in flexiblen Fertigungssystemen. Vortrag gehalten auf dem 37. Kongress der Deutschen Gesellschaft für Psychologie, September, Kiel.

Nagamachi, M., 1986, Human factors of industrial robots and robot safety management in Japan. *Applied Ergonomics,* **17,** 1, 9–18.

Nagamachi, M. and Anayama, Y., 1983, An ergonomic study of the industrial robot 1: The experiments of unsafe behavior on robot manipulation. *Japanese Journal of Ergonomics,* **19,** 259–64.

Nagamachi, M., Yukimachi, T., Anayama, Y. and Ito, K., 1984, Human factor study of industrial robot 2: Human reliability on robot manipulation. *Japanese Journal of Ergonomics,* **20,** 55–64.

Nicolaisen, P., 1979, Arbeitssicherheit. In Warnecke, H. J. and Schraft, R. D. (Hrsg.) *Industrie-Roboter* (Mainz: Krausskopf-Verlag), S. 263–274.

Nicolaisen, P., 1985, Occupational safety and industrial robots—present stage of discussions within the Tripartite Group on robotic safety. In Rathmill, K., MacConaill, P., O'Leary, S. and Browne, J. (Eds), *Robot Technology and Applications.* Proceedings of the 1st Robotics Europe Conference, Brussels, June 27–28, 1984, pp. 74–89.

Nicolaisen, P., 1986, Safety problems related to robots. Seminar on Industrial Robotics 1986, Brno, Czechoslovakia, United Nations Economic Commission For Europe.

Nicolaisen, P., 1987, Safety problems related to robots. *Robotics,* **3,** 205–11.

NIOSH, 1984, Fatal accident summary report: Die cast operator pinned by robot. Summary report 84–020. (Morgantown: NIOSH).

Noro, K. and Okada, Y., 1983, Robotization and human factors. *Ergonomics,* **26,** 10, 985–1000.

Ostberg, O. and Enqvist, J., 1984, Robotics in the workplace: robot factors, human factors and humane factors. In Hendrick, H. W. and Brown, O. (Eds), Proceedings of International Symposium on Human Factors in Organizational Design and Management (Amsterdam: Elsevier).

Parsons, H. M., 1986, Human factors in industrial robot safety. *Journal of Occupational Accidents,* **8,** 25–47.

Percival, N., 1983, Is robot technology safe? *Decade of Robotics,* 82–3.
Percival, N., 1984, Robot safety. *The Safety Practitioner,* **2,** 3, 20–4.
Rahimi, M. and Karwowski, W., 1990a, A research paradigm in human–robot interaction. *International Journal of Industrial Ergonomics,* **5,** 59–71.
Rahimi, M. and Karwowski, W., 1990b, Human perception of robot safe speed and idle time. *Behaviour and Information Technology,* **9,** 5, 381–9.
Rathmill, K., 1983, Time for a renaissance in education. *Decade of Robotics,* 88–93.
Reithofer, N., 1987, Nutzungssicherung von flexibel automatisierten Produktionsanlagen. In Milberg, J. (Hrsg.) *iwb Forschungsberichte* (Berlin: Springer), Bd. 10.
RIA (Robotic Industries Association), 1986, Proposed American National Safety Standard for Industrial Robots and Industrial Robot Systems (Dearborn, MI: RIA).
Ruppert, F., Hoyos, C. Graf., Hirsch, G. and Broda-Kaschube, B., 1990, Sicherheitsdiagnosen mit dem Fragebogen zur Sicherheitsdiagnose. In Hoyos, C. Graf. (Hrsg.) *Psychologie der Arbeitssicherheit* (Heidelberg: Asanger), S.48–59.
Schneider, J. and Diehl, G., 1988, Diagnose in automatisierten Fertigungseinrichtungen—Anforderungen, Verfahren und zukünftige Möglichkeiten. *Instandhaltungspraxis 1988.* Forum 4./5.Mai 1988 Frankfurt (Düsseldorf: VDI-Gesellschaft Produktionstechnik (ADB)).
Seeger, O. W., 1985, *Robotertechnik* (Landsberg: Ecomed).
Shah, R., 1987, Erfahrungen europäischer CIM-Anwender. *VDI-Z,* **129,** 1, 34–43.
Sheehy, N. P. and Chapman, A. J., 1988, The safety of CNC and robot technology. *Journal of Occupational Accidents,* **10,** 21–8.
Slovic, P., 1987, Perception of Risk. *Science,* **236,** 280–5.
Socher, K. and Wolowczyk, P., 1982, Gefährdungen beim Einsatz von Industrierobotern in technologischen Einheiten. *Arbeitsschutz, Arbeitshygiene,* **18,** 3, 82–5.
Sugimoto, N. and Kawaguchi, K., 1983, Fault tree analysis of hazards created by robots. In *Proceedings of 13th International Symposium on Industrial Robots and Robotics* (Dearborn: Society of Manufacturing Engineers), pp. 327–39.
Swain, A. D., 1985, The human element in systems safety: a guide for modern management. In Swain, A. D. (Ed.), Albuquerque.
Tynan, O., 1985, Change and the nature of work. Some employment and organisational problems of advanced manufacturing technology. *Robotica,* **3,** 173–80.
VDI (Verein Deutscher Ingenieure), 1989, *VDI 2854 (Entwurf): Sicherheitstechnische Anforderungen an automatisierte Fertigungssysteme* (Düsseldorf: VDI).
Vossloh, M., 1988, Wissensunterstützte Fehlerdiagnose an CNC-Werkzeugmaschinen durch störfallbezogene statistische Datenauswertung. *Industrie-Anzeiger,* **3/4,** 32–3.
Wiendahl, H. P. and Springer, G., 1986, Untersuchung des Betriebsverhaltens flexibler Fertigungssysteme. *Zeitschrift für Wirtschaftliche Fertigung,* **2,** 95–100.
Wolovich, W. A., 1987, *Robotics: Basic Analysis and Design* (New York: College Publishing).
Zimolong, B., 1985, Hazard perception and risk estimation in accident causation. In Eberts, R. E. and Eberts, C. G. (Eds) *Trends in Ergonomics/Human Factors II* (Amsterdam: Elsevier), pp. 463–70.
Zimolong, B. and Hale, A. R., 1989, Arbeitssicherheit. In Greif, S., Holling, W. and Nicholson, N. (Hrsg.) *Europäisches Handbuch der Arbeits- und Organisationspsychologie* (München: Psychologie Verlagsunion), S. 126–31.

# *Chapter 14*
# *Remote-control units for industrial robots*

**H. M. Parsons**

*Human Resources Research Organization, 1100 South Washington St., Alexandria, Virginia 22314, USA*

**Abstract.** This chapter discusses needs and problems in the design and standardization of teach pendants, a category of remote control unit (RCU). These devices are used primarily in programming industrial robots to establish the positions and orientations to which these should move in production. As miniature consoles, their design should be error avoidant and time conserving. Industry has begun to specify human factors considerations.

## *Introduction*

This chapter describes remote-control units, often called teach pendants, for the application programming of data and commands for industrial robots. Such devices are also used at times by maintenance personnel and robot operators. The generic term 'Remote Control Unit' (RCU) means that this class of devices is found also in other contexts. They are small, hand-held control/display panels manipulated at some modest distance from a computer with which they communicate by cable or radio and which in turn controls some equipment. A familiar but simpler example is a 'remote' television station selector. The robotic RCU is connected to a robot's controller, which houses a small computer and generally has an associated terminal with a monitor and keyboard. This equipment is stationed outside the robot's work envelope and any protective barrier or delineation defining it, whereas the RCU is mostly carried and used within that perimeter.

In application programming, a technician or engineer must specify the points in three-dimensional (3-D) cartesian space to which the robot's end-effector, e.g. tool or gripper, must move to operate on a workpiece, and also must specify the orientations of the robot's wrist holding it. Such positional and orientational data can be ascertained in several ways. If sufficiently precise and accurate dimensional measurements exist in some 3-D model of the robot, end-effector and workpiece, perhaps derived from Computer-Aided Design (CAD), these can be entered 'off-line' into the program at a terminal away from the factory floor. If they are not accurate enough they might be rectified, while the robot is operating, through feedback from machine vision or some other automatic sensor. But another technique is to move the robot's jointed arm so the end-effector's 'tool centre point' reaches a desired position and to move the robot's jointed wrist so the end-effector acquires the desired orientation. When the programmer then instructs the robot's computer to record the component joint angles, derived from the joints' servo-motors, its program converts these into the required data. Thus the robot is used as its own measuring device, so to speak, to establish the needed positions and orientations for a production run without any deviations that might result from using data from a model: for various reasons critical differences can occur between stored and actual values.

There are several ways to move an industrial robot to 'teach' it in this fashion; here 'teach' is a synonym for 'program'. Small manipulators can be simply pushed and pulled by hand. For programming continuous paths, as in spray painting, a technician may manually move a surrogate 'teaching' arm similar to but lighter than the real one, with some controls on it; Shulman and Olex (1985) described a human factors study of improving these. Moving large robots to desired points and orientations is the primary function of the RCU called a teach pendant. Manipulation of its numerous control elements—pushbuttons and perhaps a joystick—moves the arm and wrist joints and operates the tool, and a small display indicates status and data. Parsons (1988a,b) and Parsons and Mavor (1986/88) reported a survey of teach pendants associated with robots made by ten major manufacturers, in the first comprehensive human factors/ergonomics study of these RCUs. Much of this chapter is based on that study.

Despite their small size—no larger than an average textbook and weighing as little as two pounds—these miniature consoles can be relatively complex, and together with the applications software with which each is closely associated they constitute the critical human–machine interface in industrial robotics. That interface occurs primarily in application programming (and to some extent in maintenance) in the robot's 'system loop', so to speak, rather than in robot operation, which is largely autonomous. Due to the pendant's importance the Robotic Industries Association (RIA) has developed an American National Standards Institute (ANSI) human engineering standard for its design (ANSI, 1990).

But why the 'remote' in such devices? Why must the teach pendant be carried into the robot's somewhat hazardous work envelope, or rather, why not move the robot's arm, wrist and end-effector by means of equipment at the robot controller outside the robot perimeter, in conjunction with other programming at the computer terminal there? Though these questions are virtually unreported in the robotics literature, the reason is actually very simple. In moving the robot, the programmer must be able to see the robot's tool centre point and the workpiece at the same time and eventually to discriminate visually their precise matching before recording the arm's and wrist's joint angles. Indeed, the programmer's eyes may have to come within a few inches of the alignment, especially if illumination is poor. Despite the hazard involved in coming so close to the robot, none of the manuals reviewed in the Parsons and Mavor (1986/88) survey indicated any concern about lighting—perhaps an indication of the extent to which human factors engineering has been applied to this interface. The robotics literature has, however, pointed out a problem due to the distance between the pendant's in-house location and the computer terminal's keyboard and monitor, where the textual (line-by-line) part of a robot applications program may be composed or edited, including commands, contingencies (logic), stops and pauses and interactions with other equipment. The programmer may have to go back and forth (perhaps through a barrier gate) to use both pendant and terminal. Although apparently no systematic study has been undertaken, one time-saving solution has been to incorporate much of a program's composition and editing capability in some pendants. But that may add to the size and complexity of control elements even if menu displays and selectors replace designs requiring more buttons.

## *Functions of a teach pendant*

### Movement controls

As already indicated, the principal functions of a robot RCU—a teach pendant—are to move the arm and wrist joints so the robot's end-effector will move to some desired position with some desired orientation. An industrial robot has as many a six joints, three in an arm for positioning and three in a wrist for orientation; small robots, e.g. for assembly, are likely to have fewer. A revolute or hinged joint rotates, and a prismatic joint slides, like a telescope, as shown in Figures 14.1 and 14.2. Different robot arms have different combinations of joints.

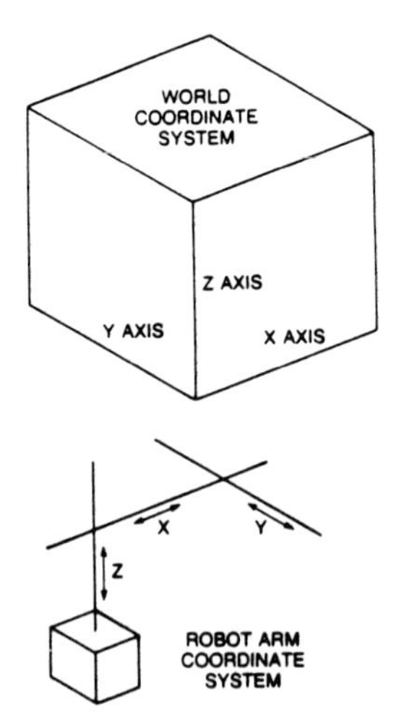


*Figure 14.1 Diagrammatic view of world and robot rectangular coordinates.*

Pendant pushbuttons or a joystick must be able to move each arm joint and each wrist joint individually in each direction, in a 'joint' mode. Though that is useful for some purposes, a programmer generally moves the robot arm in

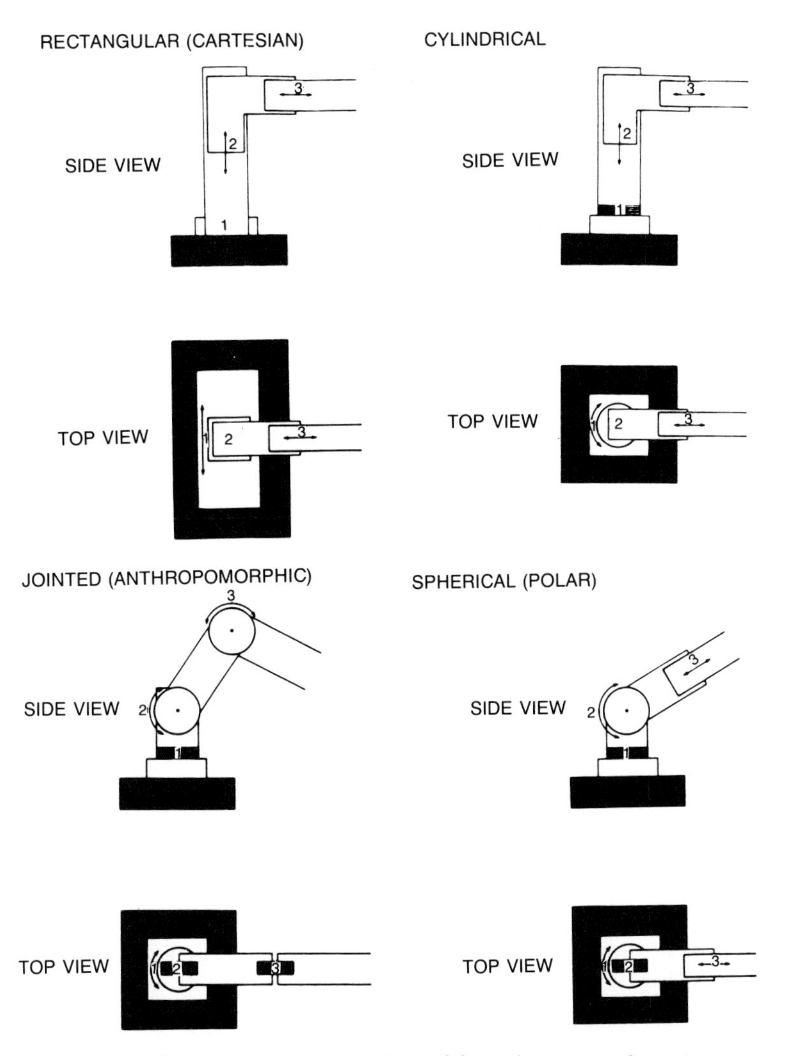


*Figure 14.2 Diagrammatic representations of four robot arm configurations.*

a 'cartesian', 'world' mode, as in Figure 14.3, in which the tool proceeds along rectangular axes (*X*, *Y* and *Z*) toward a selected point with arm joints operating together (as they do in a human arm when one points at something).

The pendant's controls must be able to move the robot in either direction along each rectangular axis. When the selected point is reached, the robot's computer calculates its three coordinate values and stores the arm's associated joint angles, to be used when the robot is moved in a programmed test or production run. The path along which the programmer moves the arm to get the tool centre point to its target is not likely to be the straight (or slightly curved) one along which the computer will move the robot in executing the program. In addition, the pendant controls must be able to orient the wrist and end-effector

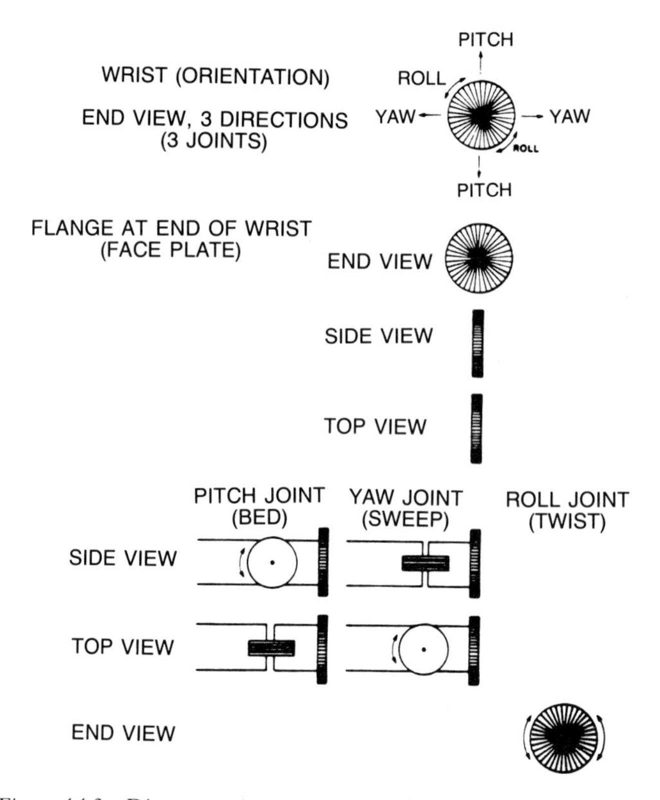


*Figure 14.3 Diagrammatic representation of robot wrist joints/movements.*

in relation to the arm in pitch, yaw and roll, effecting each of these in each direction.

A joystick can be used to accomplish all of these functions, as in the case of one robot manufacturer (ASEA, now Aséa-Brown-Boveri), with selector switches assigning the stick between joint and cartesian control and between arm and wrist/tool control. Most manufacturers have relied on pushbuttons (keys) in an array of 12, though some robots with fewer joints have a smaller number.

As an example (shown in Figure 14.4), the pendant for a GMFanuc Robotics Corporation robot with six revolute joints (degrees of freedom) has a 4 × 3 array with three pairs of buttons for arm control (by individual joints or $X$, $Y$ and $Z$ movement) and three pairs for wrist/tool control (by individual joints or pitch, yaw and roll movement); the two buttons in each pair control the different movement directions in individual joint rotation, or along an arm axis or in wrist/tool orientation. A switch assigns the buttons to the world/cartesian or joint coordinate system. Another manufacturer's pendant has twelve buttons in two columns of six each. Similarly, a relatively early but still widely used pendant (Figure 14.5) made by a company no longer manufacturing robots has six rocker switches in a column, with the direction of arm, wrist or joint

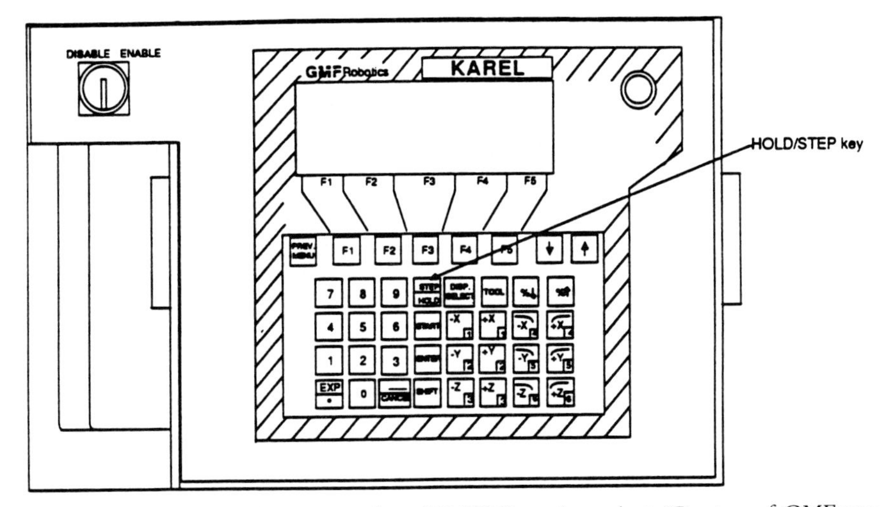


*Figure 14.4 Diagrammatic representation of KAREL teach pendant (Courtesy of GMFanuc Robotics Corporation).*

*Notes*

| | |
|---|---|
| F1–F5 | These input varying menu functions shown on the LCD display above them. There are 61 functions in 16 menus. |
| Vertical arrows | These cursor keys designate a variable on the display or scroll the display. |
| Step/Hold | Step moves the robot a single program statement or segment at a time, together with the Shift key. Hold stops all robot motion; another press releases the Hold condition. |
| Disp. Select | This brings information on to the LCD display. |
| Tool | This brings the Tool menu on to the display for tool operation. |
| % arrows | These decrease or increase increments of jog speed override values by 1% to 5% and then 5% to 100%. |
| Start | This executes a test run together with Shift key, while held down, or restarts the robot together with Shift key after Hold is released. |
| Enter | This makes a selected variable become the desired position variable. |
| Shift | This selects between another key's two functions, or it enables another key's function as indicated above or a selected menu function, e.g. recording a position. |
| Numeral pad | These 12 buttons insert numerical values into the program. |
| Disable/Enable | A switch in the upper left corner completes the connection of the teach pendant to the robot controller. |
| Emergency stop | A button in the upper right corner stops all robot motion and program execution. |
| Deadman | One of two switches on the pendant's sides must be held down for the pendant to operate. |
| XYZ/1–6 | These keys move the robot. Three XYZ keys move the arm along rectangular (cartesian) coordinates in two directions; the three other XYZ keys rotate the wrist's tool centre point around the arm coordinates, in two directions. These six keys also move individual joints designated 1–6 in two directions when a menu-associated F key establishes that mode. |

*Source:* KAREL Controller Operations and Programming Reference Manual, 1985.

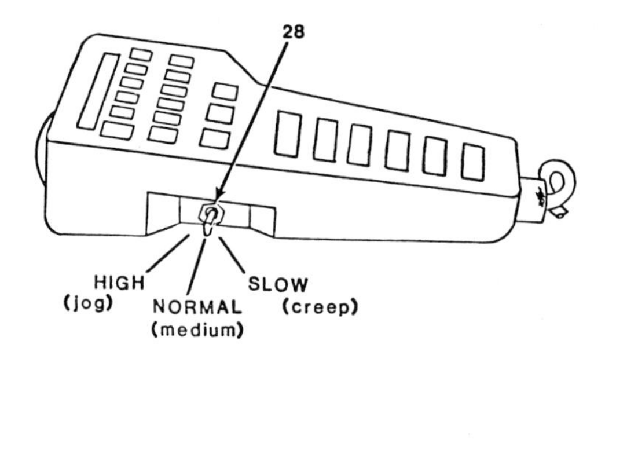


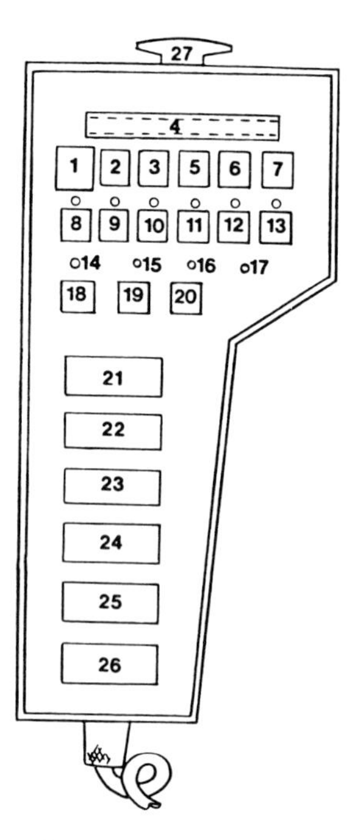


*Figure 14.5 Diagrammatic representation of a teach pendant.*

*Notes*

1. Record pushbutton. Inserts arm positions/motion instructions into the computer program.
2. In an editing mode this pushbutton selects the type of instruction to go into the program and changes the display.
3. Activates a gripper or spot welder.
4. LCD display of 22 messages.
5. Activates another gripper.
6. } These pushbuttons permit the programmer to step through the arm's programmed
7. } positions.
8. This pushbutton allows the computer to assume control of the arm.
9. This pushbutton lets 21–26 rotate the wrist or move the arm linearly in a tool coordinate system.
10. This pushbutton lets 21–26 move the arm linearly or rotate the wrist in a world coordinate system.
11. This pushbutton lets 21–26 each move a single joint.
12. This puts the arm in a 'free' mode so it can be moved manually.
13. When this is pressed, 24–26 move the wrist in very small increments or at the slowest possible speed in conjunction with 18.
14. LED indicating the teach pendant can move the arm.
15. LED indicating 20 has put the arm on 'hold'.
16. LED indicating 20 has put the arm in the 'run' mode.
17. LED indicating calibration.
18. This pushbutton moves the arm to the programmed position showing on 4.
19. This pushbutton advances the program a single step each time it is pressed.
20. This pushbutton keeps the arm on 'hold' or stops it. Pressing it again puts the arm back into the 'run' mode.

21. 22. 23. These rocker switches move the arm in plus or minus directions according to which side is pressed, linearly in tool or world coordinate systems or by rotating one joint at a time.
24. 25. 26. These pushbuttons rotate the wrist around coordinates of the tool or world coordinate system or rotate one wrist joint at a time.
27. This is an emergency stop pushbutton that puts the arm on hold.
28. This three-position switch selects between normal (medium), slow (creep) and high (jog) speed for teach programming.

movement depending on which side of the rocker is pressed. Another design has only six buttons (in a column); a selection button determines their direction of movement.

One may ask how manufacturers decided whether to use a joystick or pushbuttons. At least two initially considered a joystick but decided to use pushbuttons. Apparently ASEA was the only one to conduct an experimental test (Brantmark *et al.*, 1982); a joystick which was reported as showing a 25% advantage in speed over a pushbutton design. Subsequently the stick's gain was changed from linear to logarithmic so slower robot-to-stick movement ratios would improve fine positioning. In a still later change a guard ring was placed around the stick to prevent damage when the pendant was dropped on the factory floor. Another manufacturer (American Cimflex) changed from a joystick to pushbuttons (with a menu system) at least partly because customers complained that sticks were damaged when dropped. Though it did not incorporate a joystick, a Japanese manufacturer (Hitachi) is reported to have solved the dropping problem by equipping its pendant with a flexible cord looped around the programmer's wrist.

## Frame of reference aspects

As noted, a dedicated button may be needed to select between an individual joint and a cartesian frame of reference movement. That 3-D ('world') reference frame must have an origin ('home'), usually in the robot's base, to which it must be calibrated (perhaps at the controller) before programming or execution, and to which a special button can send it. But the programmer may want to select a different, relative origin, perhaps at the wrist flange for a tool frame of reference or in some array (e.g. pallets), and the pendant might allow this option. It may also be necessary to establish within a reference frame some software stops for safety, perhaps again through pendant manipulation.

## Movement speed

When the programmer moves the robot with a joystick, the latter's declination or extent of twist controls that movement's velocity. In one instance that variation was limited by switch settings for ten speeds. Additional control elements are needed to vary the speed of the movements initiated by pushbuttons. Different pendants have different designs. Two have three buttons, in one case to select a low, medium or high speed, in the other to select a high (50% of the maximum), medium (10%), or low (1%), whereas in still another pendant a three-position switch selects normal, slow and high speeds but a button slows each of these. On the GMFanuc Robotics Corporation pendant (Figure 14.4), two buttons increase or decrease displayed speed by 5% with each press, by 1%, or by 'fine' or 'very fine' increments. In another case a button increases or decreases speed, listing available speeds for selection or change. In one ingenious design, speed varies according to the location (with graphics guidance) where the programmer presses on one of two elongated buttons (that also initiate movement and select its direction), and a toggle switch alters speed ranges. Clearly this aspect of pendant design varies greatly. Speeds are selected by choosing alternatives, by making increments or decrements, by either discrete or continuous adjustments, by overrides, and by minimal (tick, jog, single bit) motions. Yet the speed with which the pendant moves the robot to program it (not the same as the speed at which the robot will move in executing the program) is important for two reasons. It will determine how quickly (and accurately) the programmer moves it to a desired position, with a desired orientation—and in a factory time is money. It must not exceed certain limits imposed for safety; such limits are fractions of the speed permitted in execution.

### End-effector

In addition to the robot's arm and wrist, the actions of the end-effector must also be programmed. If this is a gripper, commands must be included to open or close it, and if a tool, to activate or deactivate it. Some end-effectors, e.g. a welding gun, may require additional commands. In addition, the programmer may have to enter the end-effector's dimensions into the program, since its 'tool centre point' is what must be aligned. A gripper is often regarded as an additional degree of freedom.

### Recording positional and orientation data

When the programmer has positioned the tool centre point at the required point on a workpiece with the appropriate orientation, pressing a particular pushbutton records the joint angles in the computer's memory. As noted earlier, the computer converts them into cartesian coordinates and these are usually shown on the pendant display. Thus the programmer 'enters', 'declares', 'inserts', or 'writes' points.

### Movement commands

An application program must include commands to move the robot to the programmed points in program execution. These commands are entered into the program at a terminal or in some cases with buttons on a teach pendant, especially for testing. The pendant may also have to command the robot, during programming, to move to a point that has been already programmed either by means of the pendant or in text programming at a terminal. The point's identifier or coordinates are specified with pendant pushbuttons and another button is pressed to move the robot to the point or, alternatively, to approach it by moving to a nearby position.

### Reviewing and editing

The commanded movement during programming just described occurs when the programmer reviews or edits what he or she has just produced in moving the robot with pushbuttons or joystick. If a point's coordinates are programmed textually at a terminal, the programmer may alter (fine-tune) the data by substituting the coordinate values acquired by using the robot itself as the measurement agent. If the text program specifies only a point identifier, the programmer associates with it the coordinates thus acquired. A robot's completed application program consists of a number of steps within a cycle, repetitions of a cycle, or different cycles, perhaps repeated; the points are defined (identified and given coordinate values) in certain steps. This step/cycle structure, which includes non-point data, commands and contingencies (logic) is usually planned and written at a terminal. Thus the programmer may have to use both the terminal and the pendant to create the program. As noted earlier, the distribution of labour between the two, and thus the pendant's versatility, vary among manufacturers.

In an editing/testing mode the programmer may, with the pendant buttons, move (with or without robot motion) through a set of program steps of taught points, point by point or continuously, or may move forward or backward by a single step, or may require the robot to stop at the next step. The programmer may use the pendant to: delete or cancel a pendant input, step, point or coordinate value; erase the last keyed entry or all taught points; insert new point values, changes, or displacements; or index an array of points. Some of this editing may consist of recovery from errors. If a point has not been identified in the text program, the programmer can 'name' it numerically. For editing that involves numerical values, as much of it does, every teach pendant has a number keypad. In programming with a teach pendant, including reviewing and editing, numerous dedicated or menu-associated pushbuttons are also needed to bring stored information onto the pendant's display (see Table 14.1) and to manipulate the display otherwise, such as scrolling it, deleting items, and showing menus.

### Other control functions

Although a small number of switches at the robot controller enable a robot operator to turn power for the robot on and off, calibrate it, and start, interrupt, and stop a production run or cycle, the pendant may have pushbuttons for some similar functions, such as arm or computer power, enabling the pendant, and starting and stopping program execution. (Designs of controller panels vary among robot manufacturers, including what functions are assigned to the controller and what to its associated pendant.) Every pendant also has a prominent emergency stop button and many have a deadman's handle that will halt the robot if it is released. In addition, a pendant may have a button to put the robot's arm and wrist into a 'free' mode in which the pendant has no control; maintenance personnel may need this mode as well as the individual joint mode for testing or troubleshooting.

Various design strategies have been adopted to limit or reduce the number of pushbuttons on a pendant. Dedicated keys, for example, may have more than one function; a shift pushbutton selects which one is enabled. In fact, a subsequently discontinued version of one pendant had some colour-coded buttons with as many as five functions, along with three colour-coded shift keys on its side. Variable functions of some 'soft' buttons are designated by an associated display, a technique that usually incorporates a menu system so five pushbuttons can be shared among dozens of functions in a number of menus. A menu system, as in the GMFanuc pendant (Figure 14.4), can thus make a large number of functions available with a small number of non-dedicated pushbuttons, in marked contrast with a pendant with many dedicated, labelled keys with single or multiple functions. One pendant has alphabet buttons to spell out commands. Alternatively, some functions are relegated to the computer terminal to accompany other programming there. (Most terminals use menu systems, although the early one associated with the pendant in Figure 14.5 equipped programmers with a plastic card listing the abbreviations for 114 program commands and instructions.)

The diversity of teach pendant control elements among nine of the manufacturers surveyed by Parsons and Mavor (1986/88) is illustrated by the variation in their totals (which include soft keys and two joysticks) and in the totals of their associated functions (in parentheses) that omit menu choices: 21(89), 23(45), 25(25), 25(40), 28(43), 32(32), 42(47) and 46(86). Of particular significance, the locations and arrangements of the control elements on the pendants were also found to be diverse. As suggested by comparing Figures 14.4 and 14.5, no two have been alike. This variation in configuration, which can be fully appreciated only by examining the pictures of different pendants in the Parsons and Mavor report, has changed in its characteristics due to improvements in the state of the art of pendant design, but its extent remains considerable, and in any case earlier pendants are still being used with the robots for which they were designed.

## Information display

The programmer may receive feedback from control actuation in several ways: tactile feedback from mechanical buttons and selector switches; from some membrane buttons; and from a LED that lights up. One pendant beeps when a button is pressed, and another beeps several times when the programmer presses the record button. Pendants customarily have LCD displays above their control panels, ranging in size and thus content capability from one line of 12 characters to eight lines of 40 characters each (with scrolling in two instances). In addition to feedback from some control actuations, these displays provide information in a variety of categories as shown in Table 14.1. No one pendant, to be sure, includes all the categories. The content heterogeneity is as great as that among control elements.

*Table 14.1 Categories of information on teach pendant displays.*

| | |
|---|---|
| Status | Project number. Program number.<br>Type of unit (metric or English).<br>Default settings (parameters).<br>Electrical signals. Limit switches.<br>Software limit stops. Wrist model.<br>Close path assignments. |
| Prior action data | Number of points taught.<br>Sequences taught. Cycles completed.<br>Last 32 binary signals (I/O).<br>Last 10 errors. Last 30 errors. |
| Current action, data and feedback | Command sequence number.<br>Current program line. Step/point name.<br>Teach coordinate system in use.<br>Current mode for teach pendant.<br>Location type: single, index, approach.<br>Location: joint values, coordinates.<br>Tool at displayed point.<br>No location assigned to point.<br>Location out of reach. Error.<br>Keyed input (Feedback). |
| Future action data | Number of points remaining.<br>Next step/position.<br>Subsequent positions.<br>Next 32 binary signals (I/O).<br>Wait time. Menu options. |

# *Human factors considerations*

## Design problems

A major design issue is just what functions and information should be included in a pendant's control elements and display. The preceding section went into some detail to demonstrate the heterogeneity in these, providing prima facie evidence that some pendant designs may be incomplete in this respect. Intimately associated with this issue is the location and arrangement of various categories of pushbuttons. Here also the heterogeneity suggests some designs may be better than others. The pendants surveyed by Parsons and Mavor (1986/88) generally grouped category members together but in many instances failed to separate categories emphatically through spacing, colour or delineation. With some noteworthy exceptions, colour coding was insufficiently exploited to distinguish between category functions. Generally the emergency stop button was coded red but its location varied among pendants. The between-pendant variations in locations and arrangements could present problems to factory workers programming or maintaining robots from different manufacturers, another reason for improving category separations and coding.

Although the Parsons and Mavor (1985/88) survey did not acquire systematic data about pendant size, shape and weight, these also varied. These characteristics can affect handling capability, fatigue and accidentally dropping the pendant. Presumably some pendants are easier to grasp and carry than others. Individual differences among users, e.g. handedness, finger strength and hand anthropometry must be considered; the last should be examined among pendant users on the factory floor to make sure that spacing between adjacent pushbuttons is sufficient to prevent errors, especially in poor illumination. So must the visual discriminability of labels in such conditions. Design requirements in these respects may have to be more stringent than those for consoles, due also to the miniaturization of pendant control/display panels.

As has been pointed out, the need to conserve space on a pendant has led to giving multiple functions to pushbuttons and to using fairly complex menu systems with soft buttons. If these aspects are poorly designed they may become unexpected sources of errors, especially with less experienced users. Lessons learned about menu design elsewhere, e.g. in office environments, might be transferred to the factory. In general, skill in manipulating a pendant's control elements simply should not be taken for granted; explicit training may be needed.

Parsons and Mavor (1986/88) found that the robot manufacturers by and large failed to apply human factors engineering explicitly and systematically to the teach pendants they surveyed, though some showed more concern for users than others did in their design. Few design features have been incorporated for the express purpose of preventing errors and still fewer for recovering from them, though errors in using industrial machines can lead to accidents. Pendant designers seemed unaware of human factors studies of console design and of more recent research on human–computer interfaces. The applications software with which the surveyed pendants interact were apparently designed with equivalent unawareness of human factors issues, notably in some menu systems. However, the need to consider hardware issues in pendant design has subsequently been acknowledged by the RIA, as indicated by its efforts, noted earlier, to create an ANSI human engineering standard for teach pendants (discussed below).

## Empirical data

Except for the heterogeneity in design and the apparent human factors problems described above, little evidence other than the anecdotal exists as to the actual need for better design of teach pendants. Rate of pendant operation and error incidence remain unknown; no non-proprietary data seem to have been collected on the factory floor. In an experiment, Ghosh and Lemay (1985) obtained some time and error data from ten undergraduate engineering students with no previous experience with robots, demonstrating learning curves in using a teach pendant. Except for safety studies the only experimental test reported was the one that compared performance rates with a joystick and pushbuttons. There have been indications of feedback to robot manufacturers from robot purchasers but the contents have not been made public. Parsons and Mavor (1986/88) performed the only task analysis of pendant use—actually a task taxonomy—on which to base any systematic investigation of using pendants on the factory floor.

Safety studies have examined how different robot movement speeds might make it less likely that workers would be struck by a robot under teach-pendant control. The primary concern has been a worker's reaction time, that is, the interval between the start of robot movement and the worker's reaction of pressing the pendant's emergency stop button—in conjunction, of course, with the distance between the robot and the worker. (Conceivably the location of that button on the pendant would also be a factor; the National Institute of Occupational Safety and Health (NIOSH) has been investigating this possibility.) The faster the robot's movement and the shorter the distance, the quicker must be the reaction time—whose minimum will vary under various circumstances. Movement speeds can be established by means of pendant buttons for programming (or maintenance) purposes, at levels less than speeds during program execution, as noted above.

Sugimoto *et al.* (1984) simulated a button-pressing error and timed the interval between that pendant action and pressing the stop button, then calculating a robot movement speed such that the robot would not strike the pendant user between 20 and 30 cm from it. Helander *et al.* (1987) told their experimental participants to press the stop button if the robot arm, at four different speeds, moved beyond an expected target position. Karwowski *et al.* (1987) and Karwowski and Rahimi (1991) asked their participants to judge the maximum safe speed of one robot or two robots operating side by side. As Karwowski and Rahimi pointed out, in these studies robot movement speeds were specified in cm (or in)/s, not in deg (or min)/s. However, if the arm happens to be rotating toward the worker, the safe distance from it for a particular rotational speed would presumably depend on which part of the arm the worker is near; in the same amount of time the end of the arm would move linearly further and faster than that part near the base.

Though there have been several investigations of accidents due to industrial robots (Parsons, 1986), none of these has directly linked an accident to making an error with a teach pendant—although that does seem like a possible accident source. Nevertheless, one of two fatalities investigated by NIOSH (Etherton and Collins, 1990) involved a worker whom a robot's rotating back end crushed and pinned against a pole erected to keep the front end from hitting anything. (Illustrations in robot safety literature have not shown or warned against a hazard from the back end.) According to NIOSH Summary Report 84-020 (1984), when fellow workers discovered the victim they tried in vain to move the robot away by operating a teach pendant, but only many minutes later was the plant's director of manufacturing able to do so with a second pendant. The victim died five days later in hospital. It may never be known whether (1) further injury resulted from the extended delay in freeing him due to his associates' inability to operate the pendant or (2) their attempts to operate it may have moved the robot in the wrong direction. Although the accident report did not raise these questions, they do suggest that injuries, even deaths, might in the future be attributable to errors in manipulating pendant controls, possibly due to poor design or insufficient training, or both.

## *Standards*

The RIA has been active in developing two human-factors related standards involving teach pendants for ANSI. One first appeared as ANSI R15.06 (1986) for safety, specifying a maximum slow speed (e.g. during programming or maintenance) of 25 cm (10 in)/s, and the International Standards Organization (ISO) has endorsed this. However, the Underwriters Laboratory and the Japan Industrial Safety and Health Association have specified 14 cm (6 in)/s, which apparently was derived from the Sugimoto *et al.* (1984) study. According to Helander *et al.* (1987), the ANSI and ISO limit 'is clearly arbitrary', and these investigators developed a mathematical model relating robot movement speed and human reaction time to risk of injury by the robot.

### Human engineering standard

The other standard is ANSI/RIA R15.02/1 for human engineering design of teach pendants already mentioned. As a demonstration of concern with human factors issues it has been a welcome development. It is innovative also by incorporating in a private industry standard many of the human factors design guidelines that have evolved over several decades in the Department of Defense and are published as MilStd 1472D (1989). These cover detailed design requirements for control devices and their labels, discriminability requirements for display elements and auditory criteria; pendants may include audio warnings. Eventually current pendants may be evaluated against these guidelines, some of them mandatory.

The standard is also expected to recommend that tasks and functions be analysed as a basis for pendant design, and that designers consider pendant configurations. Whether configurations—locations and arrangement—will ever be standardized for teach pendants seems debatable, however, in view of the existing heterogeneity described earlier. For console design there are certain heuristics which, if followed, can produce some homogeneity in panel designs. For example, component control devices should be given the most accessible locations according to their criticality and frequency of use. But due to its portability and relatively small size, accessibility in terms of reach is not an equivalent problem for a teach pendant. Similarly, visual access to displays and display components on a pendant need not conform to the heuristics of placement desirability at a console or workstation. Nevertheless, standardization that implies consistency in configurations across pendants from different manufacturers may become an eventual goal so users will not make mistakes in shifting from one company's pendant to another's, especially if errors lead to accidents and consequent litigation (Parsons, 1988c).

Another consistency problem may persist with regard to the symbols for labels on pendant pushbuttons. One manufacturer has emphasized these partly because it sells robots and pendants in different nations. Symbols can help solve the language problem with RCUs much as they do on road signs. Although an ANSI standard may give less heed to international considerations than an ISO standard, ultimate agreement will be needed on symbols. From a human factors perspective important symbol designs should be selected not by a committee (e.g. of company representatives or industrial designers) but through experimental testing of representative pendant users for discriminability between potential symbols and generalization to the pendant function being symbolized.

### Feedback and error prevention/recovery

Missing for the most part from the pendants surveyed by Parsons and Mavor (1986/88) was sufficient feedback to the user about the results of control actuation (rather than just feedback that actuation occurred), especially feedback about any error that the pendant user committed. Some pendant displays do show errors, as indicated in Table 14.1, but these should be correlated with the actions that caused them. That will help programmers or maintenance personnel recover from their errors. In general, methods of error recovery, largely disregarded in pendant design, merit more attention in the future. What seems needed most is an on-the-job survey of the kinds of errors that pendant users make and careful analysis of procedures for recovering from them, including how such procedural information should be presented both on teach-pendant displays and in manuals.

It seems unlikely that error feedback and error recovery can be handled adequately in teach pendants without examining, for a human engineering standard, the software with which a pendant is associated. These RCUs are not just physical implements—hardware—but are software concomitants. Their outputs go into an application program, and what the pendant user does to produce these depends on that software. Whether or not he or she is called a programmer, the principal user engages in programming. The interactions between the pendant's controls and displays on the one hand and the applications software on the other have not yet been adequately scrutinized in developing human factors guidelines, possibly because in the past human factors engineers have concentrated on the performance, including feedback and errors, of equipment operators rather than programmers.

As has been emphasized, pendant–software interactions are epitomized in menu systems. Manufacturing engineers might benefit from human factors studies to improve menu design and other human–computer interfaces, in view of the proliferation of small computers on the factory floor including those associated with robots.

## Conclusion

The design of teach pendants for industrial robots needs, and has begun to receive, human factors scrutiny, because the pendants should be accident-avoidant in the interest of safety, time conserving in the interest of productivity and error-avoidant in the interest of both. Though their small size may imply they are simple devices, in a sense they are miniature consoles with some similar complexities, albeit with obvious differences. As programming and control devices they constitute a particular variety of remote control unit (RCU). What is learned about problems in their use and how to resolve these might be applied to other RCUs, which are proliferating in non-manufacturing settings.

## References

ANSI R15.16, 1991, *ANSI/RIA R15.02/1–1990 Human Engineering Criteria for Hand-Held Robot Control Pendants.* (New York: American National Standards Institute).

Brantmark, D., Lindqvist, A. and Norefors, U. G., 1982, Man–machine communication in ASEA's new robot controller. *ASEA Journal,* **55,** 145–50.

Etherton, J. R. and Collins, J. W., 1990, Working with robots. *Professional Safety*, March, 15–8.

Ghosh, K. and Lemay, C., 1985, Man/machine interactions in robotics and their effect on safety at the workplace. In *Proceedings of the Robots 9 Conference* (Dearborn, MI: Society of Manufacturing Engineers), pp. 19/1–19/8.

Helander, M. G., Karwan, M. H. and Etherton, J., 1987, A model of human reaction time to dangerous robot arm movements. In *Proceedings of the 31st Annual Meeting Human Factors Society* (Santa Monica, CA: Human Factors Society), pp. 191–5.

Karwowski, W. and Rahimi, M., 1991, Worker selection of safe speed and idle condition insulated monitoring of two industrial robots. *Ergonomics,* **34,** 531–46.

Karwowski, W., Plank, T., Parsaei, M. and Rahimi, M., 1987, Human perception of the maximum safe speed of robot motions. In *Proceedings of the 31st Annual Meeting Human Factors Society* (Santa Monica, CA: Human Factors Society).

MilStd 1472D, 1989, *Human Engineering Design Criteria for Military Systems, Equipment and Facilities.* (Philadelphia: Navy Publishing and Printing Office).

NIOSH Summary Report 84-020, 1984, *Fatal Accident Summary Report: Die Cast Operator Pinned by Robot.* (Morgantown, WA: Division of Safety Research, NIOSH).

Parsons, H. M., 1986, Human factors in industrial robot safety. *Journal of Occupational Accidents,* **8,** 25–47.

Parsons, H. M., 1988a, Human factors in robot design and robotics. *International Reviews of Ergonomics,* **2,** 151–76.

Parsons, H. M., 1988b, Robot programming. In Helander, M. (Ed.) *Handbook of Human-Computer Interaction* (Amsterdam: Elsevier), pp. 737–54.

Parsons, H. M., 1988c, The future of human factors in robotics. In *Proceedings of the Robots 12 Conference* (Dearborn, MI: Society of Manufacturing Engineers), pp. 3/75–3/82.

Parsons, H. M. and Mavor, A. S., 1986, *Human–machine Interfaces in Industrial Robotics.* Report for the US Army Human Engineering Laboratory (Alexandria, VA: Essex Corporation). [Also, 1988 (Aberdeen Proving Ground, MD: US Army Human Engineering Laboratory).]

Shulman, H. G. and Olex, M. B., 1985, Designing the user-friendly robot: a case history. *Human Factors,* **27,** 91–8.

Sugimoto, N. *et al.*, 1984. Collection of papers contributed to conferences held by the Machinery Institute of Japan, No. 844–5.

# *Chapter 15*
# *Implementation issues for telerobotic handcontrollers: human–robot ergonomics*

**H. N. Jacobus, A. J. Riggs, C. J. Jacobus and Y. Weinstein**

*Cybernet Systems Corporation (formerly Charles Systems), 1919 Green Road, Suite B-101, Ann Arbor, Michigan 48105, USA*

**Abstract.** Teleoperated control requires a master human interface device that can provide haptic input and output which reflects the responses of a slave robotic system. This paper addresses the design of six degree-of-freedom (DOF) cartesian coordinate force-reflecting hand controllers for this purpose.

Force-reflecting hand controllers have advantages in space-based applications where an operator must control several robot arms in a simultaneous and coordinated fashion. They also have applications in intravehicular activities (within the Space Station) such as microgravity experiments in metallurgy and biological experiments that require isolation from the astronauts' environment. For ground applications, universal, or computer-controlled hand controllers are useful in underwater activities where the generality of the hand controller becomes an asset for operation of many different manipulator types. Also applications will emerge in the military, construction and maintenance/manufacturing areas including ordnance handling, mine removal, NBC (Nuclear, Chemical, Biological) operations, control of vehicles and operating-strength and agility-enhanced machines. Future avionics applications including advanced helicopter and aircraft control may also become important.

## *Introduction: how force-reflecting handcontrollers work*

As a type of haptic (or through the sense of touch) computer input/output device, the force-reflecting hand controller has proved to be useful in teleoperation. Unlike conventional joysticks, forces on a controlled device (a teleoperator) are sensed and 'reflected' back to an operator. For example, when an active force-feedback device is used to move a teleoperated robot against a wall, the haptic input device becomes a haptic output device in the sense that it projects a 'feel like' something immovable has been encountered (by reflecting the force seen by the robot as it pushes against the wall). Back-driven motors are used to make the control resist further forward motion.

Unlike a joystick, most force-feedback devices are not hand held or console mounted. Rather, most are large bulky master robot arms (and may have shoulder harnesses which are suspended from the ceiling, such as the Argonne Arm). The notion of a compact hand-sized device is appealing, but the small-size factor has presented difficulties for mechanical design.

Force-reflecting hand controllers, like the one described in this work, are actually miniature robots (Figure 15.1). A robotic device uses multiple articulated members (similar to a human arm) to move an end-effector into an arbitrary position and orientation (arbitrary within a reach envelope). The robot senses the position of the end-effector by measuring the positions of each link (usually through measuring the angle of rotary joints, or the displacement of translational joints). In addition, most robots also measure the rate of change of position through tachometers attached to the moving joints. If the robot needs to sense forces exerted on the end-effector (externally applied or applied by the robot itself to a workpiece), it is commonly augmented by a set of force-sensing devices, normally strain gauges. Six of these can together measure any arbitrary set of torques and forces operating on the robot's end-effector.

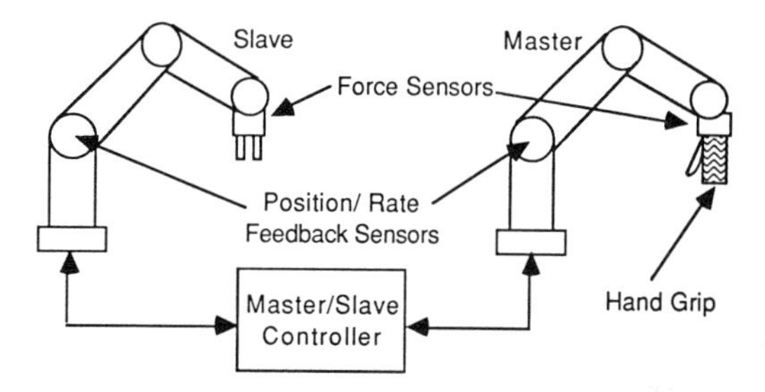


*Figure 15.1 Components of a master–slave.*

The purpose of a robot is to act on its environment to do useful work. The robotic hand controller (or master) acts as a transformer to accept actions (or motion commands) from an operator and convert them into commands to a robotic slave manipulator (or computer system simulating a robotic manipulator). In this mode the master controller controls the slave manipulator 'unilaterally' (which in turn follows the master's position). The master hand control can also accept commands from the slave manipulator's force-sensing devices which, in turn, are applied (through a scaling function) back to the operator's hands, thus communicating resistance to his/her original commands (for instance, when the slave is pushing a heavy object the controller requires more force to move it than if the slave is moving without restriction). In this mode the master controller controls the slave 'bilaterally'.

To support this active-force (or rate and motion) reflection, the controller itself must also have position-, rate- and force-sensing devices, along with motion actuators (e.g. motors) just as the slave manipulator does. In the past, the slave and master needed to be nearly kinematically similar (i.e. the joint and link structures had to be the same except for a scale factor), because this greatly simplified building electronics control systems which connected them together.

However, with modern digitally-controlled robotic servo-systems, master–slave similarity is no longer a requirement. This opens the possibility of producing very small compact force-reflecting masters. The compact size is a prerequisite for use in the cramped Space Station control console environment (as opposed to the larger cumbersome masters used in nuclear or underseas environments).

In addition, digital control allows multiple programmed modes for master–slave operation. While it is not now fully known which modes are of greatest use, the following can be supported through digital control schemes:

- Master hand controller mass and inertial forces compensation, i.e. the controller handle will appear weightless to the operator;
- Program-controlled and scaled application of forces to the hand controller from the slave manipulator;
- Program-controlled and scaled application of forces derived from the slave manipulator's inertial forces (thus making the slave's mass appear to be programmable);
- Program-controlled and scaled interpretation of hand-controller displacements (in six degrees of freedom) as commands to the slave to move to a position (position mode), move at a given rate (rate-control mode), or accelerate at a given rate (force or acceleration mode); and
- These modes will be software or operator selectable (to match specific task needs).

## *Literature review*

To derive an appropriate hand-controller design we must first be cognizant of prior efforts in the field. We have organized prior technology into a multidimensional framework of task analysis, hand-controller type (or hand-controller properties) and chronology/application area. This organization is presented in diagrammatic form in Figure 15.2. We wish especially to acknowledge a comprehensive survey (Brooks and Bejczy, 1985) and information from Honeywell, Inc., which covered a wide variety of commercially available hand controllers.

The field of telemanipulation developed in parallel with early nuclear-reactor work because of the need to handle hazardous radioactive materials (Vertut and Coiffet, 1984). Ths first telemanipulator was developed in 1948 at Argonne National Laboratories in a group led by Ray Goertz. These early manipulators were entirely mechanical, connecting the operator to a slave manipulator through a master handle which worked much like a large pair of tongs. This arrangement provided the operator with a way to position, grab and 'feel' objects in the radioactive environment. Subsequent work at Argonne, Brookhaven National Laboratories and other locations developed motorized unilateral-controlled manipulators, similar to modern industrial robots, but with simple operator controls (similar to those used in lifting and digging machines). In 1954, the Argonne group improved on these by providing bilateral force feedback through electrical servo-controls (thus providing the 'feel' available from mechanical master–slave systems without any mechanical coupling between the master and the slave manipulators). Throughout the 1950s and early 1960s this electro-

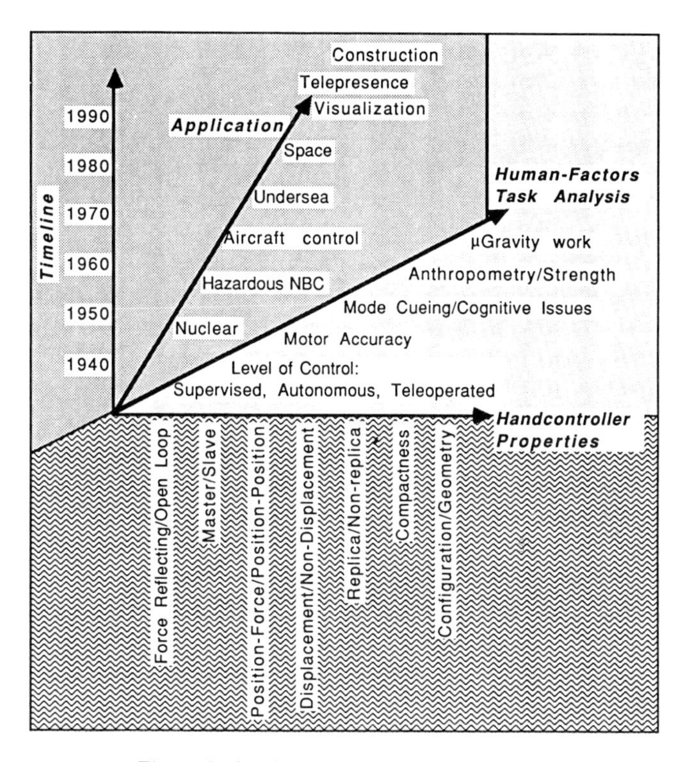


*Figure 15.2 Organization of previous work.*

mechanical master–slave concept was improved for the growing nuclear industry (with better control systems, incorporation of video/CRT monitoring systems and use of mobile platforms instead of fixed wall mountings for some systems). The Department of Energy labs (Oak Ridge, Argonne, Brookhaven, Sandia, etc.) continue to be major users and developers of teleoperated systems (Vertut and Coiffet, 1984).

In the mid-1960s to mid-1970s, relatively little new development occurred in teleoperation in the USA (probably due to the problems nuclear power was having in that country). During this period, technology development continued primarily in Europe. The US industry began a comeback in the mid-1970s with the development of a variety of teleoperators for underseas applications. A forerunner of these was built by General Mills in 1961 for the Bathyscaphe Trieste. The US Navy has operated similar vehicles (Cable Underwater Research Vehicles—CURVs) since 1966 and continues to be a major user and funder of teleoperated systems. A similar French vehicle, ERIC I, has operated since 1973 (Busby, 1976; Vertut and Coiffet, 1984). Recently vehicles for underwater exploration developed at Woods Hole (Bertsche *et al.*, 1977) publicized the technology by bringing video imagery of the Titanic to prime-time television in the USA and worldwide.

In the 1980s teleoperation technology was driven by the NASA Space Station effort. The current Space Shuttle system uses a teleoperated robot (the Remote Manipulator System—RMS) to deploy and retrieve spacecraft to (or from) the Shuttle cargo bay (Fletcher *et al.*, 1975; Lippay, 1977). This manipulator (developed by Canadian SPAR Aerospace in the late 1970s, deployed in the early 1980s) is controlled without direct operator feedback through three degree-of-freedom joysticks which operate arm position and wrist orientation independently. NASA has been a spectacular, if infrequent user, of teleoperations technology since 1967, when the Surveyor III spacecraft landed on the Moon equipped with manipulator arms for taking samples of the lunar soil. Teleoperation technology (using unilateral control) was also used on the Viking Mars landers in 1976. These manipulators were controlled open loop (without direct operator feedback) because of the long time delay from ground control to Mars and back (30 min one way). To support these space-based applications NASA has supported numerous laboratory robotics and telerobotics testbeds in-house and in the research community (Brooks and Bejczy, 1985; Szirmay *et al.*, 1987).

In the Space Station era, NASA has projects using advanced teleoperators for construction, repair and maintenance operations to avoid astronaut Extra-Vehicular Activities (EVA) (which are costly, dangerous and uncomfortable) and is funding new research and prototype development of teleoperated robotic systems and hand controllers (Saenger and Pegden, 1973; Bejczy *et al.*, 1983; Montemerlo, 1986). These new robotics devices under development (which will be controlled from an enhanced control console on the Space Station and Space Shuttles starting in the 1995–97 timeframe) will incorporate up to five robotic manipulators operating simultaneously (most likely by two crew men/women). Even if we assume that one or more of the manipulators is locked into a static position, this still requires that each crew person operate two arms concurrently (each requiring a six degree-of-freedom position, rate, or force command). To accomplish this, the crew will need easy-to-use six degree-of-freedom control mechanisms. The prior work in research, underseas maintenance and the nuclear industry indicates that these controllers must support force reflection (Herndon *et al.*, 1989), as well as position and rate control. Foreign Space Station project partners such as the Japanese space agency NASDA (with Toshiba) have also designed and built force-reflecting hand controllers and systems (Yamawaki and Sumi, 1988).

Hand controllers for the Space Station of the 1990s will be used by an increasingly wide population. In the past, space programme personnel were drawn from a small, select, and easily defined group. Anthropometric and biomechanical strength characteristics could be estimated, for example, by using Air Force pilot data. Because the environmental characteristics of the space working environment have improved, crew members can be selected from a wide range of people with emphasis on their skills and knowledge and with less emphasis on their physical conditioning. The design of a hand controller for use in the 1990s must consider crew members drawn from a civilian population with ranges (i.e. the upper and lower limits, of anthropometry, range of joint motion and strength capabilities) from 5th percentile Japanese females to the 95th percentile white or black American males (NASA, 1987, vol. I).

Effects of extended microgravity on the use of a force-feedback hand controller must also be factored into a design for the late 1990s since the Space Station will have 90–180 day length missions. There is evidence to support changes in body dynamics and posture over time during a flight (Lestienne and Clement, 1985). However, much of the data collected on long-term effects of microgravity is published in Russian (see e.g. Kasian *et al.*, 1974). Restraints, e.g. footholds, seats, may obviate these factors. But which posture should be assumed when operating a hand controller? Posture can account for large differences in human biomechanical performance (Chaffin and Andersson, 1984). Because the position (sitting, standing, or something in between) taken by the crew member when operating the hand controller will affect performance (Bejczy and Handlykken, 1981) and because of the wide range of crew-member anthropometry, it is important to ensure adjustability in the workstation area so that a variety of users can obtain the same effectiveness.

Future uses for force-feedback hand controllers include 'virtual environments' where computer-generated media can simulate an alternate environment. With a new generation of fast computers, real-time graphics can provide users with life-like three-dimensional (3-D) scenes (such as those viewed with helmet-mounted displays). With the addition of appropriate haptic input/output devices (like force-reflecting hand controllers), the user can 'explore' new spaces (for example, the terrain of Venus (McGreevey, 1989)), or the surface of a complex protein molecule (Brooks, Jr, 1977). Force-feedback from 3-D chemical models has been under development for 15 years (Brooks, Jr, 1977) but only recently has there been sufficient computation and graphical power for the technique to be useful for real-time applications. Other potential and largely unexplored applications may be found in the construction industry (for example, the control of complex excavation and earth-moving equipment; see Chapter 17, this volume.

## *Task-analysis issues in teleoperation and hand controllers*

Because telerobotics requires human-operator input for the control of distant robots, the operator must gain information about the task environment through some means of output. For instance, television output displays generated from video data are often used. As many as 90% of satellite servicing operations require visual feedback to the operator (Huggins *et al.*, 1973). Depending on the task which must be performed, degraded visual information may be used (Clement *et al.*, 1988). Some situations, while providing a direct line of sight for most work, have limitations of the line-of-sight by obstructions. In these cases where visual feedback has been limited, it is possible to augment vision by using touch, thus improving task performance. Providing force feedback may give the operator enough additional information so that the task may be performed quickly and accurately.

There are many issues to consider in the design of a compact robot hand controller with force reflection in six degrees of freedom. Surveying the existing literature clearly indicates that a multidisciplinary approach must be taken to ensure effective use of telerobotic force-feedback devices in manned space missions. The fields of anthropometry, biomechanics, cognitive psychology and motor control, among others, all have contributions to make to the effective design of hand controllers. For some examples, how great should the force reflection to the human hand be? (Current literature indicates nominally 20–4 lbs for trigger grips to about 50 lbs for wrap-around grips (Garrett, 1971; Brooks and Bejczy, 1985). An informal experiment conducted by the authors with weights confirms the 20 lbs figure.) Should it be scaled? What are the effects of weightlessness in manual control? (Roesch, 1987; Holden *et al.*, 1989). What effect do micro-g-induced-postural effects have on biomechanical strength and operator posture at a work console (NASA, 1978)?

We found that relatively few studies have directly addressed teleoperation with any type of force reflection. One survey team found only 4 such studies before 1987 (Draper *et al.*, 1987). Upon rigorous scrutiny of methodology and statistical significance, many flaws were found that make it difficult to draw conclusions. Even though much of this data was found to be dependent on the hardware and experimental conditions, some general conclusions could be drawn (Bejczy and Handlykken, 1981).

The parameters of force feedback control and metrics used in these experiments are of value. In one study, improved quality of operator performance emerged as a result of force feedback when compared with no force feedback. Even though no significant differences in mean time emerged, the supplementary force feedback allowed operators to perform better in the same time. In another study, the 'style' of operator performance changed when force feedback was removed. The operator changed from continuous motions to stepwise move, check, move motions. Small sample sizes are criticized as a statistical problem for these types of studies. Yet, much of the motor-control work which led to the development of the well-accepted Fitts' Law was performed on small sample sizes where individual subjects were studied at length (Meyer *et al.*, 1988).

The tasks used in these studies varied from remote handling tasks where the operator had to grasp a component and move it, to tasks which required grasping a bolt and then turning it to a criterion torque where the torque information was available by feel or by a visual output from a dial. Other tasks used were of the mechanical-assembly type (inserting connector ends into sockets, or peg-in-a-hole placement tasks). It is important to determine whether the presence or absence of force feedback alone was the difference between the experimental conditions of the study, or whether biases were introduced. For example, the visual information given in the dial output in the bolt-turning study may have provided specific information which allowed the subjects to meet the task criterion (Draper *et al.*, 1987). Another study showed considerable improvement with a quantified (colour bar chart) display of force–torque information (Bejczy and Handlykken, 1981).

Issues of time delay present serious problems for teleoperation over great distances, but if the hostile environment is nearby the delay may amount to only 10s of milliseconds, much of it due to computer-processing time. Some studies indicate that as little as 30 ms of time delay can negatively impact the human teleoperator, but this may be task dependent. There may be a difference dependent on whether the operator is in continuous or intermittent control. For example, other piloting studies have shown that humans can compensate for such delays when performing continuous control.

Concepts of control theory, of which there are many types, have been applied to understanding telerobotic systems. Traditional flight control of aircraft by pilots is well understood, and some of the results, along with modelling techniques, may be applicable to the teleoperator control problem for tasks in which the robot (or its end-effector) is 'flown' through paths (with rate control). However, position rather than rate control is critical for most robotic tasks, and here the analogy to flight breaks down. On the other hand, helicopter flight is analogous in some ways to the operation of a 6 DOF robot. Research interest in such integrated helicopter controls has been demonstrated (Lippay *et al.*, 1985; 1986).

Brooks has been studying the use of force feedback in the control of 3-D molecular simulation (via Argonne arm). The chemist who is 'designing' a new compound, or who is 'docking' an antigen into a virus can feel the simulated electrostatic forces between the molecules. Brooks' work is not only a compelling application area of force reflection but also proves that data from other fields must be considered in a thorough survey of teleoperation. For example, in 1976 Kilpatrick added auditory feedback to improve performance in a simulated teleoperator task. Clicks represented hard surfaces and clacks represented dull surfaces. Improvements in performance with multi-modal output may be obtained e.g. force output could be combined with sound, and/or an augmented computer display.

A compact force-reflecting hand controller is specified for the Space Station control consoles. It is not clear that empirical results using larger bulky teleoperation devices can be transferred. In summary, the relatively few studies using force-feedback devices (of any sort) have yielded inconclusive results. Different devices ranging from large and heavy master arms (Hill and Sword, 1973) to smaller joysticks have been studied across an inconsistent variety of tasks. The tasks range from actual telemanipulation (putting pegs in holes) to control of oscilloscope waveform outputs. Most of the empirical work has been performed in laboratory settings, with 'clean' laboratory tasks. It is difficult to imagine a space-based application using tasks as simple as those used in many of these studies.

Because it is possible for computerized hand controllers to vary the force feedback by scaling it in proportion to direct force, position error, combined force and position error, or in other ways (any mapping programmable in a digital computer), additional study is necessary to determine how this feature should be exploited. Prior work has typically focused on the optimality of specific control strategies (Vertut and Coiffet, 1984; Brooks and Bejczy, 1985). The potential for operator-directed switching to select a variety of control 'modes' is possible (for example, position or rate controls). Modes which lock some axes of motion while controlling others, or modes which integrate manual control of some functions and automated control of others (possibly generated by video-derived object positions and orientations) may be useful. Issues of expert use v. novice use abound (training v. retraining). Should a device be made easy to learn but then require greater time per task? Or, should a device require greater initial training time but then have quicker task execution time? Since operations will need to be cost advantageous, teleoperator controls must be efficient in a time-and-motion sense (current teleoperations data put manual operations v. teleoperations at a disadvantage ratio of 1 : 3 or 1 : 4).

Clearly, the ease of telerobotic task performance will have a large impact on future on-orbit operations for advanced missions to the Moon and to Mars. On-orbit manufacturing and experimentation will require human intervention through teleoperation or direct contact, since the processes can rarely be fully automated (Farnell *et al.*, 1989). Many dextrous tasks still require EVA with the current state of the art (Jenkins, 1987). Space-based teleoperation can permit an astronaut the option of EVA avoidance while still supporting movement (by robot) in the hostile environment outside. The ultimate test must be that an astronaut finds that he or she can complete required tasks through teleoperation more easily and safely than through direct EVA.

## *Hand controller types and features*

In their hand controller survey Brooks and Bejczy (1985) analysed the technology in terms of hand grip type (nuclear industry standard, accordion, full-length trigger, finger trigger, grip ball, bike brake, pocket knife, pressure knob, T-bar, contoured, glove, brass knuckles, door handle and aircraft gun trigger), controller configuration (switches; potentiometers; joysticks, isotonic, isometric, proportional, hybrid; replica; master–slave; anthropomorphic; nongeometric analogic and universal, control stick, floating-handle) and hand controller feedback law (rate control, direct, resolved; unilateral position control, direct resolved; bilateral position control, direct, resolved; operator-aiding control, filtering, scaling, referenced, motion constrained and motion compensated) (Figure 15.3).

### Hand controller feedback laws

*Direct control* strategies are used in replica master or master–slave systems where it is possible to relate an output from the controller to an input required by the slave manipulator under control without complex coordinate frame mappings (Johnsen and Corliss, 1971; Malone, 1973; Tewell *et al.*, 1974; Busby, 1976; Tesar and Lipkin, 1980). For instance, assume that the control master is a 6 DOF

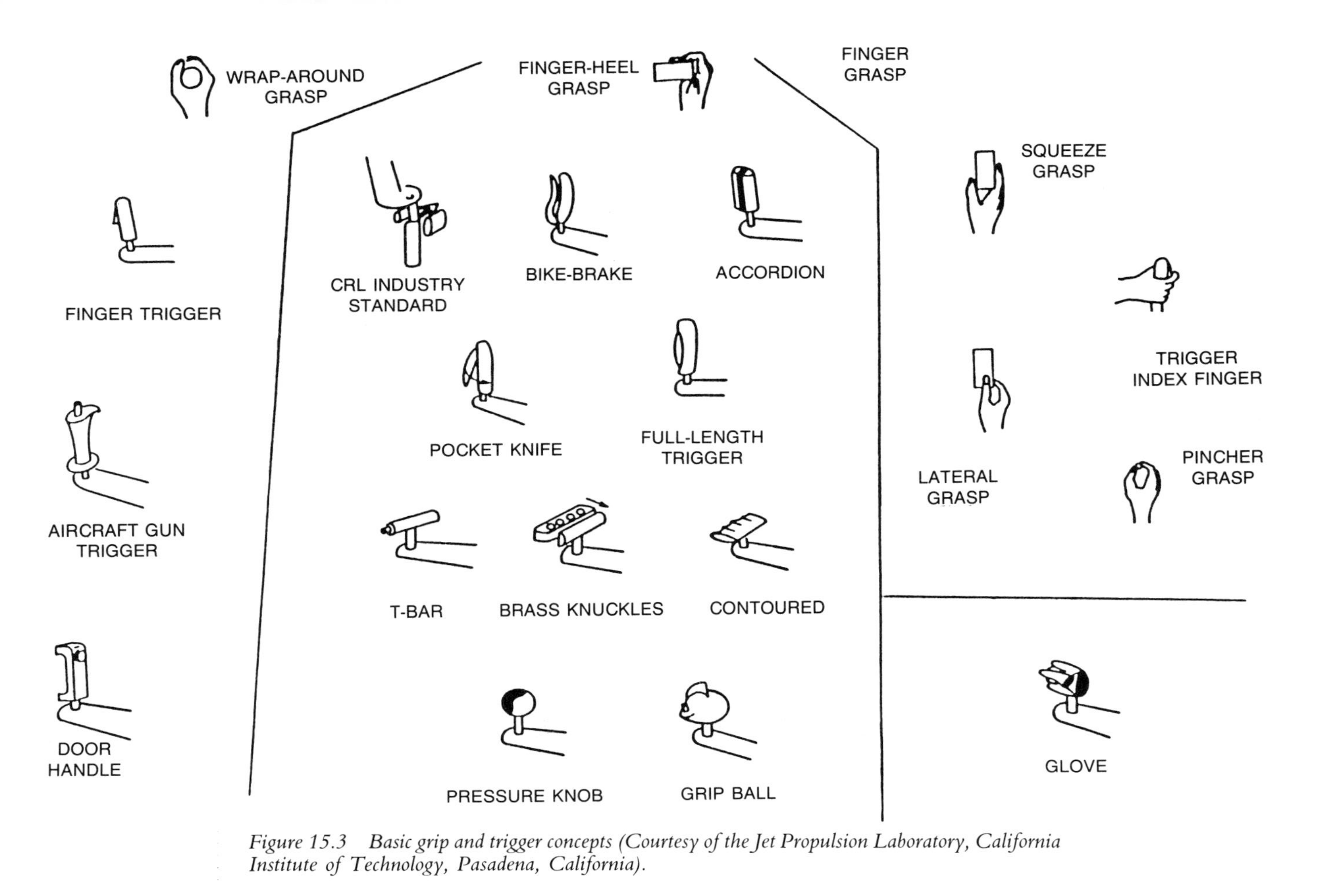


*Figure 15.3* *Basic grip and trigger concepts (Courtesy of the Jet Propulsion Laboratory, California Institute of Technology, Pasadena, California).*

sequential-link device and the slave is also, but is larger. Therefore the control directions to the slave joint actuators are identical with the feedback from each corresponding joint-sensing element in the master, scaled by the difference in master–slave size. Direct control is easiest when the master controller and the slave arm are identical (the master–slave configuration). It is still acceptable when the master is a replica of the slave (i.e. smaller or larger, but actuator and link similar). Direct control becomes progressively more difficult as the kinematic chain of the master varies from that of the slave.

*Indirect control* strategies are used when the master and the slave are different kinematically, thus operate in different coordinate spaces (Whitney, 1969; Mullen, 1973; Nevins, 1973; Shultz, 1978; Shultz *et al.*, 1979; Tesar and Lipkin, 1980). In indirect schemes, the master positions, rates and forces are transformed into a manipulator-independent space (most like a cartesian space: $XYZ$ pitch-roll-yaw). Commands expressed in this workspace frame are then transformed into the convenient space of the slave for execution (for force, rate, or position reflection the reverse process is performed). To accomplish the transformation a computer control is required. These control systems are similar to the controllers needed for conventional industrial robots. Since the controls are implemented at least in part through digital control methods, relatively fast sample and update rates are required (in the 60–200 Hz rate for different time-constant control loops). As the cost of microcomputer technology has dropped, indirect methods have gained favour over direct methods due to the added flexibility of controlling dissimilar master and slave.

*Force control* directs the slave manipulator under control to apply a force (or motor current for electric systems) directed by a force, or displacement in the master control unit (Whitney, 1977; Handlykken, 1980; Bejczy, 1983). Force alone is a difficult way to control a manipulator system due to rapid change in slave position in response to even small force commands. Direct force control is normally only used in automated assembly operations for active compliance (i.e. inserting shafts into holes, etc.). Even in this application, the slave manipulator is normally controlled for orientation (or wrist position) simultaneously with force control in $X$, $Y$ and $Z$. Force control, in robotics and teleoperation applications, is normally coupled to position control for stability. (This is not always the case, however. For instance in propulsion systems, force control is the normal mode.)

*Rate control* directs the slave manipulator under control to move at the speed directed by a force, or displacement in the master control unit (Whitney, 1969; Lynch, 1970; Mullen, 1973). This method of control has been the norm in aircraft flight controls. It makes sense because the physical action in aircraft systems is to displace aircraft control surfaces, which in turn control rate processes (due to changes in the craft drag functions) which cause aircraft orientation and direction to change. In telerobotics, rate control has been used to effect position control, and may even be the preferred control method in some cases (where the operator wants to move the end-effector rapidly from one side of a work area to another without accurate control), but does not match the requirements for most manipulation tasks. Most tasks require that an operator accurately controls position, orientation and sometimes forces applied. In these modes, rate control is inappropriate unless the rate of manipulator motion is very slow (as is the case for the RMS and for most common heavy construction machines). The major advantage of this control method is that a small hand controller can control a large manipulator without a motion scaling capability. It can also be supported by hand controls which do not reflect force or provide hand-control proportional displacement.

*Open-loop position control* directs the manipulator to move to the scaled position directed by displacement of the master control unit (Malone, 1973; Tewell *et al.*, 1974; Vertut and Coiffet, 1984). This method of control has been prevalent due to its relative simplicity (Figure 15.4). A simple displacement joystick, potentiometers, or a teach switch box (as used in industrial robot programming) can provide the control input. Early manipulators in the nuclear industry were mechanically-linked master–slave pairs. Evolution to open-loop controllers allowed these masters to be physically separated from the slaves through electronic controllers. The major disadvantage of open-loop control is that the operator has no direct feedback that the operation requested was successfully executed. Normally, open-loop position systems are, therefore, coupled with direct line of sight or television-based operator feedback. Because visual feedback may not contain high enough contrast or resolution, may not allow viewing of every area or interest and cannot 'see' effects of contacts (i.e. back forces), several methods of force reflection through the operator's master have been applied to close open-loop systems.

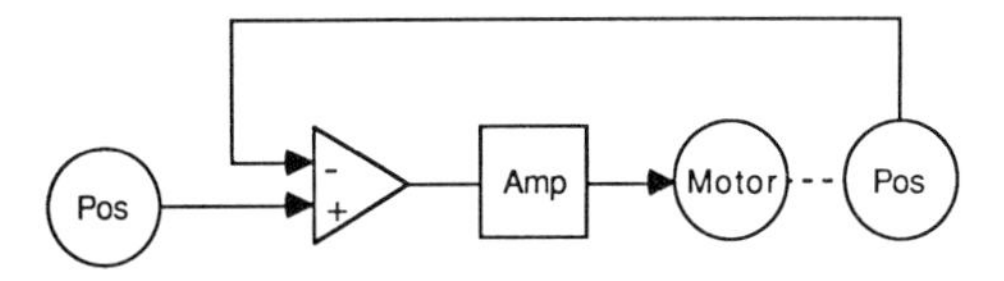


*Figure 15.4 Open-loop position control.*

*Bilateral position control* consists of two open-loop systems (Goertz *et al.*, 1961; 1966; Flatau, 1973; Brooks, 1979) coupled together (and requires an active master controller; Figure 15.5). One system directs manipulator motion from command information derived from the master control device, thus from the operator's direction. The other open-loop system connects the manipulator to the active master. This system operates on the master (and thus the operator's hand) to communicate the force and inertial state at the manipulator to the operator. Thus, the system is closed through the operator. Bilateral systems have the advantage of more information at the operator to augment partial or low resolution camera-derived data, but have the disadvantage of higher cost, higher complexity masters. The master control must become a servoed robotic device kinematically as complex as the manipulator system under control. Several specific strategies for bilateral control are discussed below.

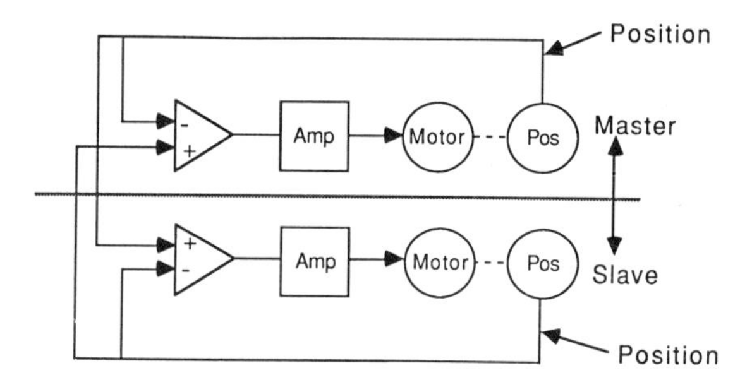


*Figure 15.5 Bilateral position–position control.*

*Position–position bilateral control* connects the master and the slave through cross-coupled position controls (Goertz *et al.*, 1961; 1966). In this method the operator moves the master which causes a position error at the slave. Thus the slave moves to make the error zero. Meanwhile, at the master, the further the operator pushes the control stick, the larger the error between the position sent back from the slave and the new master position, thus the higher the opposite centring force seen by the operator. This force is reflected to the operator and is a scaled approximation to the inertia of the slave arm. If, due to a contact in the work area, a force is exerted on the slave (for instance, the slave pushes up against an immovable object), this force slows or stops the slave. This, in turn, slows or prevents the slave in nulling the position error between the master command and its internally-measured displacement. This error, when projected back to the master, is reflected to the operator as a force. This control method is simple but it has several drawbacks. The degree of force feedback is set by a gain function, which if set too high, will cause positive feedback (an instability) between the master and the slave. If set too low it will make master–slave tracking slow and sloppy, and will back project forces inaccurately (giving a spring-like, damped feel to the operator).

*Position–force bilateral control* connects the master to the slave through a position control, and connects the slave to the master through a force control (Flatau, 1973); (Figure 15.6). In this method the operator moves the master, which causes a position error at the slave. Thus the slave moves to make the error zero. Meanwhile, at the master, the force measured at the slave manipulator is (scaled and) reflected to the operator (by pushing the control stick). If the force at the slave is measured using strain gauges it represents only object contacts, but if it is measured from the motor currents in the slave joints, it also incorporates the inertia of the slave arm. This control method appears to be optimal, but in this simple form it is very sensitive to sample rate and time delay, and can quite easily become unstable.

*Hybrid control* combines position–position characteristics and position-force characteristics, and thus can provide crisp control and feedback along with enhanced stability (Craig and Raibert, 1979). When teleoperation with force feedback is performed from a great distance, a serious problem of time delay

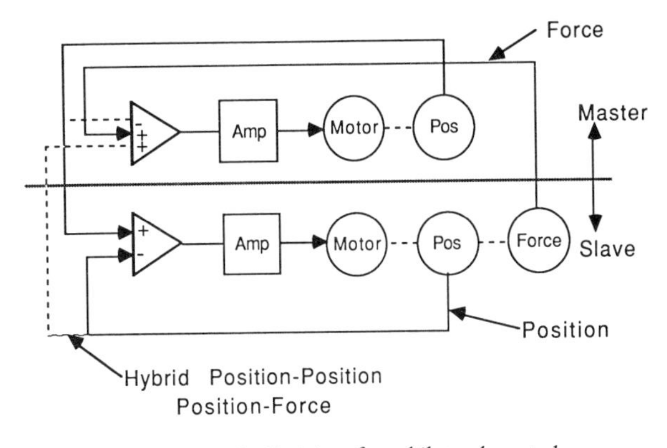


*Figure 15.6 Position–force bilateral control.*

is introduced. This was recognized in the mid-1960s (Ferrell, 1966). Further experiments showed that a system experienced instability with time delays as low as 1/10 s (Vertut and Coiffet, 1984). Anderson and Spong (1988; 1989) have also demonstrated a control which reduces stability problems and time-delay sensitivity. This system transforms the conventional feedback relations between master and slave shown in Figure 15.6 to those shown in Figure 15.7. By

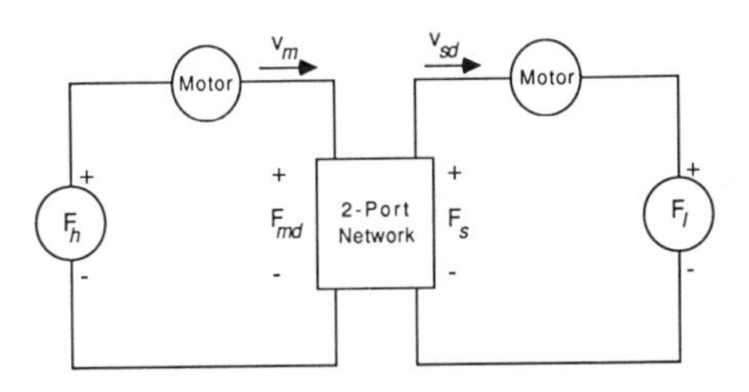


$$F_{md} = F_s \qquad v_m = v_{sd}$$

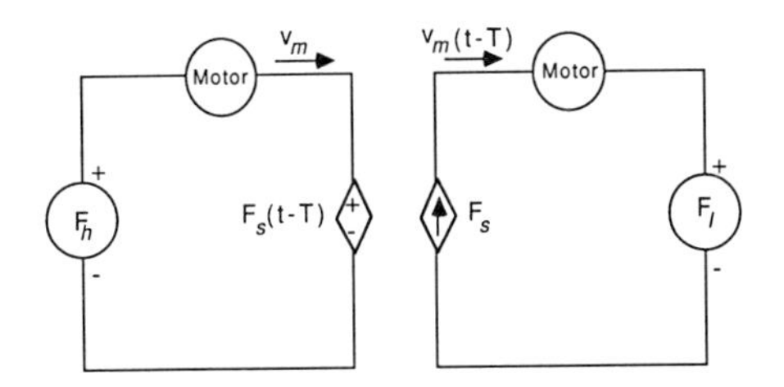


$$F_{md}(t) = F_s(t-T) + n^2(v_m(t) - v_{sd}(t-T))$$

$$v_{sd}(t) = v_m(t-T) + \frac{1}{n^2}(F_{md}(t-T) - F_s(t))$$

*Figure 15.7 Time delay stabilized force reflecting controler (from Anderson and Spong, 1988; 1989).*

adjusting the transformer terms $n^2$ and $1/n^2$ stability can be assured regardless of delay (with reduced tracking performance).

Level of autonomy of a teleoperator refers to the degree of control which is directly asserted by the operator. Alternatively, teleoperation can be fully under human operator control, fully autonomous, or an integration of the two called 'supervisory control'. In supervisory control, the computer can monitor and modify the operator's input to obtain better performance and avoid problems (Boussiere and Harrigan, 1988). To aid in control of robots over great distances, concepts of 'quickening' and 'time-braking' have been developed (Conway *et al.*, 1987).

With the advent of digital control methodologies, operator selectable, or cued control modes, are possible for task-dependent operation. For instance, we might select force control constrained to a plane perpendicular to a part insertion, while selecting position control of the insert direction and orientation. Or rate control might be selected for rapid slewing from one work area to another, with scaled position–force control selected once the new work area has been reached. Also, it is now possible to provide operator-controlled damping, velocity, force, or position envelope constraints. Greene (1973) proposes control of manipulators with more than 6 DOF through a combination of a 6 DOF controller device and position envelope constraints. These issues, while not yet well studied, should be supported in new hand-controller designs.

From this review of the various control methods possible in hand controllers, we recommend:

- providing an indirect control mechanism using digital control;
- providing control mode cueing so that multiple modes can be combined and controlled by task and operator dictated needs; and
- that bilateral control be supported (Position-Position, Position-Force, and Hybrid), along with rate control modes (for fast slewing).

## Hand controller configurations (types)

*Switch and potentiometer boxes* have been the traditional method of industrial robot control during the teaching phase (Tewell *et al.*, 1974). This method of operation mimics teleoperation but because it is done much less frequently, operator efficiency is of less concern. The main advantage of the switch box is cost and ease of industrial hardening. The major drawback is the poor level of control. For teleoperation this is more important because of the need to control not only endpoint positioning, but also manipulator dynamic path and force behaviour as well.

*Joysticks* have been used to control many multiple degree-of-freedom operations or devices (Measurement Systems; Hall *et al.*, 1970; Belyea *et al.*, 1971; Carmichael, 1979). Joysticks can be:

- isotonic, where operator action displaces the joystick handle, which generates a signal proportional to the displacement (this type does not normally have a returning force and includes 2 DOF mouse and trackball devices as well as conventional joystick configurations);

- isometric, where the operator applies a force to the joystick handle, which generates a signal proportional to that force (these sticks typically have limited or no displacement and are often implemented using strain gauges or piezoelectric crystals);
- proportional, where operator action displaces the joystick handle a small amount as a function of the force applied to the handle, which in turn becomes the output signal (these sticks usually have return springs); and
- hybrid, where some combination of the previous features are combined.

Joysticks have the advantages of being commonly available commercially, based on mature technology, relatively low in cost and simple to construct due to being purely haptic input devices (that is, they accept information from the operator, but do not provide active operator feedback, thus simplifying design). Unfortunately, as previously discussed, in teleoperation applications providing for haptic input and output is preferred.

*Replica controllers* (Figure 15.8) are active master devices having the same kinematic configuration as the slave manipulator, but are built on a different scale (Vertut and Coiffet, 1984; Schilling, 1989). They may be smaller than the slave to provide a compact control device (such would most likely be the case for space applications where the remote system may be quite large), or may be larger than the slave to provide enhanced operator control capability (as would be the case for a device to enhance a surgeon's capability to perform micro-surgery). The advantages of these devices are that they have potential for relatively small operating volume (limited by mechanical design constraints), potential for high accuracy (oversized replicas), potential for haptic output (force reflection), they can be anthropometrically well-matched to the slave. However,

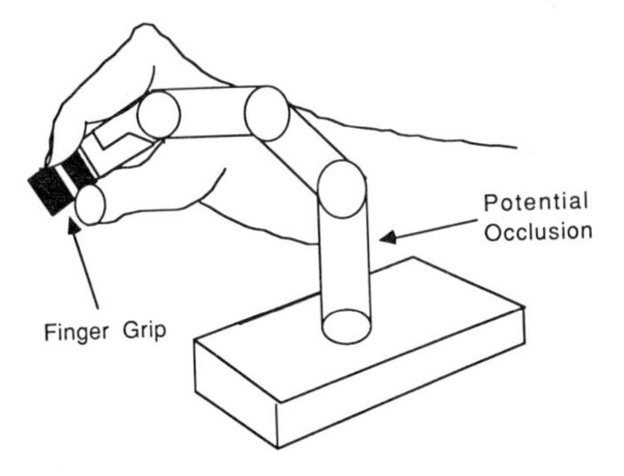


*Figure 15.8 A finger-grip replica master.*

this configuration does not represent the most compact controller arrangement possible, becomes difficult to use with unmatched slaves (i.e. becomes difficult to use if not kinematically similar to the slave) and requires the operator to be conscious of the master and slave kinematic chain configuration so as to avoid master self-occlusion (i.e. in some locations, the control will get in its own way, making use awkward).

*Master–Slave* devices have identical geometry/kinematics between the master device and its slave (Vertut and Coiffet, 1984; Central Research Labs, 1988). Thus this is a special case of replica controller and has many of the same advantages and disadvantages described previously. The key advantage of the master–slave is the simplicity possible if direct control techniques are applied. If control is direct, it is almost always more advantageous to make the master device to a convenient scale making strict master–slave undesirable.

*Anthropomorphic control* devices (Figure 15.9) are replica controllers which are configured and sized to be compatible with affixment to a human arm (Vykukal *et al.*, 1972; Bejczy, 1977). The key advantage of this approach is the ease of operator training (the operator already knows how to use his/her own arm, and thus can use the controller the same way), and the key disadvantages are potential bulk (if the operator must support the weight of the master device), stowage (storing when not in use) and that control becomes more difficult the more dissimilar the slave kinematics are from the kinematics of the human arm (and therefore the anthropomorphic control device).

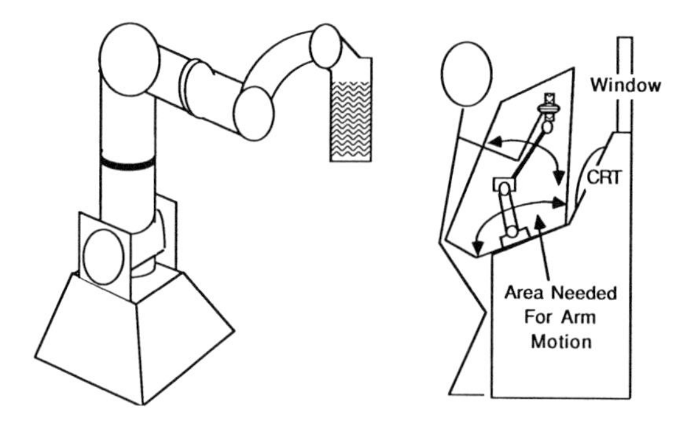


*Figure 15.9 An anthropomorphic replica (with wrap-around grip).*

Figure 15.8 shows a finger-grip replica which is grasped (at its end segment) by the operator's fingers like one would hold a pencil. This master configuration can be quite small, but is difficult to control for large motions because the operator normally rests the heel of the hand on a firm surface to get better finger control. Also the controller's own kinematic joint-link chain can occlude some motions. Figure 15.9 shows a larger serial link, anthropomorphic replica typical of masters used for underseas and nuclear applications. This configuration has proved to be versatile and reliable, but requires a relatively large operator volume (and therefore is inappropriate for the tight quarters of a cockpit).

*Nongeometric Analogic controllers* are devices which, like replica masters, have a direct mapping from the control device kinematics (joints and links) to the kinematics of the slave manipulator (Tewell *et al.*, 1974). But unlike replica masters, the nongeometric analogic controller is not kinematically similar. Therefore motions of the controller do not cause identical results at the slave. For instance, the controller may be an angle–angle joystick type device, and the slave might be an $X$–$Y$ displacement table. The control action is related to the slave motion (controller pitch to table $X$ displacement, and controller roll to table $Y$ displacement), but they are not identical. The advantage of this kind of controller is that it can control dissimilar slave devices; however the effectiveness of that capability is directly related to how dissimilar the slaves are.

*Universal controllers* take the concept of the nongeometric analogic controller one step further by incorporating coordinate transforms for translation from the controller to the slave and from the slave to the controller (Handlykken, 1980; Shultz, 1978; Shultz *et al.*, 1979; Bejczy and Salisbury, 1980). Supporting this type of controller without a digital control system is virtually impossible. The advantage of the approach is that dissimilar master and slaves can be connected effectively regardless of degree of dissimilarity. The operator orients and positions master-device hand grip (or its 'end-effector'). This position is transformed into the joint space of the slave and controls the slave to achieve the same end-effector position and orientation (even though the slave may not achieve the positioning the same way as the master device did). The advantage of the universal controller is its ability to control even quite dissimilar slave manipulators, however it requires a high performance digital control system which can perform the requisite coordinate forward and backward transformations in real-time (between 60 and 200 times/s). This type of controller is also advantageous for other applications like control of virtual environments, where the environment and objects within it are not (or cannot be) kinematically related to the controller geometry (for instance, portions of a chemical in a macromolecule simulation system).

Figures 15.11 and 15.12 show two configurations of an orthogonal universal controller. Both provide $X$, $Y$, and $Z$ motion through orthogonally organized translation stages (housed within the mechanism box). The first orthogonal controller uses a ball or finger-heel grip. This approach is usually married to a very limited travel $XYZ$ mechanism (2 in or less), and cannot be gripped as tightly as a flight-type grip by the operator. The second orthogonal configuration uses a flight stick grip for maximum operator gripping force and control. This configuration is the jump-off point for the detailed design and implementation underway in this project. From this discussion of hand-controller configurations, we recommend at least a 6 DOF universal controller to meet travel and over size constraints for specific telerobotic applications.

## Hand grips

Controller hand grips can make a joystick type device easy or difficult to use. Figure 15.10 shows some of the variability parameters for the human hand (Brooks and Bejczy, 1985; McCormick 1976; Van Cott and Kinkade 1972; Fogel, 1963; Human Scale, 1979; Johnsen and Corliss, 1971; Sheridan, 1974). As indicated in the table part of Figure 15.10, higher force levels can be controlled using wrap-around or finger-heel grips than can be controlled using finger-tip-only grips (such as those used to turn track balls). Squeezing triggers with the index finger can produce more force than can be achieved by squeezing the

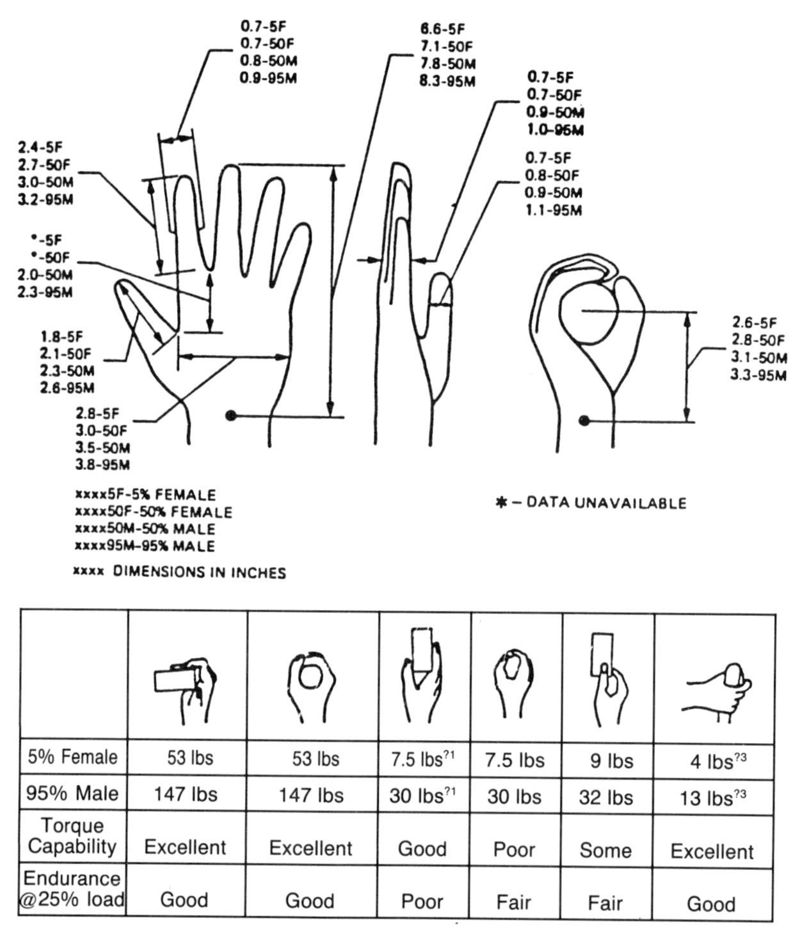


| | | | | | | |
|---|---|---|---|---|---|---|
| 5% Female | 53 lbs | 53 lbs | 7.5 lbs[?1] | 7.5 lbs | 9 lbs | 4 lbs[?3] |
| 95% Male | 147 lbs | 147 lbs | 30 lbs[?1] | 30 lbs | 32 lbs | 13 lbs[?3] |
| Torque Capability | Excellent | Excellent | Good | Poor | Some | Excellent |
| Endurance @25% load | Good | Good | Poor | Fair | Fair | Good |

?, Data unavailable; 1, Values assumed to about the same as pincher grasp but supporting evidence not available; 2, Mean value 100 male sujects; 3, Value assumed to be 1/3 of male value.

*Figure 15.10 Variability parameters for the human hand (Courtesy of the Jet Propulsion Laboratory, California Institute of Technology, Pasadena, California).*

index finger and the thumb together. Squeezing a triggering device with the whole hand is less desirable than squeezing with only the index finger because the former couples handle gripping force (i.e. the ultimate holding force which works against position and orientation forces) with holding forces (of an end-effector), while the latter does not (the index finger force can be controlled relatively independently of the gripping force applied by the lower three fingers and the heel of the hand).

Handle shape can also determine, in part, how strong a grip the operator can apply for pivoting or turning operations. An appropriately shaped grip is better than an unshaped grip. Table 15.1 shows a function of grip strength versus grip

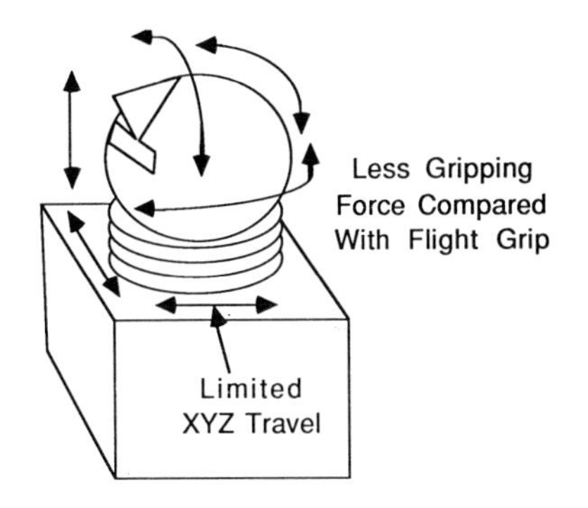


*Figure 15.11 An orthogonal controller with ball grip.*

element location. This indicates that a hand controller for use ranging from 5% female users to 95% male users will have to accommodate multiple handle shapes (thus driving a requirement for handle changeout).

*Table 15.1 Grip strength as a function of separation between grip elements.*

| Grip element separation (in) | Maximum grip strength (lbs) | Percentage loss from optimum (2.5 in) |
|---|---|---|
| 1.5 | 94 | 36 |
| 2.5 | 147 | 0 |
| 4 | 109 | 26 |
| 5 | 74 | 50 |

Figure 15.3 shows pictures of a variety of possible hand-grip types organized as finger actuated, hand-heel actuated and glove. Finger-actuated systems include track balls, switches, potentiometers and displacement balls. Hand-heel systems include wrap-around and finger-heel. In wrap-around handles the operator uses his/her lower three or four fingers to grip the handle strongly, while his/her thumb and/or index finger is free to actuate buttons or triggers. In finger-heel handles, the operator also pulls the handle to his/her hand-heel by applying finger force, but the handle is oriented horizontally instead of vertically (as in wrap-around). Because of this different orientation, the finger-heel devices are less index finger–thumb preferential for special function actuation, but also tend to provide less positive handle gripping force. The glove unit is a new innovation which will be discussed briefly below.

Table 15.2 is reproduced from Brooks and Bejczy (1985) to indicate their assessment of a figure of merit for each configuration of hand-heel grip type (finger actuated types are eliminated because hand-heel types have superior grip force performance). The Brooks and Bejczy analysis indicates that the best configuration is finger trigger. The Cybernet System handle is of this type with the addition of thumb-actuated buttons for additional mode control.

*Table 15.2 Assessment of handle designs (Courtesy of the Jet Propulsion Laboratory, California Institute of Technology, Pasadena, California).*

| | Engineering development | | | | Controllability | | | Human–handle interaction | | | Human limitations | | |
|---|---|---|---|---|---|---|---|---|---|---|---|---|---|
| | Design simplicity | Difficulty of implementation | Technology base | Cost | Stimulus–response compatibility | Cross-coupling | Secondary-function control | Force feedback | Kinesthetic feedback | Accidental activation | Endurance capacity | Operator accommodation | Total figure of merit Σ value × score |
| Value | 2 | 1 | 5 | 4 | 3 | 5 | 5 | 4 | 4 | 4 | 3 | 2 | |
| Industry standard | 2 | 2 | 3 | 2 | 3 | 3 | 1 | 3 | 3 | 2 | 1 | 2 | 97 |
| Accordian | 3 | 3 | 1 | 3 | 2 | 1 | 3 | 3 | 3 | 3 | 2 | 2 | 98 |
| Full-length trigger | 2 | 2 | 3 | 2 | 2 | 1 | 3 | 3 | 3 | 3 | 2 | 2 | 101 |
| Finger trigger | 3 | 3 | 3 | 3 | 2 | 3 | 3 | 2 | 3 | 3 | 3 | 2 | 117 |
| Grip ball | 3 | 3 | 2 | 2 | 2 | 3 | 1 | 2 | 2 | 1 | 2 | 3 | 85 |
| Bike-brake | 3 | 3 | 3 | 3 | 2 | 1 | 3 | 3 | 3 | 3 | 2 | 2 | 108 |
| Pocket knife | 3 | 3 | 3 | 3 | 2 | 1 | 3 | 3 | 3 | 3 | 2 | 2 | 108 |
| Pressure knob | 3 | 3 | 1 | 3 | 1 | 1 | 1 | 1 | 1 | 1 | 1 | 3 | 60 |
| T-bar | 3 | 3 | 3 | 3 | 2 | 3 | 1 | 2 | 2 | 1 | 2 | 3 | 94 |
| Contoured | 2 | 2 | 1 | 2 | 1 | 1 | 3 | 1 | 1 | 2 | 1 | 3 | 67 |
| Glove | 1 | 1 | 1 | 1 | 3 | 3 | 1 | 3 | 3 | 2 | 2 | 1 | 81 |
| Brass knuckle | 2 | 2 | 3 | 2 | 2 | 1 | 3 | 3 | 3 | 2 | 2 | 3 | 99 |
| Door handle | 3 | 3 | 3 | 3 | 2 | 3 | 2 | 2 | 2 | 2 | 2 | 3 | 103 |
| Aircraft gun trigger | 3 | 3 | 3 | 3 | 2 | 3 | 1 | 2 | 2 | 1 | 2 | 3 | 94 |

Glove-type devices have yet to be incorporated into a force-reflecting system, and are novel input-only devices which have been built in limited quantity (such as the Dataglove® from VPL and the SensorFrame (NASA, 1989) which convert hand gestures and positions into computer-readable forms). An obvious path for VPL was to incorporate active force-feedback into the glove itself, but this required electricity levels within the glove which could be unsafe if there were a problem. The SensorFrame (NASA, 1989) uses hand gestures within a special frame to provide computer input. For example, by detecting the hand and finger positions, an astronaut could potentially 'turn' 'virtual dials' on board the Space Station Freedom according to Linda Orr, Manager of the Graphics Analysis Facility at NASA JSC. Yet verification that the dial had been 'touched' would be unavailable through the sense of touch because this is an input-only device.

## *Hand controller requirements and design: an example*

If we consider the design of a 6 DOF force-reflecting hand controller useful in the space shuttle or space station, a set of design and performance requirements can be established (summarized in Table 15.3). Space applications require that size, weight and power consumption must be kept to a minimum. Small package size is a premium concern since the operator's console will incorporate other equipment as well, in a limited volume. This can be accommodated by incorporating the largest volume components of the hand controller, the *XYZ* translation stages, into a rectangular prism which mounts below or into the work surface/panel. This limits above-console volume to only the triple-gimbal joystick. The mechanism for a device supporting an 8 $in^3$ working volume, as shown in Figure 15.12, can be made in as little as 215 × 210 × 330 mm (8.5 × 8.25 × 13.0 in), which includes motors, sensors and drive amplifiers.

In addition to minimizing package size directly, the arrangement should be chosen which maximizes the operator's ability to see and manipulate adjacent systems (thus conserving 'valuable control and display real estate' (Brooks and Bejczy, 1985)). When we consider adjacent systems' visibility, it is necessary to take into account the impact of microgravity. In microgravity the operator's neck bends forward by 24°, altering visual line of sight by a positive 14.7° (see NASA, 1987, section 9.2–9). The magnitude of the forces and torques that a hand controller should apply to an operator has not yet been precisely determined and still requires further study. However, the indication is that between 40 N (9 lbs) and 80 N (18 lbs) continuous force capability in *X, Y* and *Z* translations is adequate. The corresponding angular torques (for pitch, roll and yaw) are 1.3 Nm to 2.7 Nm continuous about the *X, Y* and *Z* location of the handle. (NASA (1987) states forearm pronation at 17 Nm and supination at 14 Nm. Because the forearm is stronger than the wrist in rotation, the 2.7 Nm should be well within the wrist capabilities.)

For a universal hand controller it is not known what position, rate and force

*Table 15.3 Summary of a design requirements for a robotic hand controller.*

*Size related requirements*

Minimize overall size, weight and power

Minimize above-console volume

Limit occlusion of other parts of the console (i.e. switches, dials, lights, CRTs, etc.)

Limit controller movement outside its own volume (limit potential accidental interactions with other subsystems)

Minimize electronics module size

Allow physically separated locations for the hand controller and the electronics module

*Mechanical requirements*

Use brushless DC motors and resolvers to minimize maintenance

Provide very smooth motion stages with high backdriveability

Minimize use of high-tolerance machining in components

(Optional) Provide for removal of the control stick so that alternate grips can be substituted (such as the VPL Data Glove)

*Electronics requirements*

Design for small power loss in drivers (PWM)

Use PALs, microcomputers, high density memory to minimize package size/power

Provide feedback element to support accuracy requirements

Provide for control of slave manipulators

Provide for multiple fault tolerance (*** in flight/space qualified units***)

Provide adequate standard communications interfaces

*Accuracy/performance requirements*

16 bit positional feedback

10 bit or better force/rate feedback

Nominally 0.1 m (4 in) $XYZ$ translation

Servo-loop rate greater than 60 Hz

*Capability requirements*

Build with an indirect control system using digital control technology (the system will be a universal controller, i.e. can be interfaced to any slave manipulator geometry through coordinate transformations)

Support master hand controller mass and inertial forces compensation

Program controlled and scaled application of forces to the hand controller from the slave manipulator

Program controlled and scaled application of forces derived from the slave manipulator's inertial forces (thus making the slave's mass appear to be programmable)

Program controlled and scaled interpretation of hand-controlled displacements (in six degrees of freedom) as commands to the slave to move to a position (position mode), move at a given rate (rate control mode), or accelerate at a given rate (force or acceleration mode)

Make these modes software or operator selectable to match specific task needs

Support at least 6 DOF

*Human factors requirements*

Provide control mode cueing so that multiple modes can be combined and controlled by task and operator-dictated needs

Support bilateral control (position–position, position–force and hybrid), along with rate control modes (for fast slewing)

Minimize time delays by having greater than 60 Hz servo-update rates (some studies indicate that as little as 30 ms of time delay can negatively impact operator performance)

Provide positive indication of control mode changes

Provide mode change thumb buttons and trigger

*Table 15.3 (continued)*

Use a hand-heel hand grip type (finger trigger is preferred)
Provide capability for up to 80 N force steady state for force reflection
Make provision for changeable hand grips to support use by 5th percentile female users to 95th percentile male users

*Interfacing requirements*
Provide for convenient interfacing to host computer systems, slave manipulators and other equipment (Ethernet, RS232 and Parallel I/O)

*Space application requirements*
Compact design
One fault safe (i.e. hot redundancy in electronics/mechanical drives)

*Low cost manufacturing/volume manufacturing requirements*
Use common engineering materials
Build subsystems (motors, motor drivers, etc.) from common components (use identical motors, integrated circuits, resolvers, etc.); will allow volume discounts even on low volume production
Minimize high tolerances in machining of components
Use commercially available hardware wherever possible
Minimize high-wear components (use brushless motors and resolvers, etc.)
Qualify multiple sources and machine shops prior to large-scale manufacture

resolution will be necessary to support all relevant applications. We suggest, however, that the relative accuracy of the hand controller should match the relative accuracy of current state-of-the-art robots. This drives a requirement for 16 bit accuracy in the rotary position sensing system for each actuator (motor/gear drive). Because there is a 5 to 1 gear reduction from the motor shafts to *XYZ* translation, 16 bit accuracy at the motor shaft provides approximately $2^{18}$ distinct resolvable positions. Translational resolution is better than 0.001 mm (or 20 µrad for angular resolution). Controller rates and forces are digitized from analogue signals using a 10 bit A/D giving $2^{10}$ resolution elements each. These data represent the hardware limits and are beyond human capabilities. Therefore human performance will likely be the limiting factor in the overall system performance.

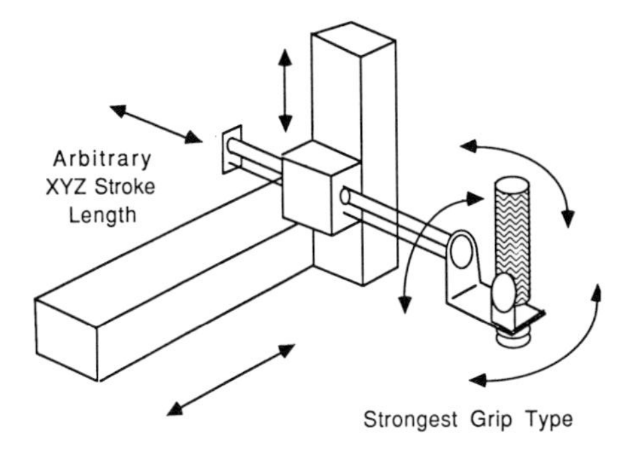


*Figure 15.12 An orthogonal controller with flight-stick grip.*

## Conclusions and applications areas

The compact orthogonal design and grip configuration has been driven by the needs of future avionics and space applications including space servicer control, advanced helicopter control and fly-by-wire aircraft control. For instance, a group of Canadian researchers reported on the benefits of an integrated 6 DOF joystick which permits helicopter flight control with one hand, increasing the potential for a safe landing by a wounded pilot (Lippay *et al.*, 1985).

In addition we expect to see a growing number of uses for computer-controlled haptic input/output devices for underwater activities, military applications, hazardous nuclear environments and advanced visualization graphics environments. These application areas will benefit from force-feedback and a greater degree of telepresence, but require universal controllers to decouple the controller coordinate frame from the coordinate system of the robotic device under control. For instance, in underwater or military operations, a common control console environment may be required to operate multiple robotic system configurations.

## Acknowledgement

The work reported here has been supported by NASA Johnson Space Center under contract NAS-9-18094.

## References

AAMRL, 1989, Anthropometry Dial-In Computerized Database. Point of Contact: Kathleen Robinette, Harry G. Armstrong, Medical Research Laboratory, Workload and Ergonomics Branch, Wright-Patterson Air Force Base, OH.

Anderson, R. and Hannaford, B., 1988, Experimental and simulation studies of hard contact in force reflecting teleoperation. In *Proceedings of 1988 IEEE International Conference on Robotics and Automation,* CH2555, Vol. 1 (pp. 584–9) Scottsdale, AZ.

Anderson, R. J. and Spong, M. W., 1988, Bilateral control of teleoperators with time delay. In Proceedings of the 27th IEEE conference on decision and control. Austin, TX.

Anderson, R. J. and Spong, M. W., 1989, Asymptotic stability for force reflecting teleoperators with time delay. In Conference on Robotics and Automation. Scottsdale, AZ.

Bejczy, A. K. and Handlykken, M., 1981, Experimental results with a six-degree-of-freedom force-reflecting hand-controller. In *Proceedings of the 7th Annual Conference on Manual Control* (pp. 465–77).

Bejczy, A. K. and Salisbury, Jr., J. K., 1980, Kinesthetic coupling between operator and remote manipulator. In *Proceedings of the International Computer Technology Conference of ASME* (San Francisco: CA) pp. 197–211.

Bejczy, A. K., Brooks, T. L. and Mathus, F., 1981, *Servomanipulator Man–Machine Interface Conceptual Design.* Final Report to Oak Ridge National Laboratories, JPL Document 5030–507. (Pasadena, CA: Jet Propulsion Laboratory).

Bejczy, A. K., Dotson, R. S., Brown, J. W. and Lewis, J. L., 1983, Force–torque control experiments with the simulated space shuttle manipulator in manual control mode. In *Proceedings of the 18th Annual Conference on Manual Control* AFWAL-TR-83-3021 (OH: WPAFB), pp. 440–65.

Belyea, I. L. *et al.*, 1971, *Design and Evaluation of Primary Hand Controllers for Fighter Aircraft* (Boeing Final Report No. AD-88-2492).

Bertsche, W. R., Pesch, A. J. and Wingate, C., 1977, *Investigation of Operation Performance and Related Design Variables in Undersea Force Feedback Manipulator Systems* (WHOI-76-47, AD-A-029344/9GA) (Woods Hole, MA: Woods Hole Oceanographic Institute).

Bicker, R. and Maunder, L., 1986, Force feedback in telemanipulators. In Morecki, A. *et al.* (Eds) *RoManSy 6 Proceedings of the Sixth CISM-IFToMM Symposium on the Theory and Practice of Robots and Manipulators* (Cambridge, MA: MIT Press), pp. 255–60.

Boussiere, P. T. and Harrigan, R. W., 1988, Telerobotic operation of conventional robot manipulators. In *Proceedings of 1988 IEEE International Conference on Robotics and Automation,* CH2555, Vol. 1, pp. 576–83.

Boussiere, P. T. and Harrigan, R. W., 1989, An alternative approach to telerobotics: using a world-model and sensory feedback. In *Proceedings of 1988 IEEE International Conference on Robotics and Automation,* CH2555, Vol. 1, pp. 1–8.

Brooks, Jr., F. P., 1977, The computer 'scientist' as toolsmith: studies in interactive computer graphics. In Gilchrist, B. (Ed.) *InfoProc* 77 (Amsterdam: North-Holland), pp. 625–34.

Brooks, T. L., 1979, 'SUPERMAN: a system for supervisory manipulation and the study of human/computer interactions', MS Thesis, MIT.

Brooks, T. L. and Bejczy, A. K., 1985, *Hand Controllers for Teleoperation—a State-of-the-Art Technology Survey and Evaluation* (JPL 85-11) (Pasadena, CA: Jet Propulsion Laboratory).

Busby, R. F., 1976, Operational equipment, navigation, and manipulators. In *Manned Submersibles* (Office of the Oceanographer of the Navy), pp. 467–535.

Carmichael, M. W. J., 1979, Elementary joystick mechanisms. In *Proceedings of the 5th World Congress on the Theory of Machines and Mechanisms,* Volume 2, Montreal.

Central Research Labs, 1988, Product literature.

Chaffin, D. B. and Andersson, G., 1984, *Occupational Biomechanics* (New York: Wiley).

Clement, G., Fournier, R., Gravez, P. and Morillon, J., 1988, Computer aided teleoperation: from arm to vehicle control. In *Proceedings of 1988 IEEE International Conference on Robotics and Automation,* CH2555, Vol. 1, pp. 590–2.

Conway, L., Volz, R. A. and Walker, M., 1987, Tele-autonomous systems: methods and architectures for intermingling autonomous and telerobotic technology. *IEEE International Conference on Robotics and Automation.*

Corker, K. and Reger, J., 1984, Human operation control of a bilateral teleoperator in part-simulation of zero gravity. In *Proceedings of the Human Factors Society 28th Annual Meeting* pp. 810–4.

Craig, J. L. and Raibert, M. H., 1979, A systematic method of hybrid position/force control of a manipulator. In *IEEE COMPSAC,* Chicago, IL.

Draper, J. V., Herndon, J. N. and Moore, W. E., 1987, The implications of force reflection for teleoperation in space. In *Proceedings of the 1987 Goddard Conference on Space Applications of Artificial Intelligence and Robotics,* Greenbelt, MD, 24 pp. CONF-870591-1.

Farnell, K. E., Richard, J. A., Ploge, E., Badgley, M. B., Konkel, C. R. and Dodd, W. R., 1989, User Needs, Benefits and Integration of Robotic Systems in a Space Station Laboratory (NASA Contractor Interim Report No. 182261, October 1987–January 1989) Teledyne Brown Engineering, Huntsville, AL.

Ferrell, W. R., 1966, Delayed force feedback. *IEEE Transactions of Human Factors,* **8,** 5, 449–55.

Flatau, C. R., Greeb, F. J. and Booker, R. A., 1973a, Some preliminary considerations between control modes of manipulator systems and their performance indices. In Heer, E. (Ed.) *Proceedings of the First National Conference of Remotely Manned Systems: Exploration and Operation in Space,* (Pasadena, CA: California Institute of Technology and NASA), pp. 189–98.

Flatau, C. R., Vertut, J., Guilbaud, J. P. and Germond, J. C., 1973b, MA22—A compact bi-lateral servo master-slave electric manipulator. In *Proceedings of the 20th Conference of Remote Systems Technology.*

Fleishman, E. A. and Quaintance, M. K., 1984, *Taxonomies of Human Performance: the Description of Human Tasks* (Orlando, FL: Academic Press).

Fletcher, J. C., Greeb, F. G., Brodie, S. B. and Flatau, C. R., 1975, *Variable ratio mixed-mode bilateral master-slave control system for shuttle remote manipulator system*, Martin-Marietta and NASA/JSC, United States Patent 3,893,573.

Fogel, L. J., 1963, *Biotechnology: Concepts and Applications* (Englewood Cliffs, NJ: Prentice Hall).

Garrett, J. W., 1971, The adult human hand: some anthropometric and biomechanical considerations. *Human Factors,* **13,** 2, 117–31.

Goertz, R. C., Blomgren, R. A., Grimson, J. H., Forster, G. A., Thompson, W. M. and Kline, W. H., 1961, The ANL Model 3 Master-Slave Electric Manipulator—its design and use in a cave. In *Proceedings of the 9th Conference of Hot Laboratories and Equipment.*

Goertz, R. C., Grimson, J. H., Potts, C., Mingesz, D. and Forster, G., 1966, The ANL Mark E4A Electric Master-Slave Manipulator. In *Proceedings of the 14th Conference of Remote Systems Technology.*

Graves, H., Bailey, A. and Mellon, D., 1972, *Study and Development of an Electric Side Stick Controller for Aerospace Vehicles* (Minneapolis, MN: Honeywell Regulator Company).

Greene, P. H., 1973, Hierarchical hybrid control of manipulators, artificial intelligence in LSI. In Heer, E. (Ed.) *Proceedings of the First National Conference of Remotely Manned Systems: Exploration and Operation in Space* (Pasadena, CA: California Institute of Technology and NASA), pp. 431–46.

Hall, H. J., Way, T. C. and Belyea, I. L. 1970, Design and Evaluation of Primary Hand Controllers for Fighter Aircraft, Technical Report AFFDL-TR-71-16, Wright Patterson AFB, OH.

Handlykken, M., 1981, Dynamic stabilization methods for bilateral control of remote manipulation. In *Proceedings of the ISSM Symposium on Mini- and Micro-Computers in Measurement and Control.*

Handlykken, M. and Turner, T., 1980, Control system analysis and synthesis for a six degree-of-freedom universal force-reflecting hand controller. In *Proceedings of the 19th IEEE Conference on Decision and Control* (Albuquerque, NM), pp. 1197–1205.

Hannaford, B. and Anderson, R., 1988, Experimental and simulation studies of hard contact in force reflecting teleoperation. In *Proceedings of 1988 IEEE International Conference on Robotics and Automation,* CH2555, Vol. 1 (Philadelphia, PA), pp. 584–89.

Herndon, J. N., Babcock, S. M., Butler, P. L., Costello, H. M., Glassell, R. L., Kress, R. L., Kupan, D. P., Rowe, J. C., Williams, D. M. and Meintell, A. J., 1989, Telerobotic manipulator developments for ground-based space research. In *American Nuclear Society 3rd Annual Conference on Robotics and Remote Manipulation,* pp. 1–9.

Herndon, J. N., Hamel, W. R. and Meintel, A. J., 1988, Robotics systems for NASA ground-based research. *Robotics,* **4,** 1, 19–25. (Note: the journal has been renamed *Robotics and Automation*).

Hill, J. W., 1979, *Study of Modeling and Evaluation of Remote Manipulation Tasks with Force Feedback* (Final Report NASA-CR-158721) (Menlo Park, California: SRI International).

Hill, J. W. and Sword, A. J. 1973, Touch sensors and control. In Heer, E. (Ed.) *Proceedings of the First National Conference of Remotely Manned Systems: Exploration and Operation in Space* (Pasadena, CA: California Institute of Technology and NASA), pp. 351–68.

Holden, K., Adam, S. and Gillan, D., 1989, 'A descriptive model of text selection performance with a control device.' NASA HCIL unpublished paper available from Marianne Rudisill, Man-Systems Division, Man-Machine Analysis Branch, NASA JSC, Houston TX, 77058.

Hollinshead and Jenkins, 1981, *Functional anatomy of the limbs and back.* (5th edition) WJB Saunders Co., pp. 160–5

Huggins, C. T., Malone, T. B. and Shields, Jr., N. L., 1973, Evaluation of human operator visual performance capability for teleoperator missions. In Heer, E. (Ed.) *Proceedings of the First National Conference of Remotely Manned Systems: Exploration and Operation in Space* (Pasadena, CA: California Institute of Technology and NASA), pp. 337–50.

Human Scale 1;2;3;4;5;7;8;9, 1979 (Cambridge, MA: MIT Press).

Jelatis, D. G., 1977, A power-assisted 45 kg grip system for master-slave manipulators. In *Proceedings of the 25th Conference of Remote Systems Technology,* pp. 158–61.

Jenkins, L. M., 1987, Telerobot experiment concepts in space. In *Proceedings of the SPIE Space Station Automation III Conference,* Vol. 85 (Bellingham, WA), pp. 92–4.

Joels, K. M., Kennedy, G. P. and Larkin, D., 1982, *The space shuttle operators manual* (New York: Ballantine Books).

Johnsen, E. G. and Corliss, W. R., 1971, *Human factors applications in teleoperator design and operation* (New York: Wiley).

Kasian, I. I., Kopanev, V. I., Cherepakhin, M. A. and Iuganov, E. M., 1974, Motor activity under conditions of weightlessness. In *Weightlessness: Medical-Biological Investigations* (Moscow: Izedatel' stvo Meditsina), pp. 218–36. (In Russian, translated abstract from DTIC No. 75A22970).

King, M. L., McKinnon, G. M. and Lippay, A., 1981, Design and development of a six degree of freedom hand controller. In *Proceedings of the 17th Annual Conference on Manual Control* (Los Angeles, CA: UCLA), pp. 455–77.

Kugath, D. A., 1972, *Experiments Validating Compliance and Force Feedback Effect on Manipulator Performance* (Final Report NASA-CR-128605) (Philadelphia, PA: General Electric Company).

Lestienne, F. and Clement, G., 1985, Postural adjustments associated with arm movements in weightlessness. In *AGARD Results of Space Expt. in Physiology and Medicine* and Informal Briefings by the F-16 working group. (see N-8531805, pp. 20–52).

Lippay, A., 1977, Multi axis hand controller for the shuttle remote manipulator system. In *Proceedings of the 13th Annual Conference on Manual Control* (Cambridge, MA: MIT), pp. 285–8.

Lippay, A. L., Kruk, R. and King, M. L., 1986, Flight test of a displacement sidearm controller. In *Proceedings of the 21st Annual Conference on Manual Control* (published as NASA Conference Publication 2428) (Moffett Field, CA: NASA/Ames), pp. 16.1–16.23.

Lippay, A. L., McKinnon, G. M. and King, M. L., 1981, Multi-axis manual controllers, a state-of-the-art report. In *Proceedings of the 17th Annual Conference on Manual Control* (published as JPL Publication 81–95) (Los Angeles, CA: UCLA), pp. 401–5.

Lippay, A. L., King, M. L., Kruk, R. V. and Morgan, M., 1985, Helicopter flight control with one hand. *Canadian Aeronautics and Space Journal,* **31,** 4, 335–45.

Lumia, R. and Albus, J. S., 1988, Teleoperation and autonomy for space robots. *Robotics,* **4,** 1, 27–33.

Lynch, P. M., 1970, 'Rate control of remote manipulation with active force feedback', MS Thesis, MIT.

Lynch, P. M. and Whitney, D. E., 1972, Active force feedback rate control of manipulators. In *Proceedings of the 8th Annual Conference on Manual Control.*

McCormick, E. J., 1976, *Human Factors in Engineering and Design* (New York: McGraw-Hill).

McGreevy, M., 1989, 'Personal simulators and planetary exploration'. Plenary address given at ACM CHI'89 meeting, Austin, TX, May.

Malone, T. O., 1973, Man–machine interface for controllers and end effectors. In Heer, E. (Ed.) *Proceedings of the First National Conference of Remotely Manned Systems: Exploration and Operation in Space* (Pasadena, CA: California Institute of Technology and NASA).

Malone, T. O., Kirkpatrick III, M. and Seamster, T. L., 1988, 'Operator interface issues in the control of telerobotic systems'. Paper MS88-273 presented at SME conference Robots 12 and Vision '88.

Measurement Systems, *Tracking Control Products for Human Operator, Catalog* (Norwalk, CT).

Meyer, D. E., Abrams, R., Kornblum, S., Smith, J. E. and Wright, C., 1988, Optimality in human motor performance: ideal control of rapid aimed movements. *Psychological Review,* **95,** 3, 340–70.

Miyazaki, F., Matsubayashi, S., Yoshimi, T. and Arimoto, S., 1986, A new control methodology toward advanced teleoperation of master–slave robot systems. In *Proceedings of 1986 IEEE International Conference on Robotics and Automation*, CH2282, Vol. 2 (San Francisco, CA: IEEE), pp. 997–1002.

Montemerlo, M. D., 1986, NASA's automation and robotics technology development program. In *Proceedings of 1986 IEEE International Conference on Robotics and Automation*, CH2282, Vol. 2 (San Francisco, CA: IEEE), pp. 977–86.

Morecki, A., Gianchi, G. and Kedzior, K. (Eds), 1987, *RoManSy 6: Proceedings of the Sixth CISM-IFToMM Symposium on Theory and Practice of Robots and Manipulators* (Cambridge MA: MIT Press).

Mosher, R. S. and Wendel, B., 1960, Force reflecting electrohydraulic servomanipulator. *Electro-Technology.*

Mullen, D. P., 1973, An evaluation of resolved motion rate control for remote manipulators, M.S. Thesis, MIT.

NASA, 1978, *Anthropometric Source Book: NASA Reference 1024, Vol. 1 Anthropometry for Designers* (Moffett Field, CA: NASA/AMES).

NASA, 1987, *Man-System Integration Standards (MSIS) Volume I, II and IV*, NASA-STD-3000 (Moffett Field, CA: NASA/AMES).

NASA, 1988, Second Annual Workshop on Space Operations Automation and Robotics (SOAR '88) (Moffett Field, CA: NASA/AMES).

NASA, 1989, *Video alignment system for remote manipulator.* In NASA tech briefs MSC-21372 technical support package (Moffett Field, CA: NASA/AMES).

Natarajan, J., Cochran, D. and Riley, M., 1984, Grasp fatigue as a function of handle size and shape. In *Proceedings of the Human Factors Society 28th Annual Meeting* (San Antonio, TX: Human Factors Society), pp. 836–40.

Nevins, J. L., Sheridan, T. B., Whitney, D. E. and Woodin, A. E., 1973, The multi-moded remote manipulator system. In Heer, E. (Ed.) *Proceedings of the First National Conference of Remotely Manned Systems* (Pasadena, CA: California Institute of Technology and NASA), pp. 173–87.

Norman, S. D., 1986, Theoretical considerations in designing operator interfaces for automated systems. In *Proceedings of the SPIE Space Station Automation II Conference* Vol. 729 (Cambridge, MA), pp. 225–30.

O'Hara, J. M. and Olsen, R. E., 1988, Control device effects on telerobotic manipulator operations. *Robotics,* **4,** 1, 5–18. (Note: the journal has been renamed *Robotics and Automation*).

Palmer, A. K., Werner, F. W., Murphy, D. and Glisson, R., 1985, Functional wrist motion: a biomechanical study. *Journal of Hand Surgery,* **10A,** 39–46.

Pepper, R. L. and Hightower, J. D., 1984, Research issues in teleoperator systems. In Heer, E. (Ed.) *Proceedings of the First National Conference of Remotely Manned Systems: Exploration and Operation in Space* (Pasadena, CA: California Institute of Technology and NASA), pp. 173–87.

Reid, L., 1975, A Preliminary Evaluation of Manual Control Problems Associated with

the Space Shuttle Remote Manipulator System. SPAR Aerospace Products, Ltd.
Roesch, J. R., 1987, *Handgrip Performance with the Bare Hand and in the Extravehicular Activity Glove* (Rept. JSC-22476) (Houston, TX: NASA Lyndon B. Johnson Space Center).
Roscoe, S. N., Highland, R. W. and Knowles, W. B., 1962, Human Engineering in Remote Handling. (AMRL-TDR-62-58). Wright Patterson Air Force, Aerospace Medical Research Laboratory.
Rouse, W. B., Geddes, N. D. and Curry, R. E., 1987–88, An architecture for intelligent interfaces: outline of an approach to supporting operators of complex systems. *Human–Computer Interaction,* **3,** 87–122.
Saenger, E. L. and Pegden, C. D., 1973, Terminal pointer hand controller and other recent teleoperator concepts: technology summary and applications to earth orbital missions. In *Proceedings of the 1st National Conference of Remotely Manned Systems* (Pasadena, CA: NASA and California Institute of Technology), pp. 327–36.
Schilling, 1989, Product literature.
Schmidt, R. T. and Toews, J. V., 1970, Grip strength as measured by the Jamar dynamometer. *Archives of Physical Medicine and Rehabilitation,* pp. 321–7.
SensorFrame, 1989, Controlling computers. *NASA Tech Briefs,* 18–9.
Sheridan, T. B. and Verplank, W. L., 1978, Human and Computer Control of Undersea Telemanipulators. MIT Man-Machine Systems Laboratory Report.
Schilling, 1989, Product literature.
Shultz, R. E., 1978, 'Computer-augmented manual control of remote manipulators', MS Thesis, University of Florida.
Shultz, R. E., Tesar, D. and Doty, K. L., 1979, Computer augmented manual control of remote manipulators. *Proceedings of the 1978 IEEE Conference on Decision and Control*, San Diego, CA, pp. 1413–7.
Stark, L. W., Kim, W. and Tendick, F., 1988, Cooperative control in telerobotics. In *Proceedings of 1988 IEEE International Conference on Robotics and Automation*, CH2555, Vol. 1 (Philadelphia, PA), pp. 593–5.
Stark, L., Kim, W., Tendick, F., Tyler, M., Hannaford, B., Barfakat, W., Bergengruen, O., Braddi, L., Eisenber, J., Ellis, S., Ethier, S., Flora, D., Gidwani, S., Heglie, R., Kim, N., Martel, B., Misplon, M., Moore, E., Moore, S., Nguyen, A., Nguyen, C., Orlosky, S., Patel, G., Rizzi, M., Shaffer, E., Sutter, M. and Wong, H., 1986, Telerobotics: problems in display, control and communication. In *Proceedings of the SPIE Space Station Automation II Conference,* Vol. 729 (Cambridge, MA), pp. 244–59.
Szirmay, S. Z., Schenker, P. S., Rodriguez, G. and French, R. L., 1987, Space telerobotics technology demonstration program. In *Advances in the Astronautical Sciences*, Vol. 63 (Keystone, CO: AAS), pp. 435–44.
Takase, K., 1987, Development of advanced teleoperation systems and international cooperation. In *Proceedings of Symposium for International Co-operation on Industrial Robots* '87 (SIGIR-87) (Tokyo: Japan Industrial Robot Association (JIRA)), pp. 303–10.
Tesar, D. and Lipkin, H., 1980, Assessment for the Man–Machine Interface Between the Human Operator and the Robotic Manipulator. Attached to Final Report on Nuclear Reactor Maintenance Technology Assessment, CIMAR.
Tewell, J., Spencer, R. A., Lazar, J. J., Johnson, C. H., Booker, R. A., Adams, D. A., Kyrias, G. M. Meirick, R. P., Stafford, R. W. and Yatteau, J. D., 1974, Configuration and Design Study of Manipulator Systems Applicable to the Freeflying Teleoperator. Final Report, Vol. II, Martin-Marietta, Denver, CO.
Turner, T. L. and Gruver, W. A., 1980, A viable suboptimal controller for robotic manipulators. In *Proceedings of the 19th IEEE Conference on Decision and Control* CH1563-6 (Albuquerque, NM: IEEE), pp. 83–7.
Van Cott, H. P. and Kinkade, R. C. (Eds), 1972, *Human Engineering Guide to Equipment Design*. (Washington, D.C.: American Institute for Research).
Vertut, J. and Coiffet, P., 1984, *Teleoperation and Robotics: Evolution and Development*, Volume

3A (Englewood Cliffs, N.J.: Prentice-Hall).

Vertut, J., Charles, J., Coiffet, P. and Petit, M., 1977, Advance of the new MA23 force reflecting manipulator system. Paper presented at the Symposium on the Theory and Practice of Robots and Manipulators, Warsaw, Poland, September.

Vykukal, H. K., King, R. F. and Valotton, W. C., 1972, An anthropomorphic master-slave manipulator system. In *Proceedings of the First National Conference of Remotely Manned Sytems: Exploration and Operation in Space*, Sponsored by NASA and California Institute of Technology, Ewald Heer, Ed.

Whitney, D. E., 1969, Resolved Motion Rate Control of Resolved Manipulators and Human Prostheses. Paper presented at the 5th Annual NASA-University Conference in Manual Control, MIT, Cambridge, MA, March.

Whitney, D. E., 1977, Force feedback control of manipulator fine motions. *Journal of Dynamic Systems, Measurement and Control*, 91–7.

Wickens, C. D., 1984, *Engineering Psychology and Human Performance* (Columbus, OH: Charles E. Merrill),

Wilt, D. R., Pieper, D. L., Frank, A. S. and Glenn, G. G., 1977, An evaluation of control modes in high gain manipulator systems. *Mechanism and Machine Theory,* **12,** 5, 373–86.

Wittler, F. E., 1975, Apollo experience report: crew station integration, volume 3: Spacecraft hand controller development (NASA-TN-D-7884).

Yamawaki, K. and Sumi, T., 1988, 'JEMRMS Hand Controller and Workstation, unpublished presentation, Space Station Robotics Working Group, Reston VA., December.

# *Chapter 16*
# *Human–robot integration for service robotics*

**K. G. Engelhardt and R. A. Edwards**

*Center for Human Service Robotics, Carnegie Mellon University, 4616 Henry St., Pittsburgh, PA 15213, USA*

**Abstract.** The quest for innovative technological answers to unmet human needs drives scientists and technologists to explore previously uncharted domains. The creation of intelligent robotic systems to address physical and cognitive human needs raises new and complex issues of human–system integration. Human–robot interaction for robotic technology which works in close proximity to human beings demands new methods for relevant research and development. The emerging field of 'service robotics' challenges system designers to address human compatibility issues, safety and utility in innovative ways. System research and development for humans with a range of abilities provide critical opportunities for examining ways in which robotic technologies can augment human capabilities and identify key parameters in human–robot integration. Research agendas encompassing state-of-the-art service robotics are discussed and projected for the coming decade.

## *Introduction*

The convergence of various demographic trends and technical capabilities during the past decade have established a mandate for a new class of human-made tools: 'service robotics'. The distinguishing characteristic of service robotics—their close proximity to humans in service sector roles—creates new demands for both the human and the robot from a human–robot interaction perspective. Human populations with particular needs (e.g. people with disabilities) provide a unique resource for examining ways in which robotic technologies can augment human abilities across the spectrum of functions. Since no human being is normal in every respect and consequently levels of ability vary widely, the concept of universal design is an important component of designing useful and safe service robotic tools.

## *Background to service robotics*

Robots have been defined as systems that can 'sense, think, and act' (Reddy, 1986). Service robotics encompasses 'systems that function as smart, programmable tools, that can sense, think, and act to benefit or enable humans or extend/enhance human productivity' (Engelhardt, 1987). Service robotics represent a unique classification of robots. Their utilization is in close proximity to humans in shared workspaces as opposed to industrial robots, which operate primarily behind safety barriers. Service robotics demand higher levels of human system integration than have previously been addressed.

Based upon a decade of research efforts into human needs, as well as robotics, new approaches to technological innovation and human–system integration were needed. The challenges to the management of service delivery to elderly and disabled individuals motivated exploration into new options for addressing human independence and alternative health care delivery. The fundamental hypothesis of this work is that human-, not technology-driven research and development can help create systems that are more appropriate to serve identified human needs. Potential task sets that may be amenable to robotic intervention are derived from utilization of classical task analysis paradigms in concert with socio-medical research methods. Sensitivity to human dignity and independence are also critical considerations in conducting this type of research. Such human-oriented technological development increases the relevance and utility of emerging technologies, such as robotics.

Service robot evolution is analogous to biological evolution. Different forms will evolve to fill different niches best suited to their particular range of skills. No less than in the case of their human counterparts, robotic species' differentiation and successful survival are a result of taking advantage of appropriate environmental niches. The more successful ones will work in optimal settings best suited to their individual strengths and which minimize their weaknesses (Engelhardt, 1984; 1986; 1988). An analogy to different kinds of vehicles for specific uses offers an illustration of this point. For instance, even though a car and a pick-up are both motor vehicles and function in roughly the same ways, they are generally utilized for somewhat different purposes. Further, a piece of construction equipment such as a crane or earth mover is also a vehicle, but again, the 'jobs' they perform for humans are different from a family minivan or sports car. Likewise, robotic applications in health and human service domains will have a wide spectrum of forms and functions. Whatever the niche, the more interactively a robot works with a physically limited human, the more demand will be placed on its technological capabilities. Increasingly complex tasks that require direct human contact will demand ever more sophisticated sensory integration and a higher degree of machine intelligence. Thus, we can begin to work toward developing 'hybrid vigour' for these unique integrated systems. The fusion of defined user needs/desires with the appropriate application research and development holds promise for spurring the creation of new, flexible forms and ever more versatile systems to serve human demands.

From a simple, programmed robot arm that can perform repetitive tasks in a structured workspace, to the more sophisticated systems that have integrated sensory and communication capabilities which can be programmed off-line using high-level languages and perhaps coupled to Computer-Aided Design/Computer-Aided Manufacture (CAD/CAM) and voice systems, the potential designs seem almost limitless. However, as robotic systems are designed, they will need to evolve to fill particular niches. Investigation is required regarding the optimal designs for (a) ability type/level, (b) diverse settings ranging from work environments for non-disabled individuals to in-home care for very mildly impaired individuals to long-term institutional care of severely disabled individuals and (c) the best mix of sensory and robotic capabilities to address human needs in specified domains.

The origins of service robotic technologies can be traced along several pathways (Figure 16.1) with varying amounts of crossover among the multiple lineages which primarily evolved in parallel: industrial manipulators (robotics), teleoperators, computers, automated office equipment (the use of technology in non-manufacturing), durable medical equipment and consumer/medical electronics. The convergence of these areas and visionary research and development along with rapidly evolving miniaturized, lower cost, flexible microprocessor-based technologies have provided an array of tools and resources for creating a new class of hybrid 'smart' systems. Some of the contributing technologies to service robotics are briefly described below.

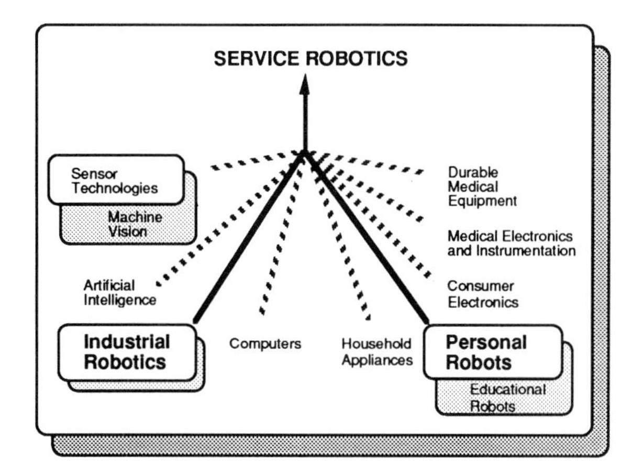


*Figure 16.1 Origins of service robotics.*

*Sensors*

Both sensor integration and end-effector design will play significant roles in successful user-acceptable designs. Versatile, high performance end-effectors with fine control capabilities, generalizable functions and increased sensor-augmented intelligence will be needed to perform jobs in close proximity to humans. Sensors in the manufacturing environment have provided the additional 'critical mass' requisite for ensuring the viability of robotic handling of parts, inspection and quality control. The identification of these sensory capabilities is only one side of the challenge. Creative and thoughtful synthesis of the technologies to meet health and human service requisites is the other, especially considering the approximately $2.65 \times 10^{32}$ possibilities for combining existing sensory capabilities (Engelhardt, 1985).

### *Durable medical equipment*

Service robotics, especially 'medical' robotics, also are derived from existing health service technology roots, especially in terms of the regulatory compatibility requirements for durable medical equipment. Medical devices are regulated by the Food and Drug Administration (FDA) based upon the 1976 amendments to the Food and Drug Administration Act. The FDA assigns devices to one of three classes with each class subject to different regulations. For example, 'Class 3' devices are subject to 'pre-market' approval because satisfactory performance of a particular device has not been demonstrated and, therefore, it is not possible to establish an adequate performance standard for the device. Additionally, data to support effective and safe use of this class of device do not exist as the design is often experimental. The 'risks to health' for Class 3 devices include bodily injury if the device malfunctions. If robotic assistants are perceived as Class 3 devices, their diffusion process could be substantially affected by the FDA approval procedures.

### *Consumer/medical electronics*

The growth of the consumer electronic industry during the past twenty years also contributed to the proliferation of microprocessor-based technology. Calculators, entertainment equipment, kitchen appliances and office equipment that incorporated computer technology into their devices provided mechanisms for users to become familiar with the operation of 'programmable' technologies. Concurrently, the evolution of electronic durable medical equipment, primarily in the areas of diagnosis and therapy, enabled medical personnel to acquire familiarity with electronic technology in their work environments. The recognition that technology, in general, could improve caregiving of ageing individuals was accelerated by a NATO sponsored conference focusing on technology and ageing (Robinson *et al.*, 1984).

### *Personal robots*

The 1980s witnessed the growth of 'educational robots' and toy robots for two main purposes: training of industrial robot users; and coursework in primary, secondary and community/technical college programmes. The 1980s also spawned much interest from hobbyists. The popularity of Star Wars in the motion picture media and other positive portrayals of robotic technologies encouraged the utilization of 'promotional robots' to entertain. The first International Personal Robot Congress held in April, 1984, recognized the potential for robotic technology to assist in caregiving domains (Engelhardt, 1984).

*Industrial robots*

'With the market considerably smaller today than 3 years ago, the Robotics Industries Association (RIA) reported that orders dropped precipitously from $483 million in 1985 to $364 million in 1986. Orders booked in 1987 totaled 4063 units at $341 million, down 29 percent and 6 percent from their respective 1986 levels. Foreign and domestically produced products shipped in 1987 totaled $299 million, down 32 percent from $441 million in 1986. On a unit basis, RIA stated 1987 shipments as 3949 units, a 37 percent decline from the previous year. Again, based on RIA data, 1988 orders fell approximately 20 percent and shipments by nearly 25 percent. In 1987, robot imports totaled 5479 units, an increase of 1572 or 40 percent over 1986. Imports declined 40 percent in 1988 to approximately 3332 units, valued, with parts and accessories, at $78.6 million. Japan accounted for 55 percent of the total followed by West Germany with 18 percent, Sweden with 9 percent, and Canada with 6 percent.' (Robotic Industries Association, 1989).

Since 1984, the number of industrial robot companies has fallen from 72 to 56. The decline and its magnitude indicate the severity of the industry's shake-out over the past four years. However, the industry's contraction is making the survivors more profitable. 'Current orders placed with the smaller number of firms moved some firms into the black for the first time while others returned to profitability.' (*US Industrial Outlook*, 1989, 20–4). Robot producers and vendors are serviced by a larger number of accessories manufacturers. Accessories include, but are not limited to, vision recognition systems, sensing and proximity devices, end-of-arm tooling, interface modules and compliance devices. Many lessons have been learned by the mistakes of the industrial robotics companies and the industry is on the rebound.

In the USA, federal funding of robotics has been concentrated in nonhuman service related areas. The Department of Defense spent approximately $200 m in 1986, NASA spent approximately $23 m in 1986, and the National Science Foundation (NSF) spent approximately $2.6 m. By comparison, human service robotics is not even recognized as a funding category. Isolated grants, primarily through the SBIR programmes, are estimated to total less than $1 m (Robotic Industries Association, 1989). The most primitive forms of this new breed of assistive robotics existed in research laboratories in the 1970s and 1980s. The Veteran Administration Rehabilitation Research and Development Program has supported rehabilitative robotics projects which might be considered precursors to or a subset of service robotics.

## *Potential user populations and need*

The mandate for focusing our attention and resources on health and human services is broad. The National Health Interview Survey (1983–5) indicated that 32.5 million persons have activity limitations in the US; (14.1% of the non-institutions population). More than 4% (7.7 million) of the age 15 and over non-institutionalized population in the USA need assistance in activities of daily living (ADL) and instrumental activities of daily living (IADL). ADL include bathing, dressing, eating, walking and other personal functioning activities. IADL cover preparing meals, shopping, using the phone, doing laundry and other measures of living independently. Two and a half million people need assistance in ADL and 5.1 million people need assistance in IADL (National Center for Health Statistics, 1989). In 1985, there were 1 318 300 residents in nursing and personal care homes in the United States. Forty-five per cent (597 300) were age 85 and over which represents a rate of 220.3/1000 population in that age group (US Department of Health and Human Services, 1989). 'By 1990 there will be 6.2 million elderly Americans with one or more basic disabilities, up from almost 5 million in 1984, according to estimates by the Urban Institute, a research organization.' (US News and World Report, 1989, p. 22).

Disability has implications for the overall state of the nation. For decades severely disabled individuals were often not considered employable. 'Those between the ages of 35 and 60 are four times more likely to become disabled than die.' (Roth, 1989). Of the population aged 18–69, 11.5%, or 17.4 million persons, report some degree of limitation in working at a job or business due to chronic health conditions. This includes 6.6% or 9.9 million persons, who are unable to work altogether. An additional 3.4% of the total population are limited in activities other than working at a job or business. It has been well documented that involvement of these potential users in the innovation process is the critical factor to successful technological creation and utilization. Consequently, the methodological approach conceived and implemented to create service robotic systems focused on how best to incorporate potential end users at the onset of the innovative process. 'Interactive Evaluation, (I/E)' is a need-driven method for creating service robotics.

'Need' can be considered from several perspectives. Appropriate technological solutions to human need can be characterized by clearly defining areas of need and systematically specifying technological systems that offer the best answers to the problem areas. First, individuals with temporary or permanent physical limitations have needs for replacing lost functional capabilities. These 'needs' are based on the assumptions that independence is a universal human goal and that maximum functional independence is currently not being achieved through existing measures. A person's control (both perceived control and functional control) of his/her personal space is an important component of human dignity and quality of life. One of our research goals has been to identify and elaborate these areas of need and to research robotic technologies which might be used to address these 'shortfalls' in human performance functioning. Our basic assumption is that both the humans and the machines have 'disabilities' and that an appropriate performance match of both the abilities of the human and the capabilities of the robot are essential for successful human–system integration (Engelhardt, 1985).

Accurate information about the incidence and prevalence of disabilities is difficult to quantify because of problems inherently involved in collecting descriptive data. Since disabilities are generally a result of trauma or disease, patients are classified by the aetiology of their disability rather than by functional deficits or, better yet, remaining functional abilities. Presently, functional needs can only be approximated, particularly in ageing populations. 'The National Center for Health Statistics reports that 81% of all persons over age 65 have chronic conditions that impair their ability to function independently. Further, 46% are limited in their activities because of chronic conditions, while 39% are limited in major activities.' (American Health Planning Association, 1982; Holt, 1987).

Second, need can be considered from the professional caregiver's perspective. Informal caregivers (family, friends) play a significant role in caring for and maintaining the disabled and/or older person at home. Their burden is significant (Chappel, 1983; Sangl, 1983). Their need for assistive technology is as great as, or greater than, the formal caregivers'. As numerous informal caregivers become frail themselves, and further tax existing health care delivery mechanisms (Brody, 1985; US Senate Special Committee on Aging, 1985), the need for alternative methods of care provision, such as assistive technology, will intensify. Formal caregivers and health-care professionals, as well as informal caregivers, have stated that by replacing odious, routine, demeaning, or boring tasks they could be relieved of the less desirable caregiving responsibilities, and could be freed to deliver care that humans can uniquely provide. In this way, robotic technology holds potential for augmenting their quality caregiving by reducing some of the burdensome aspects of it. Fetch and carry tasks, lifting and transferring, vital signs monitoring, feeding, bowel and bladder care are all example tasks in which caregivers would like 'a hand' from technology (Engelhardt, 1984; 1986).

Third, need can be considered from the perspective of medical administrators who are presently facing enormous challenges in an increasingly competitive health-care environment. Technology that can demonstrate cost savings while maintaining or improving existing quality of care and/or productivity can address one of the most pressing needs in today's cost-conscious industry. Health-care administrators will need to consider the cost/benefit/risk relationships that exist for potential robotic devices. Multipurpose tools can provide increased cost justification and increased functional capabilities. This class of technology also holds the potential for answering previously unsolvable, labour intensive challenges, such as patient wandering or meal preparation and delivery.

Legislation and regulation of employee safety and acceptable working conditions are inherently part of all industries. The health-care milieu is a unique 'industry' in that patients with significant, often life-threatening, vulnerabilities also require protection along with employees. In some respects the health-care environment can be considered a 'hazardous environment' (exposure to communicable disease, handling of contaminated materials and x rays are three examples of hazards) which is traditionally the type of environment for which robots are well suited. Successful introduction of robotic technology into health-care settings will require knowledge of existing regulations for particular health-care environments and for specific device designs.

'Health services are very labor intensive. In 1987 there were 3,114,000 full time equivalent (FTE) personnel or 401 FTE/100 adjusted patient census in community hospitals in the United States. Labor represented $80,992,000 (53%) of the total expenditures of $152,585,000.' (American Hospital Association, 1988). 'Many hospitals try to solve the nursing shortages by relying on overtime for nursing staff. More than 40 percent of hospitals employ temporary or agency nurses. Many hospitals in New York, New Jersey, and Pennsylvania hire nurses from foreign countries such as the Philippines, England, Canada and Ireland.' (*US Industrial Outlook,* 1989).

Fourth, need can be considered from the macro, societal perspective. Needs exist in a larger framework. It is important to note that there is no monolithic health care industry. This 'industry' is a mosaic of socio-political entities which coexist in the 'non-system' of health care delivery in the United States. These entities include (but are not limited to) federal, state and local governments, as well as religious and proprietary corporations, private non-profit organizations, consumer groups and professional associations. Health care is a more than $400 billion industry comprising nearly 11% of our Gross National Product for the past several years.

Fifth, need can be considered from the global perspective. As nations evolve greater mutual interdependencies and advances in communications technologies provide avenues for increasing interaction, innovative mechanisms for addressing universal human needs can be shared. Such sharing has the potential for helping each of us, with our own national perspectives and biases, to envision and create new tools for addressing common challenges facing all humans. Our uniting denominator is that we demand that technological advances serve human needs and work in concert with us to enhance and enable our abilities. For most elderly individuals, some residual capabilities remain (though often declining) that can and should be utilized as long as possible in order to maximize the individual's functional abilities. Technologies that: (a) support these remaining abilities, (b) compensate for partially lost capabilities, and (c) are flexible enough to adapt to varying demands are needed throughout the world. Projects designed to attain these goals are a fertile area for research as technological progress continues to provide new options. The concept of coping must give way, in a new decade and a new century, to an expanded understanding of rehabilitation. (Engelhardt, 1989).

Needs assessment is required, but how does basic research occur? From the perspective of the robotic domain, basic research is requisite in fundamental needs and potential applications. In this case, it requires the combined knowledge and experience of persons with both robotic and health-care backgrounds. This might be accomplished through the use of multidisciplinary teams. However, too often the technical person, with little or no experience in health-care delivery, will find it difficult to imagine or visualize application areas in this unfamiliar field. On the other hand, the clinicians' reluctance to accept robotic technology, or their unfamiliarity with or misconceptions about robots, may prevent them from envisioning relevant uses for a device they can only barely start to imagine. Involving potential users, who can clearly articulate their needs and who demand technological utility, can be the driving force behind the direction of causality in discovering appropriate application areas (Engelhardt, 1985).

*Service robotics: research and methodology*

The interactive evaluation approach to researching and developing robotic systems provides the framework for systematic introduction of human factors research into the design of robotics. Most technology development occurs in isolation from its intended end users regardless of the application domain. In the case of robotics for health services the discrepancies among technically-trained developers, health professionals and patients is large. Health-care technologies have often been inappropriate for the individuals or consumers for whom they are designed (Office of Technology Assessment, 1982). In robotics, end users were only included 'after the fact'. Benchmark testing (evaluation) occurred in isolation because designers assumed they were the 'experts' in their fields. Interactive evaluation (IE) is a process for delineating needs and assessing the feasibility of developing and disseminating prototype robotic systems. Evaluative research conducted within this framework is dynamic, iterative, interdisciplinary (Figure 16.2) and includes basic, baseline and applied research.

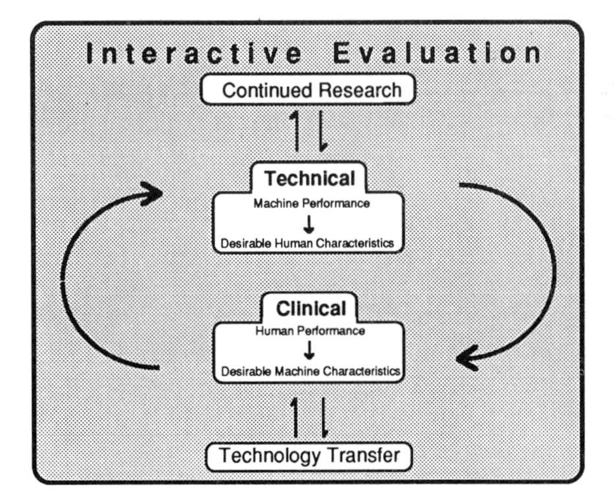


*Figure 16.2 Model of interactive evaluation.*

Innovative research approaches, which work to blend the protocols of different fields, are required to examine human–system interactions and integration (Engelhardt and Edwards, 1986). The fields of psychology, medicine, human factors, computer science, engineering and design all offer paradigms to help us gain basic knowledge required to build more intelligent systems. Multidisciplinary research protocols allow a wide variety of allied professionals and disabled persons to be involved in the lifecycle of a potential product. Evaluative research includes potential end users as research and development partners throughout all stages of potential product creation, from idea conceptualization to technology transfer. This method facilitates the definition, design, development and dissemination of appropriate state-of-the-art technology in a potentially cost-effective manner (Walsh *et al.*, 1985). This type of research and development process is energy intensive requiring more 'effort' on the part of professionals unaccustomed to working together. This may be one reason why there have been few, if any, evaluative methods that focused on the interactive approach. The benefit of this approach in terms of a higher quality, more usable system and research data generated far exceeds the cost, particularly as the complexity of the system increases.

The successful and interprofessional community 'team' was predicated on the following factors: (1) we identified and reached agreements on their problems; (2) we gathered the necessary baseline data; (3) we established measurable goals; (4) we obtained and sustained top-level support; (5) we allocated quality staff, finance and other resources/support needed to follow through on initiatives; and (6) we continually assessed results, strategies and, where necessary, remained flexible. These factors have been identified as critical to successful teams (US Department of Labor, 1986).

The first step in the IE approach is to conduct basic research into human needs. The outcomes of the needs research provide task taxonomies that can then be analysed in a stepwise, systematic mode from the various technical and clinical perspectives. Hypothetical scenarios are formulated to test assumptions and stimulate discussion of potential design trade-offs. Human performance determines the desirable machine characteristics. If technological capabilities exist which meet the criteria of the desirable machine characteristics, then feasibility research can begin. If technological capabilities partially exist, the human performance provides the justification for development of new technological capabilites. Likewise, machine performance can help define the desirable human characteristics. Together these methods provide the mechanism for screening out inappropriate technologies and inappropriate use of technologies for particular user subgroups. Another capability derived from this approach is that the division of labour between the intelligent human and the smart system can be examined and optimized. In this way both the human and the machine can perform the portions of the task which each does best. The IE approach is one in which the sample population is more representative of existing and/or future demographics.

## *Agenda for research in service robotics*

In order to focus on service robotics and their specific research needs, the Center for Human Service Robotics (CHSR) was created. It is an outgrowth of work begun in the Health and Human Services Robotics Laboratory for exploring and designing human service systems based on robotic and artificial intelligence technologies. The major goals of this research are: to gain basic knowledge of human–system integration; to facilitate development that will contribute to the evolution of robotics in the service of humans; to help improve the quality of, and option for, human services through the application of advanced technologies; and to create innovations that will extend human productivity for individuals with a spectrum of abilities. Current activities focus on research associated with the design, development, and interactive evaluation of hybrid robotic systems.

## Basic command and control research

In our centre, one of our basic research projects, Mobilanguage, is focused on the development of command and control vocabularies for mobile and remote robots. It includes research on human preference and natural language interfaces. The goal of this area of investigation is to optimize effectiveness and efficiency of voice input/output in the operation of mobile robots. An optimal control vocabulary will have sufficient complexity to allow the user to perform complicated task sequences and yet have enough simplicity that it does not burden the user's cognitive workload. The types of meaningful utterance which people choose to control and command smart systems will contribute to our basic knowledge regarding human perception of machine movement.

The necessity of command and control research projects has derived from past experience with a stationary voice-controlled robotic system, the first generation Veterans Administration/Stanford University robotic aid. While developing and standardizing training procedures for this newly evolving robotic system, exploratory observational data showed that as the orientation of the robot changed, confusion increased regarding the movement of the robotic arm based on particular commands. The labelling of the robot arm with colour coded directional indicators (arrows) helped decrease this confusion. More detailed discussion of this observation is described in Engelhardt and Edwards, 1987.

### *Baseline research: interactive errors*

Our past research had demonstrated the importance of examining the interactive error rate of voice recognition systems. With the human in the loop, the errors are not just machine-generated but are often human errors in combination with machine errors (Edwards *et al.*, 1985). This experience suggests that any mobile robot which is voice operated might be more effectively controlled if the command vocabulary fits the perceived movement pattern, and if visual markers, such as movement direction indicators, are placed on the robot so they are useful as cues to the human operator. Such markers need to be visible when the robot is within the visual range of the user and when it is seen on a remote monitor.

The Robotic Vocational Workstation (RVW) allows users to operate the robot in real-time and preprogrammed motions with both voice and keyboard entry.

The directional markers have assisted naive users in learning the robot's motions. The markers have provided feedback regarding the name of the joint as well as the direction of motion. Colour coding of the markers with the corresponding markers on the keypad enhanced the process of learning which keys cause which motions. Simulations of the robotic arm on the screen assist in teaching the corresponding voice commands.

Our experience also suggests that, to develop better mobile robot training materials, more attention needs to be paid to the human factors variables such as interactive error avoidance strategies for commanding robots. Perception of control (or one's locus of control) is expected to increase if users learn that they can easily and quickly recover from errors. The strategy of choosing only commands that are meaningfully familiar and comfortable to the user will be important as a robot safety (error avoidance) strategy. This type of research is critical because the gap between machine-appropriate communication and meaningful human communication is wide. Historically, the gap was initially bridged by punch cards and assembly-type languages; and subsequently by 'high-level' programming languages and menus. Natural language interfaces, such as voice control, hold promise for reducing the burden of learning 'computerese' in order to access particular applications or customize off-the-shelf software for a special user need (Reddy, 1986). Voice control of robotic systems requires special training. Learning to use one's voice for commanding operations of an advanced system is significantly different from manual manipulation of controls. The command vocabulary for a vocational workstation is being evaluated and upgraded during testing at CHSR. The overall objective is to decrease user frustration and increase ease of use.

Changes in the vocabulary are based upon the following constraints and needs: (a) phonetic dissimilarity, (b) natural language and (c) new commands needed to add new functional features. The actual process of changing the vocabulary includes the following procedures: (1) tallying the types and sequences of mis-recognitions through laboratory use of the first generation RVW; (2) utilization of the voice recognition system's template differentiability score to grade new alternative commands; (3) best guess choices of words which might be less likely to be confused based on experience or training with voice command and control; and (4) group discussions of experienced users and consensus prior to any changes in the vocabulary. Researchers and developers need carefully to consider methods that can decrease development time and minimize design errors that occur from limited involvement with the actual challenges of real-world, non-simulated environments. Simulation of many parameters is desirable and works well for specific areas of automation. However, for human–robot teams, the understanding of the best way to make a robot adapt to a human assumes greater importance. The work envelopes under consideration are no longer robot work envelopes but rather more complex interactive envelopes in which both the human and the robot exist as a cohesive team. With increased mobility, from tracks to autonomous navigation, the envelope may be an entire room or a whole building. This interactive envelope and the behaviours of humans must be studied in settings that provide (realistic) information for robotic and artificial intelligence designers, developers and manufacturers as well as clinicians seriously considering utilizing these technologies. We must begin to understand boundary conditions and baseline performance characteristics of both humans and robots better. Feasibility is not simply technical feasibility but also clinical feasibility. Evaluation of 'feasibility' from the human user point of view is also critical (Engelhardt *et al.*, 1989).

### *Baseline research: voice command and control*

Humans have demonstrated extensive adaptability when communicating with living entities different from themselves. For example, we train animals to respond to a subset of verbal commands most successfully communicated in a firm tone in the imperative tense. We simplify our sentence structure and speak more slowly when communicating with children in order to increase the effectiveness of our communication with them. We listen more attentively, employ more gestures and often speak more slowly when communicating with someone who does not have the same native language as we do. We also generally alter our communication style when communicating through a telephone or other mass media such as radio, television, film, or video. Likewise, humans are likely to adapt their communication patterns successfully to interact with mobile robots. They are also likely to tolerate a certain amount of communication error as we already do when we interact with other human and non-human entities.

Six pilot subjects have navigated structured, mazelike environments from both a regular view and a remote view. Their verbal commands have been analysed from a functional perspective. One objective has been to classify the commands in terms of the roles they play in conveying desired movement to an intelligent mobile entity. A secondary goal has been to collect baseline data on specific verbal and gestural commands a wide age range of individuals utilize to direct an 'intelligent' mobile entity. This pilot research is being conducted as part of the 'Mobilanguage' project. This is especially important in complex service environments which mobile robots will share as public spaces with humans (Engelhardt and Edwards, 1987). Existing mobile robot bases provide the necessary technological tools for exploring the appropriate voice commands. Several are being investigated as part of this research.

Baseline studies of human-to-intelligent human, human-to-intelligent machine (robot), within structured and unstructured environments will help to formalize anecdotal and sparse evidence regarding the interaction of mobile intelligent robots with humans of varying ages and abilities.

### *Training on prototypes: how to gather baseline data on human-robot interaction*

A critical component to successful implementation of the IE process is the training of potential users actually to work with the prototypes. The prototype-system training procedures were based on andragogical (adult-learning) principles (Knowles, 1978) and were introduced systematically with a set of increasingly difficult tasks. The training task sets had been chosen for 'meaningfulness' as well as for their increasing control complexity. For instance, the first problem that users had to solve was using the robot under step-by-step voice commands to pick up a cup and straw and bring them to his/her mouth. Giving oneself a drink of water was the highest priority need identified by our quadriplegic users. This need request was then translated into performance parameters for a first level task set with the prototype robotic system. Subsequent tasks required increasingly difficult control of the robotic movements, such as moving ('piloting') the robot from one spatial plane to another (Engelhardt, 1984; 1989).

Research into appropriate and successful training procedures provides multiple types of human factors and safety information. It allows researchers to incorporate user input into the design and development of the prototype. Training enhances the developers' ability to discover how the system handles unexpected situations that occur with actual use by actual users. Finally, it provides information about training procedures which have a significant impact on acceptance, understanding and use of a complex assistive device like a robot.

User training consists of the development and standardization of procedures to teach naive individuals how to use the RVW. Baseline performance information, successful human–system profiles and a user training manual are also generated during this process. The development of training procedures plays a major role in assessing actual usage, and, therefore, utility of any sophisticated assistive technology. Examination of the role of training in increasing the users' perception of control over his/her environment by using assistive robotic technologies and generation of these data help illuminate the barriers to user acceptance and technology diffusion.

Standardized training procedures are selected based upon research conducted on user responses to most frequently requested tasks. Training is being standardized to 4–6 sessions, each lasting approximately 2 h. During each session, the user reviews existing expertise, acquires new information, practises the lessons in both structured and unstructured situations and finishes the session with a task or tasks which help the user assimilate the entire lesson. The sessions also incorporated increasing levels of task complexity. For example, later sessions required the user to perform tasks in multiple planes which required utilization of more complex robotic movements. These exercises helped the user become more familiar with real-time operation of the robot for non-routine tasks as well as understanding how to instruct the robot to perform preprogrammed task sequences. This andragogical or 'adult-learning' method includes such concepts as self-directedness, learning readiness, immediate applicability and problem-centred learning tasks. This type of training procedure has been successfully implemented in this work at Carnegie Mellon University, and our previous work. Over 250 users ranging in age from 3 to 90 years have been able to accomplish useful tasks with voice-controlled prototype robotic systems (Engelhardt *et al.*, 1984; 1989).

## Applied research: safety and service robotics

In the medical community, efficacy is closely linked to safety. Efficacy may be thought of as the probability of benefit from use of a medical technology. The Office of Technology Assessment (OTA) defines safety as 'a judgement of the acceptability of relative risk in a specified situation' (Office of Technology Assessment, 1978). Stringent clinical demonstration of the expected benefits in relation to the relative risks will be a requirement for 'medical' service robotic systems.

Safety is, and continues to be, the number one issue that transcends all other questions. Safety for the human and safety for the robot (and other humans and objects in its environment) is of paramount importance in the acceptance and diffusion of robots into health and human service settings. Safety is always relative. It is dependent on the desires and values of the individuals developing, designing, manufacturing, purchasing and using the technology. The priorities of the organization and the society also impact on safety. We will often spend great sums trying to 'band-aid' a fix for accommodating human safety after the system is designed and developed. Our research makes optimization of human factors and human safety with robots our highest priority as is consistent with a human-driven approach.

The safety requirements for the spectrum of care delivery settings from medical to human service are being examined. For instance, our research has demonstrated that the rapid robot manipulation and motion speeds required in the manufacturing environment appear to be undesirable in health and human service domains. Passive safety features, which already exist in various forms of industrial robot system, can be built into the electromechanical design of service robotic manipulators. For example, fire resistance, manual/passive movement capability, elimination of sharp edges and protrusions, padding, speed restrictions, low inertias of moving parts to limit the effects of collisions and maximum static forces as well as switch strips, mats and vests all offer increased potential safety for human users, the robot itself and their shared environment. Safe service robotic systems are expected to contain lower level supervisory system managers which monitor critical parameters such as forces, velocities, accelerations, checkpoints, deviance into 'off-limits' areas, self turn-off and start up, diagnostic error messages and informed, meaningful feedback to the user regarding system status. Such built-in system safety managers, combined with passive safety design features, form the basis for the application of the conceptualization of 'inherent safety' for service robotics (Engelhardt, 1987).

At a recent conference ten deaths were reported in Japan only, associated with robots in manufacturing environments during the last ten years. All of these were reportedly caused by human override of machine safety design, not robot error relative to safety standards (Nagamachi, 1988). How safe is safe? What is the efficacy? Who must share the risks as well as reap the benefits of technological innovations? How do we protect both employees and patients in health-care delivery domains? The first prerequisite for safety is to have the robot 'KNOW MORE ABOUT HUMANS and HUMAN ENVIRONMENTS'.

Intelligent hybrid systems will be mandated to acquire understanding of human physiological parameters, functional capabilities, needs, behaviours and perceptions. This is essential if we are to build systems that successfully respond with appropriate actions to human-defined needs.

Rendering the robot more autonomous by fusion of capabilities such as force, tactile and proximity sensors should enable a significant reduction in the cognitive workload of the robot user. Study of multiple sensors and their integration will also allow us to determine redundant and overlapping safety mechanisms to be implemented along with voice, manual overrides and emergency stops. Such fusion and redundancy further supports the concept of fail-to-safety, or 'safety architecture', for the human–robot system.

As end-effectors become more sophisticated, they will be able to be used in more human service areas. The need for sensitive and delicate touch, not necessarily required for many tasks in a manufacturing environment, will become essential to robotic systems that help lift or transport a frail patient whose body or garments must be handled with a confident but gentle touch. This is, indeed, the requisite domain for applications of both 'hi-tech and hi-touch' and these constraints can serve to push smart technology toward more utilitarian configurations.

Again, the most critical issue in service robot development is safety research. This is one of the most neglected areas of the innovation lifecycle, with safety 'features' often built in as an afterthought. The need for applied safety and human factors research *before* the prototypic technologies take product form cannot be overstated. We have the know-how and the technological capabilities to build more 'inherently' safe systems; that is, robotic systems that make human safety the number one priority for design. Issues such as speed, motion control, sensor integration, sensor hierarchy for safest usage and human safety training are but a few of the factors that can be addressed from the onset of system design. University-based research with a received emphasis on applied and baseline experimentation for technical innovations can play a pivotal role in guiding the creation of safer, more useful systems. The educational environment provides the structure for exploring safety and human factors questions in-depth.

### *Smart homes*

Technology that is designed to enhance remaining capabilities and to enable the individual with adapted, new, replaced and/or additional capabilities may be thought of as 'enabling technology' or 'augmentative technology'. Astronauts who operate manipulators in space for satellite maintenance are utilizing technology which augments their existing human capabilities; that is, they are not limited by their own physical strength or proximity. Likewise, augmentative technology can be utilized by abled as well as disabled persons to increase or expand the limits of what they can accomplish. The remote controller for a television enables the viewer to operate his/her television without physically touching it. The automobile with wheels which can 'run' faster than human legs extends the human's mobility capability whether abled or disabled. The caregiver who must help care for a disabled, but non-institutionalized person, as well as the carereceiver, could also benefit greatly from enabling technology that would augment caregiving goals and carereceiving needs.

An integrated system of such enabling technologies might well be a 'Smart Home' (Engelhardt, 1987). Working toward the use of this class of technology for enhancing a person's living space is a relatively new concept, but well-suited to the needs of older individuals and their quest for least-restrictive environments and self-sufficiency. A 'Smart Home' can provide a caring, forgiving environment that engenders a therapeutic, rehabilitative and enabling life space by: optimizing the use of residual abilities (including retraining) and compensating for lost capabilities, as well as enhancing, augmenting, or extending typical human capabilities. Such a smart home creates a new level of human–system integration in which robotic technology is working in concert with human beings. In fact, if one applies the definition of service robotics in the broadest sense, microwave ovens are primitive 'amoeba-like' robotics already existing in many homes. For example, the temperature probe provides a sensing capability when it identifies that a piece of poultry has achieved a certain degree of being cooked. The thinking ability is the comparing of what has been sensed with the prescribed instructions input into the microwave by its operator. The action is the very simple activity of sounding a buzzer. Microwaves enable a human (such as a child or frail older person) with a range of abilities safely to cook something because of the absence of an open flame or a hot element. Although this example is primitive compared with the potential, it does illustrate the gradual emergence and incorporation of smart technology into our daily lives.

One goal of the CHSR's 'Smart Home' project is to develop a demonstration home environment (which could be mobile), based on the premise that artificial intelligence and robotic technology can be utilized to create a forgiving environment for individuals with severe to moderate impairments (Engelhardt, 1987). Creating a functional, aesthetic, smart environment for this subpopulation will lead to a design that will appeal to young and middle-aged persons without apparent disabilities, and thus, to economies of scale. Such a forgiving home incorporates not only the convenience and efficiency components of Smart House™ efforts, but also, socio-medical, human factors and product design expertise as well.

Since many health-care professionals feel that the kitchen environment is the 'interface between independence and institutionalization', we are focusing our initial efforts on meal-related tasks and on expert nutrition systems including medication reminding. Such a smart kitchen is relevant to addressing the meal-related needs of other 'constrained' individuals, such as astronauts. Therefore, this advanced application development is significant across the service domain from the general consumer to the homebound senior to residents of the Space Station.

An extensive network of human services has developed during the past decade which focuses on providing assistance to this population outside institutional settings. Particular emphasis has been placed on expanding home-based care for the elderly, who are the fastest growing subset of long term care clients.

In tandem with this publicly-funded commitment, the assistive role of families as care providers for older citizens is considered a major factor in delaying or inhibiting premature institutionalization for at-risk older individuals (Comptroller, 1977; Chappell, 1983; Sangl, 1983). Research shows that where these informal support systems are available the basic physical and instrumental self-maintenance needs of the elderly can be met (Shanas, 1979; Feller, 1983; Sangl, 1983).

Over the next two decades the historical capacity of both elements of this dual caregiving system will be challenged due to the dramatic increase in the older population (Taeuber, 1983; US Senate Special Committee on Aging, 1985). These growing numbers of older Americans will require increased publicly-funded long-term care services. In addition, they will place increased demand on ageing spouses and families who will also be ageing, and, in some instances, infirm themselves (Brody, 1985). The mobility of today's society as well as the increase in employed women and the stresses of family, financial and housing concerns are also making family caregiving more difficult.

Such need requires us as a nation to explore new ways for addressing these challenges. Service robotic technologies are one important area of investigation as well as the roles other recent technological advances can play in daily living and in easing the caregiver burden. Consequently, Carnegie Mellon's CHSR is conducting a Technology Assessment Study in conjunction with Pennsylvania Family Caregiver Support Project as implemented by the Southwestern PA Area Agency on Aging. The Family Caregiver Support Project includes: caregiver assessment, carereceiver assessment and environmental assessment in all four regions of Pennsylvania. Our region has also included an additional component designed to gather baseline data on profiles of existing household technologies and human systems integration parameters. The three-year study incorporates the following objectives:

- relationship of technology assessment results to environmental, caregiver and caregiver assessment results;
- relationship of technology assessment results to general attitudes toward technology and toward innovation adoption;
- identification of needs for existing and future technological interventions; and
- training materials and video tapes for caseworkers regarding the potential of technological interventions for improving independent living.

This research seeks to address the assistive role an advanced technological intervention can play in extending the self-care capacity of older persons and thereby to expand the caregiving capabilities of both formal and informal care providers. In the long term, the resultant robotic technologies used in the smart home kitchen could be implemented in the homes of disabled elderly people and could significantly reduce their need for either formal or informal caregiving assistance with meal tasks, permit them to function more independently for longer periods of time and enable formal caregiving resources to be focused on more severe caregiving problems.

Highlights of the results from a pilot analysis of 58 households in rural Pennsylvania are as follows.

- 71% had 2 or more televisions (100% had at least 1 TV)
- 22% had 1 VCR
- 57% had 1 or more remote controllers
- 62% had 2 or more telephones (100% had at least 1 telephone)
- 53% had 1 microwave oven
- 69% had 1 electric coffee maker
- 69% had 1 electric blender
- 93% had 1 or more vacuum cleaners

Technology that has been available within the last 40 years and that has been perceived as useful can be found in these households. Based upon a 113 client sample of Southwestern PA's Family Caregiver Support Program, 51% of the clients have a degree of upper-limb dysfunction that impairs their activities of daily living such as meal-related tasks (Bergwyn and Savage, 1989).

The market for a smart home kitchen is a national market that is expandable to international arenas. As mentioned earlier, a large number of Americans are living in a community setting but are unable adequately to conduct their own meal preparation activities. Approximately 10–30% (100–300 000) of these individuals are so limited in their meal-related capabilities that they would benefit from a smart home kitchen. Once the medical efficacy can be demonstrated as to improved health status resulting from better nutrition, the system could be prescribed by physicians, thereby rendering it eligible for third-party reimbursement.

There are additional younger and disabled individuals who could benefit from a smart home kitchen. According to the 1978 Survey of Disability and Work, approximately 878 864 Americans aged 18–64 were severely disabled and homebound. It is likely that a high proportion of these individuals would also derive utility from a smart home kitchen (US Department of Health and Human Services, 1982). Furthermore, this kitchen is appropriately aesthetic and practical for use by upwardly-mobile individuals without apparent disabilities.

As our technologies evolve into almost limitless numbers of configurations, the forms of robots that are part of our trained consciousness will reach new understandings. For instance, as our buildings become smarter the small manipulator arms or mobile wheeled vehicles will be only part of the integrated components incorporated into truly 'smart' systems that will become robotic offices and homes. The robots will become part of a system of distributed intelligence throughout a network of rooms and spaces and will become 'transparent' to the end user. For instance, local control of basic heat and light for persons working in dense or multi-use spaces will be controlled by sensors that will 'know' about the individuals and their needs and preferences and will be able to adapt accordingly, much as our own bodies' homeostasis mechanisms operate.

*Figure 16.3 Photograph of CHSR's prototype smart home kitchen.*

## *EDUCATION: a critical research priority*

First, the focus for our educational system generically should be the concept of lifelong learning, not just education for the first 18 to 25 years of our expected 75 year lifespan. This becomes an increasingly important issue when it is estimated that up to 50% of our technological knowledge base becomes obsolete every four years. The OTA states that 'experience, training and education may be rendered useless by new information' (Office of Technology Assessment, 1987). Furthermore, we must accord more recognition to interdisciplinary and applied principles. Technological knowledge is evolving much more rapidly than the academic disciplines expected to teach it. Indeed, some of the most innovative milestones have emerged from cross-disciplinary efforts. One way this can best occur is through applied research which is (1) directed toward more than 'widget building' or 'technology for the sake of technology', and (2) incorporates a research base that prioritizes human factors as well as technical factors.

We have created a basic robotic curriculum for children in conjunction with Carnegie Mellon University's Children's School and the National Pilot Program for Developmental Advances in Science, Health and Technology. Our goal has been to begin to teach concepts and vocabulary for robotic technologies to pre-school and elementary school age children. We also sought to teach how the technology relates to human abilities because complex biological and technical concepts could be related to everyday experiences.

We are developing a sequential and integrated science, health and technology curriculum to help stimulate children's interest and increase their familiarity with this important part of their future. We conducted additional assessments with the children to try to understand better what their opinions of robotic technologies were. We utilized a hands-on andragogical approach to motivate participation of students and teachers. We also tried to see how we could use robotic technologies to teach other basic science concepts in mathematics, physics and biology. For instance, the use of infra-red and sonar sensors in navigation can be used to teach children about distance and wayfinding. Adult training and retraining programmes based upon this curriculum are also under development with various local community organizations.

Education and literacy are serious issues for Americans of all ages. In a technologically-oriented society, it is imperative that our citizens be afforded the opportunities to learn and relearn pertinent information throughout their lives if we are to maintain our technological viability. 'An older less adaptable workforce will face a job market that requires increasingly flexible skills, and many workers changing jobs five or six times during their worklives.' (Bennett and McLaughlin, 1988). We feel that robotic technologies offer the tools to allow creative learning experiences. Again, it is important that we not delay technical education until high school or college but incorporate this information into early childhood curricula and in ongoing teacher education.

### *The Robotic Vocational Workstation: human–robot integration issues*

The Robotic Vocational Workstation II (RVWII) is a second-generation prototype voice-controlled robotic system that has been researched and developed using an iterative, stepwise refinement methodology. The RVWII is targeted at individuals with diminished, if any, use of their upper limbs. Persons with these severe disabilities were previously considered unemployable. The goal has been to create a robotic system which will allow a person with upper-limb physical disabilities to be gainfully employed full time (eight hours). The process of integrating an intelligent human and a smart advanced technological device, such as a robot, requires careful consideration of factors from multiple perspectives. Development of the RVWII from two of those perspectives (the design of the system's environment and the design of the system's software) is described below. The first generation Robotic Vocational Workstation (RVWI) was implemented in 1988. This system was tested at the Health and Human Service Robotics Laboratory (now the CHSR), Harmarville Rehabilitation Center and the Vocational Rehabilitation Center. The subjects were 9 males and 3 females ranging in age from 20 to 45 years. These subjects represented a wide variety of disabilities including spinal cord injury, congenital amelia, spinal tumours and muscular dystrophy. In addition, the first generation RVW was placed in Magee Women's Hospital where it functioned 8 hours a day, 4 days a week, for 7 months. This extensive testing of the system demonstrated weaknesses in the system. A major weakness was in the robotic arm chosen for the system. The arm (a UMI RTX SCARA arm) proved inappropriate and unreliable. One of its major functional deficiencies was its inconsistent calibrating and returning to programmed positions. Also, it was found that an increased degree of system cost-effectiveness and increased application acceptability could be introduced by completely redesigning the system environment. The results from the first generation RVW guided the redesign of the uniquely flexible RVWII. It was these considerations that prompted the development of the RVWII system. It includes a better robotic manipulator, improved system workspace environment and more advanced and flexible software.

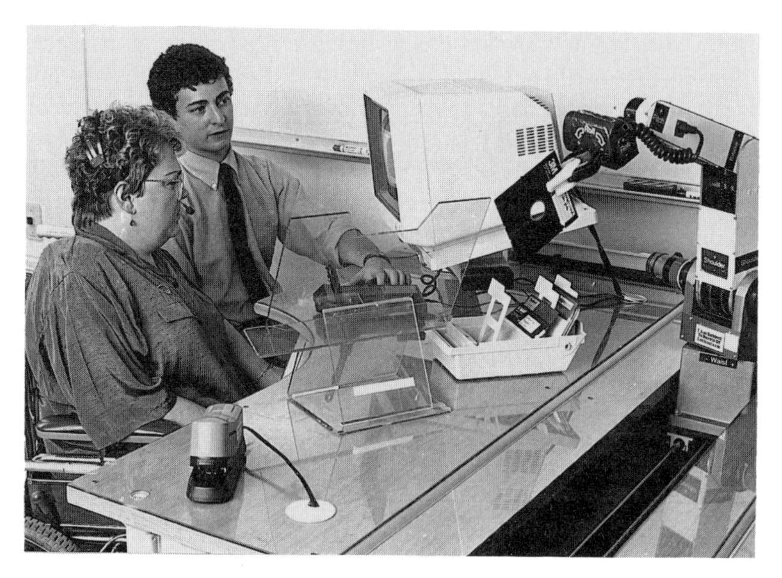

*Figure 16.4 Photograph of CHSR's Robotic Vocational Workstation II.*

## The RVWII: system environment design perspective

The design principles used in the creation of the RVWII system have been derived from classical human factors, product design and ergonomic guidelines. These have been integrated with health service information regarding persons with a wide range of disabilities. The concept of accessibility for the human, the robot and the workpieces in the environment directed our design effort. The system environment design consists of three modules.

### *User station*

The first module the user station acts as the primary interface tool to the other two modules, but can also be used as a stand-alone voice-operated computer workstation. This module contains the computer and all its necessary peripherals, a telephone and a readerboard. The keyboard is mounted on a three-dimensionally adjustable tray to account for various users preferences for keyboard angle or mouthsticking positions. Research gained from the first generation workstation showed that many users with keyboard ability preferred to mix some keystrokes with their voice commands. The broader design of the modules—guided, in part, by the choice and configuration of the robot arm and accompanying track—allowed for more leg and chair clearance. The auto boot-up makes further added improvement in docking methods needed for users to approach the first-generation workstation. The users can approach the computer from a variety of angles without the encumbrances of hitting table legs or undercarriages. The user station also acts as a barrier to keep the user from inadvertently entering the robot's work envelope, yet allows the user the space to interact safely with it.

*Robot station*

The second module in the system contains the robotic manipulator and controller. The robot chosen for the RVWII was the CRS M1A manipulator. This is a five DOF articulated arm which proved to be highly reliable and robust. It could also lift up to 4 lbs, far more than any comparably priced robot. The CRS robot is small so as not to overshadow the user (as the UMI arm did) and is very quiet in its operation.

Another component in the robot station is a horizontal track on which the arm is mounted. The track chosen for this was a Daedal horizontal table. This track is recessed into the table for three reasons. The first is that the added height obtained from the track's thickness translates to vertical work envelope lost by the robot. The second reason is the aesthetic, less industrial, more office-like feel gained from enclosing the track within the table. And the final reason is cleanliness. The mechanics of the track are protected by a retracting sheath that moves with the robot. The track gives the added work envelope of the entire RVWII work surface.

In a particular task domain, the successful interaction of a robot with its workpieces is largely dependent on the capabilities of the end-effector. In-depth research for office workpiece parameterization has been conducted with models and simulations.

A key component of our first year's research with the RVWI system was the workpiece inventory. The project team analysed the workpieces that the robot must handle and characterized them according to several dimensions such as size, shape and weight. These workpiece characteristics have served as the performance specifications for overall system design, including our specialized environment and custom-designed end-effectors.

To understand better the force requirements necessary safely to hold and manipulate the variety of office workpieces, a device was assembled to measure specific forces for different test gripper faces. With this device the project team has assessed the holding forces imparted on the various test gripper faces, and calculated the best leverage and motor torque requirements to fulfill specified office tasks.

Information gathered from the workpiece inventory helped specify three primary criteria for the new end-effector: 1) Opening width, 2) Approach profile, and 3) Closing strength. Other criteria important to the end-effector design are: closing speed v. torque (motor noise became a factor here), aesthetics (overall mass and approachability of design concept and how it blends with the robot arm and overall office environment). Also designed into this end-effector are interchangeable subcomponents. The joint between the fingers and the enclosed drive system is designed to shear should the robot crash or accidentally hit an object with excess force. This feature further enhances the inherent safety we have worked to design into the system.

*Equipment station*

The last module contains many of the peripherals with which the robot works and interacts. These peripherals include such items as: filing cabinet, file folder holder, laser printer, bookshelf (books, manuals, periodicals, etc.) and stapler. Other peripherals may be included depending upon user preference and job description. Each component of this station is modular in design. For example, the file cabinet is easily removable for exchange or transportation. Not every peripheral need be on the equipment station. The readerboard in the system is located on the user station. Additional peripherals, such as a paper shredder or waste basket, may be included on the floor next to the system.

*System versatility*

Because of the modular design of the system environment, five different configurations of the individual stations are possible.

- User station—stand alone
- Robot station—stand alone
- User station and robot station
- Robot station and equipment station
- User station and equipment station.

One possible scenario is the following: during the user's normal 8 h workday, he or she will use the entire system with all three stations. After hours, the robot station can be removed to perform an autonomous task while the user station and/or the equipment station can be used by another operator (with or without disabilities).

Furthermore, each of the three stations is independently adjustable in height. This allows for accommodating the characteristics of a wide variety of user ergonomics and preferences. The system components are also easily transportable. Each station has casters for mobility and latches for system integrity while the stations are attached. This modular concept is integrated into the power and data transmission between system stations. A single connector for power and a single connector for data are all that is required between the user station and the robot station (Sample *et al.*, 1990).

## The RVWII: system software design perspective

The RVWII software consists of a specially designed program that controls a robotic manipulator, a horizontal track, a customized voice recognition system and a telephone management system. Through this program, a disabled user gains full control over all of the systems' functions through voice control. The design of this software had to:

- be user friendly and user customizable;
- have on-line help;
- have a simple, easy to understand command structure;
- be accessible from within other software packages; and
- be easy to learn.

The software also had to satisfy a number of hardware system constraints concerning:

- choice of host computer system;
- integration of robot/track/computer/voice system; and
- small program size (occupying less computer memory) versus maximum functionality trade-offs.

By following these constraints and gaining end-user input during the development process, we have been able to create a powerful and flexible voice-controlled robotic system.

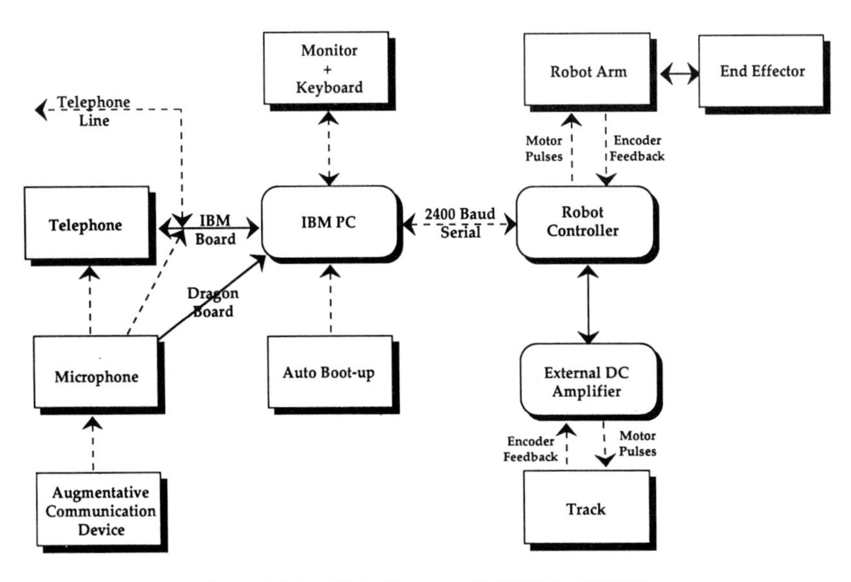


*Figure 16.5 Block diagram of CHSR's RVWII.*

*System functionality*

The basic system functionality is broken down into four sections: the voice recognition system, the robot control, the telephone manager and application specific software packages.

The voice recognition system handles all of the voice training and recognition for the RVWII. The voice system (Dragon Systems Voice Scribe 1000) will recognize utterances and convert them to various keystrokes to be interpreted by any other piece of software. The voice system is a speaker-dependent system, so each person who uses the system must first take a few minutes to train the system to his or her voice (approximately 10 min per 100 words trained). It also has the ability to retrain or redefine and voice command 'on the fly' without any knowledge of the underlying programs. The voice system makes the entire RVWII system operate completely hands-off; every aspect of the robot and the telephone (described below) can be controlled by the user's voice.

The robot control portion of the software handles all user interactions with the robot. The user can playback 'preprogrammed' tasks that are either created by the user (by recording series of robot positions in real-time mode) or are supplied with the system. Preprogrammed tasks include such daily activities as manipulating file folders, retrieving output from a printer and inserting disks into a disk drive, to name a few. Careful task analysis has shown that many system users would like the robot to perform such personal tasks as handing out a business card or giving out a flower. Any such personal task can be preprogrammed and operated with simple, natural language commands. A user can say GET FLOWER to hand a flower to someone else.

Another aspect of the robot control is the 'real-time' control of the robot. Real-time control is the commanding of the robot's motions directly. The user can directly control each of the axes of the robot: the track, waist, shoulder, elbow, pitch, roll and gripper. By simply naming the axes he or she wishes to move and using a plus or minus command, the user can move each axis in any direction. The user may also at any time change the speed at which the robot is moving, and control whether the robot should move in small, discrete steps (called discrete mode), or in continuous motion (called continuous mode). In discrete mode, each axis will move a preset, user-determined increment each time the user gives a command to do so. In continuous mode, the robot will move the select axis in the selected direction until the user gives the command *stop*. Having the two methods of motion and complete control over the speed and increment of motion gives the user a complete and flexible method of commanding robot motion.

The telephone manager gives the user control over a variety of telephone functions. The user has the ability to answer the phone and/or dial any phone number. There is a built-in Rola-dex™ style phone book which allows direct speed dialling of commonly used phone numbers. For example, the user may say CALL MIKE, and the system will automatically look up Mike's phone number for quick and easy dialling. The user may add or alter names in the phone book, and may supply both a home and business phone number for each entry. The storage capacity of the phone book is limited only by the disk space available on the system. The telephone manager also has the ability to record and playback phone conversations or personal memos in digitized form. This ability of the system gives the user a 'note-taking' capability in a completely hands-off fashion.

The application specific software packages handle voice recognition and voice keyboard input for a variety of commercially available software packages, such as Lotus 123 and DBASE III. Each application specific software package is tailor-made for a specific commercial software package to optimize voice recognition and voice keyboard entry by containing a customized set of words specific to that application.

*User interface*

The user interface for the RVWII is designed to be simple and functional. Most user choices are offered as menus of commands. By offering menus, or lists, of commands, the user is clearly informed of possible choices at any one time. The simplicity of this interface not only benefits the user, but is also simple and economical from a programming point of view.

While the user interface is generally clear about the possible choices of commands, there is always the potential for confusion. It is for this reason that on-line help is available in every aspect of the RVWII program. The user may, at any time, say HELP and the system will immediately display some helpful text based on the context in which the user said HELP. This context-sensitive help system gives the user concise information on what he or she is currently doing, without the need to wade through pages of irrelevant information on other parts of the system.

Another aspect of the user interface unique to a voice-controlled program is the vocabulary chosen to command the system. It is important that this vocabulary be as close to natural language as is practical. Command words should be long enough to be descriptive and to avoid confusion with other words in the system. For example, CALL JOHN AT HOME is a valid sequence of voice commands to the system. The best way to decide upon appropriate commands is to gain the input of a variety of end users during the development process. Their questions and comments about the interface and functions provide an incalculable asset to the development of the software.

A comprehensive training manual is being developed concurrently with the software. This manual includes in-depth information about the entire system on subjects such as hardware and software installation, user training, software descriptions and tips for more efficient use. This manual is also being developed with the aid of end users and specialists to ensure that it will be a proper teaching tool.

*Asynchronicity*

In order for the RVWII to be useful on an IBM PC under MSDOS, the RVWII software must be accessible at any time. Because of the absence of multi-tasking capability in MSDOS, the RVWII software was written to be a Terminate and Stay Resident program (TSR). As a TSR, the RVWII software is loaded once at system start-up time, and is thereafter in the memory of the computer. The software then waits for the proper command from the voice system to activate it. Upon activation, any currently running program is halted and the RVWII window is displayed on the screen. To the user, this is all transparent. The proper software is loaded automatically at boot-up. If the user is using Lotus 123 and the telephone rings, he or she needs only say TELEPHONE to activate the telephone manager and answer the call. There is no need to first exit Lotus 123 and return to DOS. However, because the program is totally memory resident at all times, it must be kept small. In order to keep the program size small, many size/speed trade-offs have to be made. The program is also very disk intensive, using disk space rather than RAM to do many of the space-greedy tasks.

### *Hardware*

While there is a large number of different computers used in small offices in this country, perhaps the most prevalent seems to be the IBM PC. The IBM PC has a very large array of software and hardware enhancements available for a personal computer, including voice recognition, telephone management and other hardware. It was for these reasons that IBM PC technology was chosen first for the RVWII. A Macintosh-based version of the RVWII is also under development.

The robot chosen for the RVWII was the CRS M1A manipulator. This is a five DOF articulated arm which proved to be highly reliable and robust. It could also lift up to 4 lbs, far more than any comparably priced robot. The CRS robot is a complete, self-contained robotic system with its own controller. The robot has its own programming language (called RAPL) and can communicate with a host computer through a serial line.

The final component in the RVWII system is a horizontal track on which the arm is mounted. An off-the-shelf track was chosen, modified and integrated with the CRS arm for smooth coordination of the track and arm. The track has two possible methods of control. The first is via a printed circuit board designed for the IBM PC. This method gives the PC direct control over the track's functions. The second is via an external DC power amplifier attached through the CRS robot. Using this method, the track is under the direct control of the robot, which is in turn under the direct control of the PC. Either of these two methods of control may be used with the RVWII system.

### *Hardware integration*

Integrating the various aspects of the RVWII system required that each individual component be completely controlled by the IBM PC. By designing the software to accommodate this, the system gives the user absolute control of all aspects of the system. This is consistent with the project's objectives of providing as much independence to the disabled user as possible.

Communication between the PC and the CRS is through a 2400 baud serial line. This speed must be kept slow due to hardware constraints in the robot. Because of this speed limitation and the necessary control we require of the robot by the PC, we limit what the CRS controller is allowed to do to small, finite commands whose progress can be closely monitored and controlled. By doing this we can give the user ultimate control to stop or modify whatever the robot is doing at any time with minimal time delay.

The robot has a specially designed connector for attaching various end-effectors. The end-effector may then be controlled via the robot. Both gripping force and distance of grip may be adjusted. Any external end-effector sensing is reported either to the robot or directly to the PC, depending on the needs of the software.

To maximize the speed and efficiency with which the robot performs its preprogrammed tasks, the tasks are stored completely on the IBM PC and sequenced to the robot in succession. The PC adjusts gripper position, track position and robot position for each step of the program. The specific robot positions, the absolute position of each axis on the arm, are stored on the robot while executing a task to minimize time delay required to transfer that information from PC to robot. These robot positions can be transferred to and from the PC's hard drive giving a virtually unlimited number of potential robot programs.

Other application-specific hardware can easily be added to the system. Such hardware might include a Fax board sending and receiving facsimiles or an ERMA board for communication with mainframe computers (Sandrof *et al.*, 1990).

Even though development of the RVWII represents the most current considerations of human–system integration, our work is still nascent in terms of what can and what must be accomplished in this area. This system incorporates the most advanced contemporary products, methods and thinking. Yet this stage of development is comparable to the early phases of flight or automotive transportation. The current RVWII represents a significant leap forward from previous generations of robotic workstations for people with disabilities. It is still a primitive tool relative to advances which can occur in the next decade. By addressing the design constraints required by persons with high-level disabilities, we can design systems that are more accessible and more appropriate for humans who have no apparent disability.

## Conclusion

Rapidly evolving technologies and changing populations with their widening spectrum of abilities and characteristics present new challenges to researchers and developers of service robotics. Service robotics are hybrid systems incorporating the repeatable and programmable flexibility of robotic and sensor technologies with the adaptability, judgement and decision making of humans and the emerging capabilities of artificial intelligence. The hybrid vigour achieved by such integration exceeds the capabilities of any single component of the human–robot system. The creation of such systems requires interdisciplinary, intergenerational teams of individuals from various perspectives and disciplines.

As manufacturing continues to change and service delivery emerges as a dominant economic and social force in the coming decades, the utilization of smart technologies in the service sector will become increasingly important. By replacing odious, routine, demeaning and boring tasks, adoption of technologies that augment human productivity for humans of any ability level serves not only to provide access to a larger potential labour pool for the employer, but also to make systems easier for all users. Smart technologies can offer new ways to facilitate autonomy and safety much like the telephone, television, airplane and automobile afforded new opportunities for communication and transportation for previous generations. The systems that will be evolving in the next quarter century will require our best human thinking to bring the creations of our imaginations to utilitarian fruition.

Our burgeoning older population with its wide spectrum of abilities and characteristics presents new challenges to human factors, robotics and ageing specialists regarding augmentation of human abilities and reduction or prevention of hazards. The concept of utilizing advanced technologies as tools for intervening in human independence is relatively new. Little effort has been directed toward understanding technology as intervention. Based on existing and emerging innovations, technological interventions utilized to increase human safety, security, autonomy and careproviding capabilities are critically in need of research.

Prospects look favourable in terms of identifying potential applications in health and human service areas and the technology that will be amenable to these needs. Utility and cost-effectiveness will be among the primary issues. Successful dissemination will be incumbent upon researching and designing reliable, appropriate systems. Cost justification may be expanded to include cost and care justifications. Thoughtful, careful consideration, evaluation and stepwise refinement toward smarter systems must occur at every stage of the research, design, development and dissemination process.

The challenging question: how do robotic scientists begin to research in a domain that is ill-defined and rapidly changing and for innovations that are not yet developed but rapidly evolving? Our work has taken a process approach. First, we work in an interdisciplinary manner to identify needs by working directly with potential end users and simultaneously to examine whether there are existing or future planned technologies that can fill/address these needs. This method also serves to illuminate human and technical performance and functional capabilities that will be required, and allows us to delineate and specify robotic system designs based on identified human driven needs. '. . . Many aspects of human life can be understood only by studying man's functioning with all its complexities and in the responses that (s)he makes to significant stimuli.' (Dubois,

1968, p. 133). If robotic technologies are, indeed, going to serve human beings, then the human–robot interaction must focus on the human–robot team as a synergistic whole complete with multiple and varied complex interactions.

## Acknowledgements

Thank you to Whitney Sample and Michael Sandrof and rest of the team that has made this chapter possible. Special thanks to Sanford Blatt.

## References

American Hospital Association, 1989, *Hospital Statistics 1988* (Chicago, IL).

Bennett, W. J. and McLaughlin, A., 1988, *The Bottom Line. Basic Skills in the Work Place* (Washington, DC: USGPO).

Bergwyn, E. and Savage, B., 1989, Caseload Study—Internal Report. Southwestern Pennsylvania Human Services, Inc. (Monessen, PA).

Brody, E. M., 1985, Parent care as a normative family stress. *Gerontologist,* **25,** 1, 19–29.

Building a Quality Workforce, July, 1988, US Department of Labor, US Department of Education, US Department of Commerce.

Chappel, N. L., 1983, Informal support networks among the elderly. *Research on Aging.* 5, 77–9.

Dubois, R., 1968, *So Human an Animal* (New York: Charles Scribner's Sons).

Edwards, R., Engelhardt, K. G., Van der Loos, H. F. M. and Leifer, L. J., 1985, Interactive evaluation of voice control for a robotic aid. *Journal of the American Voice Input/Output Society,* **2.**

Engelhardt, K. G., 1984a, High technology and its benefits for an aging population. Expert testimony. *Hearing Before the Select Committee on Aging, House of Representatives*, 98th Congress, 2nd Session, USGPO Comm. Pub. No. 98–459 (Washington, DC: USGPO).

Engelhardt, K. G., 1984b, Robotic concepts and applications in health and human services. *Proceedings of the 2nd International Robot Conference.*

Engelhardt, K. G., 1985, Applications of robots to health and human services. *Proceedings of Robots 9 Conference, Robotics International of SME* (Dearborn, MI).

Engelhardt, K. G., 1987, Innovations in health care: roles for advanced intelligent technologies. *Pittsburgh High Technology Journal,* **2,** 5, 69–72.

Engelhardt, K. G., 1988, Robots in the service of humans: concepts and methods. *Proceedings of the IEEE International Workshop on Intelligent Robots and Systems* (Tokyo: Japan).

Engelhardt, K. G., 1989, *International Journal of Technology and Aging,* **2,** 1,

Engelhardt, K. G. and Edwards, R., 1985, Robots in longterm health care. *8th Annual Conference, Rehabilitation Engineering Society of North America* (Memphis, TN), pp. 10–2.

Engelhardt, K. G. and Edwards, R., 1986, Increasing independence for the aging: robotic aids and smart technology can help us age less dependently. *Byte: The Small Systems Journal,* **11,** 3, 191–6.

Engelhardt, K. G. and Edwards, R., 1987, Mobilanguage. In *Cognitive Engineering in the Design of Human–Computer Interaction, Volume II* (New York: Elsevier), pp. 415–22.

Engelhardt, K. G., Awad, R. E., Leifer, L. J. and Perkash, I., 1984, Interactive evaluation of prototype robotic aid: a new approach to assessment. *Proceedings of the 2nd International Robot Conference*

Glastris, P., 1989, The mixed blessings of a movement. *US News and World Report*, 22,

Holt, S. W., 1987, The role of home care in long term care. *Generations,* **XI,** 2, 9–12.
Knowles, M., 1978, The Adult Learner: A Neglected Species, 2nd ed.
Kraus, L. E., 1989, Disability by the numbers. *Worklife,* 10
Nagamachi, M., 1988, Ten Fatal Accidents due to Robots in Japan. In Karwowski, W., Parsaei, H. R. and Wilhelm, M. R. (Eds) *Ergonomics of Hybrid Automated Systems I* (Amsterdam: Elsevier), pp. 391–6.
National Center for Health Statistics, 1988, *Disability Data Book.* NIDRR (Washington, DC: USGPO).
Norwood, J. L., 1988–9, *The Occupational Outlook Handbook.* US Department of Labor (Washington, DC: USGPO).
Office of Technology Assessment, 1978, *Assessing the Efficacy and Safety of Medical Technologies.* (Washington DC: USGPO).
Office of Technology Assessment, 1982, *Technology and Handicapped People.* (Washington DC: USGPO).
Pennar, K. and Mandel, M., 1989, Economic Prospects for the Year 2000. *Business Week,* **25 September,** 158.
Reddy, D. R., 1986, *Annual Review.* (Pittsburgh, PA: The Robotics Institute, Carnegie Mellon University).
Robinson, P. K., Livingston, J. and Birren, J. E. (Eds), 1984, *Aging and Technological Advances.* Vol. 24, NATO Human Factors Series, (New York: Plenum).
Robotic Industries Association, 1989, *Industry Report.* (Ann Arbor, MI: RIA).
Roth, W. C., 1989, Let us work. *Parade Magazine.*
Sample, W., Sandrof, M., Edwards, R. and Engelhardt, K. G., 1990, Robotic office workstations. *Proceedings of Fifth International Service Robotics Congress,* (Detroit, MI)
Sandrof, M., Sample, W., Edwards, R. and Engelhardt, K. G., 1990, Development of a robotic vocational workstation. *Proceedings of Rehabilitation Engineering Society of North America* (Washington, DC), p. 123.
Sangl, J., 1983, The family support system of the elderly. In Vogel, R. and Palmer, H. (Eds) *Long Term Care: Perspectives from Research and Demonstrations.* Health Care Financing Administration (Washington, DC: USGPO).
Schneider, W., Schmeisser, G. and Seamone, W., 1981, Computer-aided robotic arm/worktable system for the high level quadriplegic. *IEEE Computer,* 41–7.
Sterns, H. L., Barrett, G. V. and Alexander, R. A., 1985, Accidents and the aging individual. In Birren J. E. and Schaie, K. W. (Eds) *Handbook of the Psychology of Aging* (New York: Van Nostrand Reinhold), pp. 703–24.
Swartzbeck, E. M., 1983, The problems of falls in the elderly. *Nursing Management,* **14,** 12, 34–8.
Thibadeau, R. H., 1986, Artificial perception of actions. *Cognitive Science,* **10,** 117–49.
US Bureau of the Census, 1986, *Statistical Abstract of the United States: 1986.* 107th edition (Washington, DC: USGPO).
US Department of Health and Human Services, 1982, *1978 Survey of Unstabilized Work* (Washington, DC: USGPO).
US Department of Health and Human Services, 1989, *Health United States, 1988* (Washington, DC: USGPO).
US Department of Labor, 1988, Creative Affirmative Action Strategies for a Changing Workforce. *Opportunity 2000* (Washington, DC: USGPO).
*US Industrial Outlook*, 1989, Prospects for over 350 Industries, 30th Edition, pp. 22; 20–4; 51–3.
US Senate Special Committee on Aging, 1985, *Developments in Aging: 1984*, Vol. 2. Appendices (Washington, DC: USGPO).
Walsh, P. J., Edwards, R., Engelhardt, K. G. and Perkash, I., 1985, A formulation of the interactive evaluation model. *Proceedings of the Ninth Annual Symposium on Computer Applications in Medical Care* (Baltimore, MD)
*Webster's New World Dictionary*, 1975 (New York: The World Publishing Co.).

# *Chapter 17*
# *Robot implementation issues for the construction industry*

**M. J. Skibniewski**

*Division of Construction Engineering and Management, School of Civil Engineering, Purdue University, West Lafayette, IN 47907, USA*

**Abstract.** This chapter addresses significant implementation problems associated with robotics application in the construction industry. Problems associated with the industry's economic and financial environment and with the human factors such as labour attitudes towards robot application are discussed. Several robot equipment management issues are described along with current developments in research on optimum robot assignments to construction tasks and on efficient robot management.

## *Introduction*

Robotics application in construction differs from that in manufacturing due to ill-structured work environments, a relatively low repetition of tasks, the need for equipment mobility and a constantly changing job site (i.e. the structure being erected) as the work progresses. In contrast, the equipment in manufacturing plants is stationary, the work subject usually moves along the production line and the work environment is well-structured. Thus, the need for a machine to have cognitive abilities in a large portion of manufacturing tasks is reduced or eliminated. The reverse is true in construction, which has been a source of major difficulties in the development and implementation of computer-controlled equipment.

For these reasons, the construction industry experience with industrial robotics is relatively small. The application of robots in construction follows successful experiences in manufacturing, mining, nuclear plant maintenance and space exploration. Lessons learned in those domains have been applied to construction industry robotics in similarly executing structured tasks. A number of robots have been developed and tested on various construction sites throughout the world (see Tables 17.1 and 17.2). Some of these robots are presented below.

*Table 17.1 Example construction robotic prototypes developed in the United States.*

| System description | Application | Research centre |
|---|---|---|
| John Deere 690C | teleoperated excavation machine | John Deere, Inc, Moline, IL |
| Laser-Aided Grading System | automatic grading control for earthwork | Gradeway Const Co and Agtex Development Co, San Francisco, CA; Spectra-Physics, Dayton, OH |
| Automatic Slipform Machines | placement of concrete sidewalks, curbs and gutters | Miller Formless Systems Co, McHenry, IL; Gomaco, Ida Grove, IA |
| Micro-Tunnelling Machine | teleoperated micro-tunnelling | American Augers, Wooster, OH |
| Robotic Excavator (REX) | autonomous excavation, sandblasting, spray washing and wall finishing | The Robotics Institute, Carnegie Mellon University (CMU), Pittsburgh, PA |
| Autonomous Pipe Mapping | mapping subsurface pipes | The Robotics Institute, Carnegie Mellon University, Pittsburgh, PA |
| *Terragator* | autonomous navigation | The Robotics Institute, Carnegie Mellon University, Pittsburgh, PA |
| Remote Core Sampler (RCS) | concrete core sampling for radiated settings | The Robotics Institute, Carnegie Mellon University, Pittsburgh, PA |
| Remote Work Vehicle (RWV) | nuclear accident recovery work, wash contaminated surfaces, remove sedimets, demolish radiation sources, apply surface treatment, package and transport materials | The Robotics Institute, Carnegie Mellon University, Pittsburgh, PA |
| Wallbot | construction of interior partitions metal track studs | Massachusetts Institute of Technology, Cambridge, MA |
| Blockbot | construction of concrete masonry walls | Massachusetts Institute of Technology, Cambridge, MA |
| Shear Stud Welder | weld shear connectors in composite steel/concrete construction | Massachusetts Institute of Technology, Cambridge, MA |
| Automated Pipe Construction | pipe bending, pipe manipulation and pipe welding | University of Texas at Austin, Austin, TX |

*Table 17.2 Example construction robotic prototypes developed outside the USA.*

| System description | Application | Research centre |
|---|---|---|
| *(a) Concrete* | | |
| Shotcrete Robot | Spray concrete for tunnel liner | Kajima Co and Obayashi Co, Japan |
| Slab-Finishing Robot | Finish surface of cast-in-place concrete | Kajima Co, Japan |
| Automatic Laser Beam-Guided Floor Robot | Finished surface of cast-in-place concrete | Obayashi Co, Japan |
| Horizontal Concrete Distributor (HCD) | Place concrete for horizontal slabs | Takenaka Komuten Co, Japan |
| Automatic Concrete Vibrator/Tamper | Vibrate cast-in-place concrete | Obayashi Co, Japan |
| Automatic Concrete Distribution System | Carry concrete from batching plant to the cable crane | Obayashi Co, Japan |
| Concrete Placing Robot for Slurry Walls | Place and withdraw tremie pipes and sense upper level of concrete as it is poured | Obayashi Co, Japan |
| *(b) Non-concrete spraying* | | |
| Fireproof Spraying Robot | Fireproof structural steel | Shimizu Co, Japan |
| Paint Spraying Robot | Paint balcony rails in high-rise buildings | Shimizu Co, Japan |
| *(c) Structural members* | | |
| *Auto-Claw, Auto-Clamp* | Erect structural steel beams and columns | Obayashi Co, Japan |
| Structural Element Placement | Place reinforcing steel | Kajima Co, Japan |
| Automatic Carbon Fibre Wrapper | Wrap existing structures with carbon steel | Obayashi Co, Japan |
| Structural Element Welding | Weld large structural blocks for cranes and bridges | Mitsubishi Heavy Industries Co, Japan |
| *(d) Inspection* | | |
| Wall Inspection Robot (KABEDOHDA) | Inspect reinforced concrete walls | Obayashi Co, Japan |
| Wall Inspection Robot | Inspect facade | Kajima Co, Shimizu Construction Co and Taisei Co, Japan |
| Bridge Inspection Robot | Inspect structural surface of the bridge | University of Wales, Institute of Science and Technology, Cardiff, UK |
| GEO Robot | Finish facade/surface | Eureka, France |
| *(e) Tunnelling* | | |
| Shield Machine Control System | Collect and analyse data for controlling tunnelling machine | Obayashi Co and Kajima Co, Japan |

Table 17.2 (continued)

| System description | Application | Research centre |
|---|---|---|
| *(f) Excavation* | | |
| Super Hydrofraise Excavation Control System | Excavate earth | Obayashi Co, Japan |
| Tunnel Wall Lining Robot | Assemble wall liner segments in tunnels for sewer systems and power cables | Ishikawajima-Harima Heavy Industries, Tokyo Electric Power Co, Kajima Co, Japan |
| *(g) Other* | | |
| Clean Room Inspection and Monitoring Robot (CRIMRO) | Inspect and monitor the amount of particles in the air | Obayashi Co, Japan |
| Integrated Surface Patcher (ISP) | Hot resurfacing on highways | Secmar Co, France |
| Material Handling | Pick and distribute construction materials eg. prefabricated concrete materials and pipe | Tokyo Const Co, Ltd, and Hitachi Const Co, Ltd, Japan |

## *Examples of construction robots*

The following is a short description of some of these robotic systems (Skibniewski and Russell, 1989).

The John Deere 690C excavator is a teleoperated machine, controlled by a human from a remote site. A six-cylinder, four-stroke turbo-charged diesel engine propels the excavator at a travelling speed of approximately 16 km/h. The arm on this excavator has a lifting capacity of approximately 52.5 kN over side and 48.5 kN over end. The rated arm force is 72.1 kN, while the bucket digging force is 114.4 kN. The machine has been implemented by the United States Air Force for unmanned repairs of aircraft runways damaged during bombing raids and under enemy fire. Other areas of implementation, including heavy construction work and combat earth moving in forward areas, are under investigation.

A Dayton, Ohio based firm developed a microcomputer-controlled, laser-guided soil grading machine. A laser transmitter creates a plane of light over the job site. Laser light receptors mounted on the equipment measure the height of the blade relative to the laser plane. Data from the receiver are then sent to an on-board computer, which controls the height of the grader's blade. An automated soil-grading process implemented by this machine relieves the operator of manual positioning and controlling the grading blades, thus increasing the speed, quality and productivity of grading.

Miller Formless Systems Co. developed four automatic slipform machines for sidewalk curb and gutter construction. All four machines are able to pour concrete closer to obstacles than with alternative forming techniques. They can be configured to order for the construction of bridge parapet walls, monolithic sidewalks, curbs and gutters, barrier walls and other continuously formed elements commonly used in road construction. All the slipforming machines have the capability of operating in a playback mode while following a preset and precleared path of work. With lower labour requirements than traditional forming techniques, the labour saving potential on large volume projects is substantial.

The primary task of Carnegie Mellon University's robotic excavator (REX) is to remove pipeline sections in areas where explosive bases may be present. This robot is an autonomous machine able to sense and adjust to its environment. The manipulator arm can lift a 1.4 kN payload at full extension and over 4.5 kN in its optimal lifting position. REX uses two primary sensor modes, tactile and acoustic, allowing for three-dimensional (3-D) imaging.

Carnegie Mellon's *Terragator* is designed for autonomous, vision-based outdoor navigation in the conditions frequently resembled on construction sites. This machine is extremely survivable and powerful, thus preventing problems that inhibit machines designed for interior use. By changing gearing, the *Terragator* can be configured as a low-speed, high-torque or a high-speed, low-torque machine. The *Terragator* is a six-wheel-drive vehicle designed to ensure that it will not become entrapped.

Massachusetts Institute of Technology developed two robots for the construction of drywall interior building partitions, a *Trackbot* and a *Studbot*. Circumventing the need for a complex navigational system, the *Trackbot* is guided by a laser beacon aligned by a human. The machine is separated into two parallel workstations, an upper station for ceiling track and a lower station for the floor track. Detectors are mounted on the ends of the effector arms to ensure that the laser guidance system achieves the necessary precision. When the *Trackbot* completes a run of track, the *Studbot* begins the placement of studs. Location assessment is made by following the track and employing an encoding wheel or electronic distance measure instrument. The *Studbot* then references a previously stored floor plan to ascertain locations of studs to be placed. The stud is removed from its bin, placed into position, and spot welded into place by the positioning arm.

The *Blockbot* robot developed at the Massachusetts Institute of Technology is intended for automation of masonry block wall erection. To facilitate construction, the blocks are stacked upon each other with no mortar between the levels. The wall is then surface-bonded with a fibre glass-reinforced bonding cement. This process produces a wall with strength comparable to a traditional mortar-bonded wall.

The Construction Industry Institute of the University of Texas at Austin in cooperation with Bechtel Group and Grove Manufacturing Co. developed a remotely-controlled robot manipulator for placing piping sections into a complex piping structure on the construction site. The manipulator is joystick-controlled by a human operator located either on the ground next to the manipulator vehicle or in a cabin near the end-effector. This device enables better access for piping sections to their placement location and accelerates the piping construction process, thus considerably increasing work productivity.

Traditionally, in tunnelling work, a skilled operator is needed to regulate the amount of shotcrete sprayed on a tunnel surface and the quantity of the concrete hardening agent used. The Tokyo-based Kajima Construction Company developed and implemented a semi-autonomous robotic applicator by which high-quality shotcrete placement can be achieved.

A robot designed by Kajima Construction for finishing cast-in-place concrete slabs is mounted on a computer-controlled mobile platform and equipped with mechanical trowels to produce a smooth, flat surface. By means of a gyrocompass and a linear distance sensor, the machine navigates itself and automatically corrects any deviation from its pre-scheduled path. The robot is able to work to within 1 m of walls and replaces at least six skilled concrete finishers.

The Horizontal Concrete Distributor developed by Takenaka Construction Company of Tokyo is a hydraulically-driven, three-boom telescopic arm that cantilevers from a steel column of a building. The boom can extend 20 m in all directions over a 1000 $m^2$ of surface area. A cockpit located at the end of the distributor houses the controls for an operator to manipulate the boom direction and flow of concrete. The robot weighing approximately 5 tons can be raised along the column by jacks for the next concrete pour. On the average, the relocation procedure takes only 1.5 h.

A Tokyo-based Shimizu Construction Company developed several versions of a robot for spraying fireproofing material on structural steel. The robot uses the same materials as in conventional fireproofing, works sequentially and continuously with human help, travels and positions itself. Sufficient safety functions for the protection of human workers and of building components are provided in the machine. The robot can spray faster than a human worker, but requires time for transportation and set-up.

*Auto-Claw* and *Auto-Clamp*, two robotic devices for steel beam and column erection have been developed by Obayashi Corporation of Japan. Both robots speed up erection time and minimize the risk incurrred by steel workers. The *Auto-Claw* consists of two steel clamps fastened to the beam while in motion to the erection location. The clamps are automatically released by remote radio control once the beam is securely in place. Fail-safe electronic circuitry prevents the accidental release of the clamps during erection. The *Auto-Clamp*'s purpose and mechanics are essentially the same as those of the *Auto-Claw*, except that the *Auto-Clamp* uses a special electro-steel cylinder tube to secure and erect columns. The steel cylinder is electrically inserted and locked into the hole by remote control, whereupon the column can be erected. The *Auto-Clamp* has a rated lifting capacity of 136 kN and is also equipped with a fail-safe system preventing the cylinder from retracting from the hole during erection.

Secmar Company of France developed a robotic Integrated Surface Patcher, which consists of a carrier with rear wheel steering, an emulsion tank, an aggregate container, a built-in spreader and a compaction unit. The machine uses a hydraulic system driven by an additional motor to operate its functional modules. The electronic valve controls are operated with power supplied by the vehicle battery. This unit is used primarily for hot resurfacing road repairs, including surface cutting, blowing and tack coating with emulsion, as well as for repairs requiring continuous granular materials. It is also suitable for deep repairs using aggregate/bitumen mix, cement-bound granular materials, untreated well-graded aggregate and for sealing wearing courses with granulates.

In summary, there are three basic types of construction robots in use: teleoperated equipment, hard automation-based devices and semi-autonomous machines. The teleoperated equipment has the capability to perform tasks with a direct and continuous feedback from human operators. Thus it is only an extension of *human* handling capabilities into an inaccessible or hazardous work environment. As an example, a drywall panel handling device has been developed by the Taisei Corporation, and a ceiling panel handling fixture is undergoing testing and refinements at the Shimizu Corporation. Hard automation-based machines are applicable in the handling of high-volume, repetitive and somewhat simple tasks that are cumbersome or otherwise unattractive for human workers. In this case, no substantial sensory feedback or information processing is required from the machine. Examples of such machines include fresh concrete surface-finishing machines developed by the Takenaka Corporation, and a concrete-smoothing robot from the Obayashi Corporation. Autonomous construction equipment should be capable of handling most or all of the essential work functions currently handled by human workers, including proximity and tactile feedback and processing, navigation, elementary judgement and decision making. Such equipment is currently a subject of intensive research and development, both in academic and commercial institutions.

## *Challenges of implementation*

Construction is the largest single industry in the United States with the annual production volume of over $400 billion in the early 1990s. Over the last decade, the overall construction productivity has been in a decline, whereas manufacturing productivity has been steadily improving during this time (see Building Research Board *et al.*, 1986). In comparison with manufacturing firms, most US construction contractors operate with low profit margins (usually about 3 to 4%) and they are highly leveraged (i.e. finance their business operations by borrowing money). Therefore, virtually no funds are available for in-house research and development. Only a few large construction firms, such as the Bechtel Corporation in San Francisco, California and the Stone and Webster Corporation in Boston, Massachusetts, conduct large-scale in-house research and development activities. Any investment in new equipment and technology is expected to produce substantial short-term economic results to satisfy creditors. This makes a long-term investment in research and development extremely difficult for small and medium-sized firms. Robotics implementation in an environment as challenging as a construction site requires such a long-term investment of funds and effort.

A decision logic incorporating a variety of company and project factors is necessary for responsible decisions regarding investment in construction robotics implementation. A summary of these factors and their relationships is presented in Figure 17.1 (see also Skibniewski, 1988b).

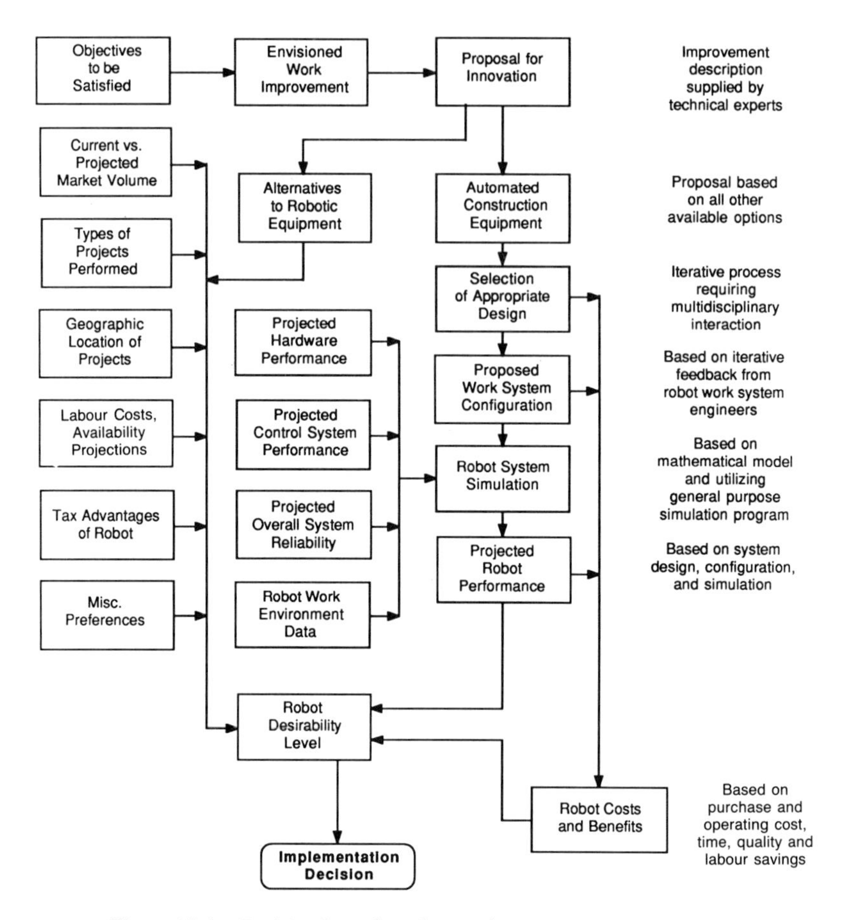


*Figure 17.1 Decision logic for robot implementation in construction.*

Another problem of robot implementation on construction sites pertains to the labour concerns. Skilled construction labour is to a large extent organized in sometimes militant, rigidly-controlled craft unions, exercising rigid work rules which are traditionally blamed for their contribution to the overall low labour productivity in the industry. A removal of these rules, which was necessary to introduce new construction methods and technologies, has in some cases encountered craft resistance. The issue at stake is a perceived loss of job security for the craft constituency should the new technology result in a reduced demand for human labour, skilled or unskilled. Thus, it is necessary to point out that the application of robots on construction sites will most likely lead to eliminating only those human-performed tasks that are currently considered undesirable. Most of the tasks eliminated by the developed robots have experienced prior labour shortages, primarily for work safety or health reasons, or due to the lack of job satisfaction with the performance of such work (Construction Labor Relations, 1988).

Historically, labour unions in the manufacturing industries have not consistently resisted automation efforts on part of management (Ayres and Miller, 1983). Instead, they have been undergoing transformations and promoting retraining of their members to enable better adaptation to the new work conditions. Labour displacement due to the introduction of robotics has been minimal. In fact, robot implementation frequently led to an upgrade of prior job description for formerly unskilled employees to that of a robot operator, and to an overall increase in employment due to the creation of new jobs in robot technical support. With this positive experience in the manufacturing sector, there is no reason to expect a negative impact of robotics on labour in the construction industry. However, a few research establishments are involved in sociological studies related to the impact of construction robots on the future workforce in this industry. For example, a comprehensive research study aimed at determining the impact of robotics on the workforce in the Australian construction industry is currently underway at the Commonwealth Scientific and Industrial Research Organization in Melbourne, Australia.

## *Economics of construction robotics*

The most decisive factor for a robotic application in construction will be its impact on the overall cost of the construction process (Skibniewski, 1988a). The promising areas of application are in tasks, where the work volume, high repetition and simple control requirements result in a promising robot automation potential. Such tasks include surface treatment (e.g. cleaning, painting, sandblasting, etc.), inspection (e.g. non-destructive concrete and steel testing, ceramic tile adhesion assessment, piping installation mapping, etc.) and excavation.

Robot-related costs to be reconciled during the analysis include capital costs and operating costs. The capital costs include research and development expenditures (hardware and software, work system engineering, calibration and field hardening. The operating costs include energy, maintenance, downtime, repair, tooling, set-up, dismantling, transportation, operator and other related expenditures. Tables 17.3 to 17.6 show example costs for a mobile surface treatment robot in both categories.

Robot-related benefits include construction labour and material savings, improved work quality, extension of work activity into additional locations and time periods, and possibly improved productivity. Table 17.7 lists example benefits expressed in monetary terms for a surface-application robot accrued on a typical construction project site.

Table 17.3 *Cost breakdown of R&D on Japanese construction finishing work robot.*

| Description | Cost ($) |
|---|---|
| *(a) Prototype stage* | |
| Manpower for R&D (engineering, system/machine design, planning, scheduling, coordination, etc.) 408 man-days | 74 100 |
| Product testing: | |
| work surface treatment | 15 000 |
| teaching | 1 000 |
| travelling | 1 500 |
| navigation wiring | 900 |
| Purchase of industrial robot hardware | 90 350 |
| Purchase of tractor and traveller (for robot mobility) | 25 000 |
| Subtotal (approximate) | 208 000 |
| *(b) 'Mark 2' system* | |
| Manpower for R&D (engineering, system/machine design, planning, scheduling, coordination, etc.) 254 man-days | 45 550 |
| Product testing: | |
| position sensor | 1 800 |
| surface treatment by robot | 600 |
| material feeder | 1 100 |
| Conversion from prototype to 'Mark 2' system (improved hardware, controller, software) | 40 000 |
| Subtotal (approximate) | 89 000 |
| Total (approximate) | 300 000 |

Table 17.4 *Estimated cost of sandblasting robot component.*

| Item | Cost ($) |
|---|---|
| Base of robot | 270 000 |
| Horizontal linkage of manipulator | 71 000 |
| Vertical support | 107 000 |
| Surface preparation end-effector | 34 000 |
| Applicator effector | 30 000 |
| Surface preparation subsystem | 400 000 |
| Applicator subsystem | 40 000 |
| Control systems hardware: | |
| computer servo-loops | 40 000 |
| optical collision avoidance | 40 000 |
| man–machine interface | 14 000 |
| Software (including control of system elements) | 600 000 |
| Total (approximate) | 1 650 000 |

Details of the economic analysis summarized in the above tables can be found in Skibniewski and Hendrickson (1988) and in Skibniewski (1988a).

## Construction robotic equipment management

As can be expected from the conditions in which the US construction industry operates, the most important decision factor for robot implementation will be

*Table 17.5 Estimated cost of sandblasting system components.*

| Component | Cost ($) |
|---|---|
| Automatically guided vehicle (AGV) | 50 000 |
| Robotic manipulator | 40 000 |
| Control systems: | |
| sand hopper controller | 1 000 |
| sand flow controller | 900 |
| air pressure controller | 1 200 |
| power supply controller | 1 800 |
| Guidewires and guidepaths | 1 000 |
| Custom-built sensors | 2 000 |
| Graphic displays and communication | 5 000 |
| Total (approximate) | 100 000 |

*Table 17.6 Projected operating cost of sandblasting robot.*

| Benefits/savings | Value/project ($) |
|---|---|
| Operator labour | 7 500 |
| Scaffolding elimination | 3 000 |
| Health and safety | 3 000 |
| Work quality | 750 |
| Productivity gain | 0 |
| Extension of activities | 1 425 |
| Total (approximate) | 15 500 |

*Table 17.7 Estimated benefits from example roboticized sandblasting project.*

| Item | Cost/project ($) |
|---|---|
| Supervision cost (one technician) | 1 250 |
| On-site reprogramming and adaptations | 300 |
| System re-set-up (three technicians, one day) | 600 |
| System dismantling (two technicians, one day) | 400 |
| Electric energy (battery and power line) | 300 |
| Transport to new work site | 500 |
| Maintenance and repair | 400 |
| Total (approximate) | 3 500 |

a short-term profit potential resulting from labour savings through productivity improvement and possibly through increased construction quality. Unlike the Japanese engineering construction firms, the US companies will be very unlikely to invest in robot research and development and will rely on robotics technology developed by commercial systems houses. Such robots will then be either sold or leased by commercial vendors operating in the construction equipment market.

A major difficulty that construction firms will initially face is that of estimation of robot costs and benefits, as described in the previous section. This will improve as more experience with particular robot applications is gained. Detailed information on the cost and benefit items for various robot applications in typical job site settings can accelerate the pace of robotization if it is made available to all interested construction firms. Future construction robot equipment vendors will be well positioned to fulfill this function in cooperation with robot system developers and manufacturers.

Once more robotics become available for implementation on construction sites, significant challenges to both management and technical staff will emerge. For the reasons outlined above, robots must be managed wisely in order to insure maximum economic benefits for the contractor's firm.

Despite a number of advantages over traditional methods of performing construction tasks, robots are currently, and will continue to be in the near future, in short supply in comparison with other construction equipment. Thus they should be regarded as a scarce resource and their use should be maximized to their full operating potential. With maximized robot utilization on as many construction projects as possible in the contractor's portfolio, economic benefits of robot use can be easier to attain. Consequently, robot development costs can be recovered faster and robot use can spread to other applications and types of construction tasks.

The Construction Robotic Equipment Management System (CREMS) was developed as a response to this need (Skibniewski *et al.*, 1989). The principal objective of CREMS is to provide a comprehensive construction robot management capability for construction field personnel. CREMS is intended as an aid to decision making regarding the implementation of robots on the job site. Such a system should prove highly desirable when managing a fleet of diverse single-purpose robots and increase the application efficiency of robots available to a construction company. The discussion of CREMS presented below is based on its description in Russell *et al.* (1990) and Skibniewski *et al.* (1990a). Detailed technical information on the system architecture and implementation can be found in these source papers.

CREMS, as shown in Figure 17.2, consists of four basic modules: Construction Task Analysis Module (CTAM), Robot Capability Analysis Module (RCAM), Robot Economics Evaluation Module (REEM) and Robot Implementation Logistics Module (RILM).

Construction Task Analysis Module (CTAM). This module acquires and analyses all pertinent data regarding a given construction task. The data analysed include manpower and productivity requirements, achievable work quality and characteristics of the work environment. The information derived is both a qualitative and quantitative evaluation of task suitability for robotic and non-robotic performance.

Robot Capability Analysis Module (RCAM). This module examines the capabilities of a given robot or group of robots for the performance of a construction task(s). The data analysed include robot weight, power supply, payload, work envelope, available tools and control and sensory systems.

Robot Economic Evaluation Module (REEM). This module performs a detailed comparative analysis of non-robotic versus robot job site work alternatives

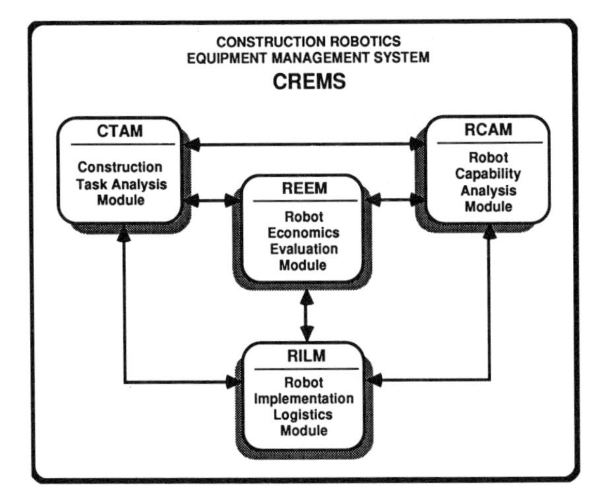


*Figure 17.2 The principal modules for CREMS.*

relevant for the considered job site location. Factors for analysis include task performance cost with non-robotic work techniques, operational costs of robotic equipment, resulting labour and time savings and quality improvements.

Robot Implementation Logistics Module (RILM). This module performs robotic equipment scheduling functions to ensure optimal use throughout the company projects. These functions include robot work-time determination, tooling and manpower allocation, robot maintenance schedules, robot site set-up and disassembly and transportation schedules. During the scheduling and resource levelling procedures, the robots are treated simply as another project resource and are subject to similar management approaches as other conventional construction equipment.

CREMS has been implemented within a *Hypercard*™ programming environment on an Apple Macintosh II microcomputer.

## CREMS submodules

Figure 17.3 provides an overview of program submodules and their functional relationships. They consist of two categories: (1) input submodules and (2) analysis submodules. Each is described below.

*Construction Input Submodule* is structured to allow the input of principal design and site-specific parameters to determine the degree of robot applicability for a given task. It receives two main types of data: information from the plans/drawings for the task and information derived from specifications that control the task.

The submodule processes information from plans/drawings for the operation to obtain the relevant and necessary parameters required to perform the robot applicability analysis. A submodule designed to perform such an analysis is described below. Specific examples of data required by this submodule include

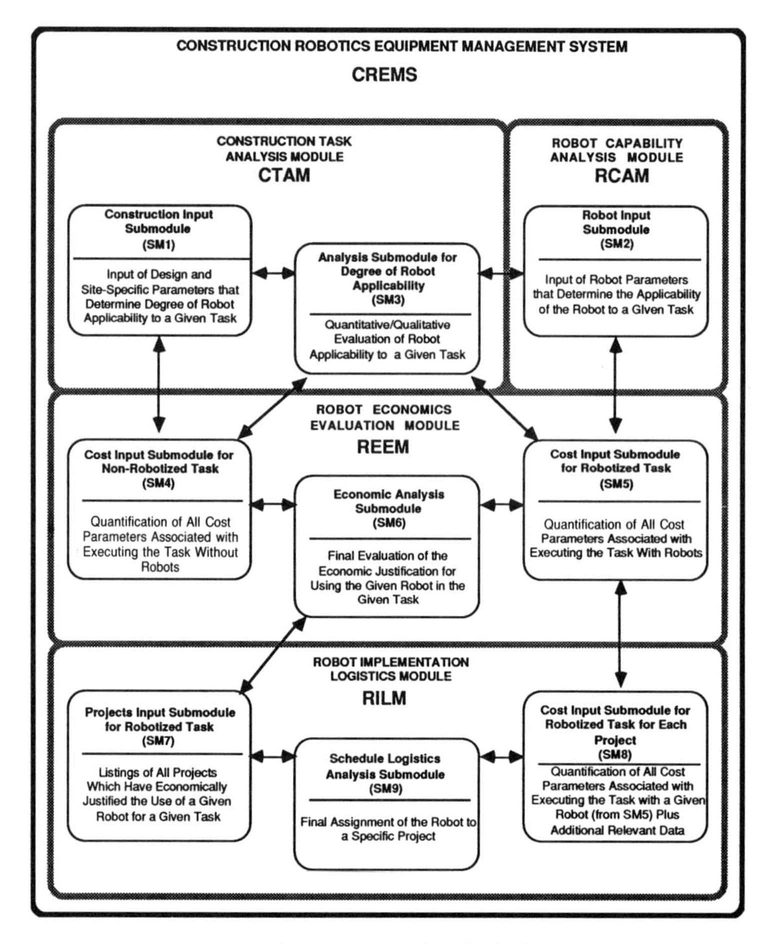


*Figure 17.3 Submodules of CREMS.*

areas, perimeters, borders, obstacles, clear heights, weights of objects, paths and corners. Using the project specifications, SM1 extracts all relevant information about quantities, materials, construction methods and other specific issues that are required to perform the robot applicability analysis.

*Robot Input Submodule.* With a similar objective, this submodule is structured to allow input of the principal parameters that determine the robot's applicability to execute the given task. Although this submodule is self-contained (i.e. works on a stand-alone basis), as shown in Figure 17.3, it is a part of the RCAM module.

This submodule receives three types of information: (1) physical characteristics of the robot, (2) robot execution capabilities for the operation, and (3) performance constraints that control the robot's operation. The submodule extracts all relevant information regarding the type of robot and its functions required to perform the robot applicability analysis.

*Analysis Submodule for Degree of Robot Applicability*. The analysis determining whether or not a task can be accomplished with a robot, and to what degree it can be executed is performed by this submodule. The analysis compares the design, site-specific parameters, and quality requirements of the given operation to the robot's characteristics and capabilities. This provides a qualitative and quantitative evaluation of the robot applicability to the specified task.

This submodule includes both qualitative and quantitative elements since numerous stages of the analysis require the application of judgement from field personnel. Examples of calculations involved in this module include paths, percentage of effective area (i.e. where the robot can actually perform the operation), expression of obstacles and constraints in terms of effective areas.

The submodule contains the knowledge necessary to establish whether or not to continue the evaluation of robot use for a given task from an economic perspective. If the decision is not to continue, the evaluation is terminated at this point, identifying reasons for not recommending the robot's use and providing guidelines for increasing the degree of robot applicability. This may involve altering the design and site-specific parameters of the task or modifying the robot configuration.

*Cost Input Submodule for Non-robotized Task* is structured to allow input of the principal cost components associated with the conventional, non-robotic execution of a given task. This submodule evaluates alternative non-robotic approaches and resource configurations and selects the most cost-effective alternative for a given task. It identifies principal cost components associated with each method of task execution accepting three types of information: (1) the direct cost component of the task (i.e. labour, permanent materials, equipment allocated specifically for the task), (2) indirect cost component of the task (i.e. overhead, equipment distributed over different operations, temporary materials), and (3) other associated costs of the task (i.e. preparation tasks, transportation, clean-up). The submodule extracts all relevant data regarding the cost of executing a task without a robot, as required to perform the economic analysis.

*Cost Input Submodule for Robotized Task* is organized to allow the input of principal cost components associated with the execution of a given task using robots. This submodule identifies principal cost components associated with the execution of the task using a given robot for a specific task and receives four types of information: (1) the direct cost components of the task (i.e. labour, permanent materials, equipment allocated specifically for the operation), (2) indirect cost components of the task (i.e. overhead, equipment distributed over different tasks, temporary materials), (3) other associated task costs (i.e. preparation tasks, clean-up), and (4) all costs associated with the robot (i.e. transportation, set-up, calibration, operation, maintenance, dismantling).

*Economic Analysis Submodule* performs the final qualitative and quantitative evaluation of the economic factors associated with using the given robot for a specific task. The result of the complete analysis is a statement regarding the economic justification of robot use. The possibility to override this decision due to subjective considerations will be embedded with this submodule.

*Project Input Submodule for Robotized Task* is structured to allow the input of data on all projects which have met the economic requirements previously determined. It processes information to obtain the necessary parameters to perform the scheduling logistics analysis. Specific data relevant to this submodule include: (1) project name, (2) project location, (3) project owner, (4) project type, (5) estimated project cost, and (6) estimated construction task cost.

*Cost Input Submodule for Robotized Task for Each Project* is structured to allow the input of principal cost components associated with the execution of a given task using the robot. It contains a complete list of these data items. Additional data include (1) impact of implementing the robot on the project's critical path, (2) the effective usage of the robot (i.e. robot work time), (3) availability of necessary manpower to maintain and operate the given robot, (4) availability of necessary construction equipment to off-load and place the robot at the desired work location, and (5) willingness of the project personnel to implement robotics on the job site.

This module extracts all data regarding costs and other considerations for executing the task with a robot.

*Schedule Logistics Analysis Submodule* performs the final quantitative and qualitative evaluation for scheduling the robot for given tasks across all projects currently in the contractor's portfolio. The result of this analysis is the selection of a project for which the robot should be implemented. The possibility of overriding this decision can be embedded within the submodule.

It is envisioned that all the input submodules will contain knowledge that establishes whether or not the information input at this stage of the analysis is sufficient to perform the evaluation (either of robot applicability, cost, or robot scheduling). If the information is incomplete, the submodule identifies additional data to be input and provides guidelines for data-collection procedures in the event that these data are not readily available.

## Robot applicability analysis

To formalize the contents of the system's modules and submodules further, three analysis frameworks can be distinguished within CREMS. The first framework determines a robot's applicability to a given construction task (i.e. CTAM and RCAM). The second analyses the economic attractiveness of executing a given task by robot versus competing non-robotic, conventional construction methods (i.e. REEM). The third framework determines, from a group of applicable projects, a specific project in which to implement the considered robot.

CTAM input begins with a given construction task. This is the initial input to the module. The next stage of analysis requires the input of task-specific parameters. These parameters include the description of the task, its principal characteristics and general constraints such as weather, congestion, accessibility, hazard, etc. The result of this initial input is a set of task parameters used to determine, based on a set of pre-established rules or criteria, whether there is a robot available that can execute the task given its characteristics. If the determination is positive, two parallel processes are generated.

One process selects the robot, automatically generating a generic default set of robot characteristics and parameters stored in a data base. Based on this information, an additional process is launched to determine the robot's capabilities once it has been adjusted and calibrated for the given task. The output of this process consists of a set of robot parameters (both physical characteristics and performance) to be used in the next stage of the CTAM analysis.

Another process determines project parameters such as physical layout, work constraints, quality requirements and, in general, a description of what needs to be accomplished to execute the task successfully. The output of this process is a set of project parameters to be used in the next stage of the CTAM analysis. These outputs are provided to subsequent processes to determine the robot effectiveness for a given task and the robot productivity.

Next, the portions of the task which can be assigned to the robot are determined. The output of this procedure, the new task layout plan, provides an input to a complex intelligent query that determines the quantity of work which the robot can execute. Subsequently, robot productivity in executing the task is determined. The program will automatically stop if the values of both the quantity of robot work and robot productivity are determined as low. Conversely, the process continues to the performance of cost analysis.

### Cost analysis

The cost analysis of using a robot for a given task begins with a process determining all non-robot costs (e.g. the cost components necessary to execute the task without using the robot such as labour, material and equipment). Another process determines the robot costs related to a given task (e.g. cost components necessary to execute the task with the use of robots, such as robot set-up, dismantling and operation). The process is followed by a query determining if additional non-robot costs are required to execute the given task completely (e.g. labour, equipment and material necessary to execute completely the task).

The final process of the cost analysis utilizes the quantities of the costs associated with the task performance with and without the use of a robot. The output of this process is the final robot implementation decision, based on performance cost comparison, determining whether or not to use the robot in executing the task.

### Schedule logistics

Once the suitability of a given robot for the performance of a specific construction task is determined, the robot can be assigned to that task and dispatched. However, as may frequently be the case with a large construction company, more than one project site will compete for the robot assignment at any given time. Thus, considering the robots as scarce resources, sound and timely dispatching decisions based on a careful weighing of each assignment's advantages and disadvantages must be made. The variety of decision issues is quite complex and includes task work volumes, task durations and sequences, job site locations and associated travel distances, robot set-up, operation and dismantling resources needed, and others.

The complexity of the robot assignment problem precludes the use of one purely classical method such as a dynamic linear programming assignment model in the assignment decision making. A hybrid decision model containing both classical optimization approaches and heuristic knowledge is necessary. Such a model can encompass the specific challenges facing construction equipment managers when dealing with relatively new and unfamiliar robot equipment.

The conceptual framework of schedule logistics relative to the use of a given robot for specific tasks and projects is presented in detail in Skibniewski *et al.* (1990b).

The schedule logistics determination commences upon the completion of cost analysis and begins with determining whether the contractor currently needs to implement the same robot on other sites. If this determination is positive, several decision processes are generated.

The first process compiles a listing of all project sites on which the robot could be implemented. Prior to being included on the list, all projects and their associated tasks must have been screened by the cost feasibility analysis procedure described above. The compilation includes a list of project names, locations and types which will be stored in a data base.

The subsequent process uses the robot cost as well as additional data such as the robot effective usage and its impact on the critical path of the project schedule. The result of this process is the determination of the robot implementation logistics cost and additional data necesssary for evaluation.

Robot costs and additional data for each competing project are evaluated by the next process. This evaluation includes combining both quantitative and qualitative data to arrive at the project-specific robot implementation decision. At this step, a prioritized list of projects on which the robot should be implemented is determined. The final process schedules the robot for the selected project using a scheduling program.

## *Conclusions*

Robot applications in construction still constitute a difficult technical and managerial challenge. However, potential benefits may be significant in this industry, which is the largest single sector of the US national economy (over $400 billion in the annual project volume in the early 1990s, including the support industries). Due to the lack of investment in research and development by construction firms in combination with other factors, the industry experienced an overall actual decline in productivity over the past decade, while manufacturing experienced productivity improvements in the same period.

To facilitate more comprehensive impact of robotics and automation on the construction industry in the future, further research and development is essential. In particular, more attention should be focused on the redesign of construction sitework environments to enable direct technology transfer from other industries to construction, rather than on development of customized automated construction equipment that would closely resemble the human-like performance of traditional construction tasks (Skibniewski and Nof, 1989). Sound methods for systematic technology transfer and evaluation are also necessary (Gaultney *et al.*, 1989; Skibniewski, 1988b). Better quality and lower price of the construction product will increase the competitiveness of the construction firms and can ultimately lead to greater demand for the services of this large industry.

## *References*

Ayres, R. and Miller, S., 1983, *Robotics: Applications and Social Implications* (Cambridge, MA: Ballinger Publishing Co.).

Building Research Board, Commission on Engineering and Technical Systems, National Research Council, 1986, *Construction Productivity* (Washington, DC: National Academy Press), p. 2.

Construction Labor Relations, 1988, 130,000 More Concrete Masons Needed in Next 10 Years, **33,** Feb., 1391.

Gaultney, L., Skibniewski, M. and Salvendy, G., 1989, A systematic approach to industrial technology transfer: a conceptual framework and a proposed methodology. *AIT Journal of Information Technology,* **4,** 1, 7–16.

Russell, J., Skibniewski, M. and Vanegas, J., 1990, Framework for construction robot fleet management system. *ASCE Journal of Construction Engineering and Management,* **116,** 3, 448–62.

Skibniewski, M., 1988a, *Robotics in Civil Engineering* (Southampton-Boston-New York: Computational Mechanics Publications and Van Nostrand Reinhold), p. 233.

Skibniewski, M., 1988b, Framework for decision-making on implementing robotics in construction. *ASCE Journal of Computing in Civil Engineering,* **2,** 2, 118–201.

Skibniewski, M. and Hendrickson, C., 1988, Analysis of robotic surface finishing work on construction site. *ASCE Journal of Construction Engineering and Management,* **114,** 1, 53–68.

Skibniewski, M. and Nof, S., 1989, A framework for programmable and flexible construction systems. *Robotics and Autonomous Systems,* **5,** 2, 135–50.

Skibniewski, M. and Russell, J., 1989, Robotic applications to construction. *AACE Cost Engineering Journal,* **31,** 6 (Morgantown, WV: American Association of Cost Engineers), 10–8.

Skibniewski, M., Vanegas, J. and Russell, J., 1989, Construction equipment management system (CREMS). In Tucker, R. (Ed.) *Proceedings, 6th International Symposium on Automation and Robotics in Construction,* San Francisco, CA, June (Austin, TX: Construction Industry Institute, University of Texas), pp. 404–11.

Skibniewski, M., Russell, J. and Cartwright T., 1990a, Construction Robotic Equipment Management System (CREMS) Software, Technical Report, Division of Construction Engineering and Management, Purdue University, February.

Skibniewski, M., Tamaki, K. and Russell, J., 1990b, Construction robot implementation logistics. *Proceedings, 7th International Symposium on Automation and Robotics in Construction*, Bristol, England, June, pp. 543–55.

Yoshida, T. and Ueno, T., 1985, Development of a Spray Robot for Fireproof Treatment. Shimizu Construction Co. Technical Research Bulletin No. 4, Tokyo, Japan, pp. 48–63.

# SUBJECT INDEX

Page numbers in bold denote chapters primarily concerned with that subject